STUDENT SOLUTIONS M[ANUAL]

for use with

NELSON

Introduction to
Probability and Statistics
Second Canadian Edition

Mendenhall · Beaver · Beaver · Ahmed

Prepared by Adamic/Gillis/Caron

NELSON EDUCATION

Student Solution Manual for use with
Introduction to Probability and Statistics
Second Canadian Edition
Prepared by Peter Adamic, Dan Gillis, and Sylvain Caron

Vice President and Editorial Director:
Evelyn Vetch

Editor-in-Chief, Higher Education:
Anne Williams

Executive Editor
Jackie Wood

Marketing Manager:
Sean Chamberland

Developmental Editor:
My Editor Inc.

Content Production Manager:
Christine Gilbert

Manufacturing Coordinator:
Ferial Suleman

Design Director:
Ken Phipps

Managing Designer:
Katherine Strain

COPYRIGHT © 2011
by Nelson Education Ltd.

For more information contact Nelson, 1120 Birchmount Road, Scarborough, Ontario M1K 5G4. Or you can visit our Internet site at www.nelson.com

ALL RIGHTS RESERVED. No part of this work covered by the copyright hereon may be reproduced or used in any form or by any means—graphic, electronic, or mechanical, including photocopying, recording, taping, web distribution or information storage and retrieval systems—without the written permission of the publisher.

Table of Contents

Preface v

Chapter 1: Describing Data with Graphs 1

Chapter 2: Describing Data with Numerical Measures 23

Chapter 3: Describing Bivariate Data 47

Chapter 4: Probability and Probability Distributions 67

Chapter 5: Several Useful Discrete Distributions 89

Chapter 6: The Normal Probability Distribution 103

Chapter 7: Sampling Distributions 123

Chapter 8: Large-Sample Estimation 137

Chapter 9: Large-Sample Tests of Hypotheses 153

Chapter 10: Inference from Small Samples 171

Chapter 11: The Analysis of Variance 205

Chapter 12: Linear Regression and Correlation 229

Chapter 13: Multiple Regression Analysis 257

Chapter 14: Analysis of Categorical Data 273

Chapter 15: Nonparametric Statistics 297

To the Student

In the tradition of the previous Canadian edition, Introduction to Probability and Statistics, Second Canadian Edition text exercises are graduated in level of difficulty; some, involving only basic techniques, can be solved by almost all students, while others, involving practical applications and interpretation of results, will challenge students to use more sophisticated statistical reasoning and understanding. The variety and number of real applications in the exercise sets is a major strength of the text. New to the Second Canadian Edition are "big picture" projects, or mini cases, added throughout the text. These provide an opportunity for students to build on knowledge gained from previous chapters and apply it to big picture projects. Rather than working with problems based only on the individual sections, students will be using almost all of the concepts, definitions, and techniques given in that chapter, thus bolstering students' success rate. More examples and exercises have been added to selected chapters and a number of new and updated real data sets from applications in many interesting fields.

The *Student's Solutions Manual to accompany Introduction to Probability and Statistics*, Second Canadian Edition, was prepared to assist the student in mastering the skills required for an understanding of probability and statistics. The selected questions have been chosen by the authors of your text to allow you to discover the range and depth of your understanding.

This *Student's Solutions Manual* contains the worked-out solutions to odd-numbered in-text exercises, and the solutions to the case studies and projects at the end of each textbook chapter.

Peter Adamic, Dan Gillis, and Sylvain Caron
Laurentian University

Chapter 1: Describing Data with Graphs

1.1   **a**   The experimental unit, the individual or object on which a variable is measured, is the student.
        b   The experimental unit on which the number of errors is measured is the exam.
        c   The experimental unit is the patient.
        d   The experimental unit is the azalea plant.
        e   The experimental unit is the car.

1.3   **a**   "Population" is a *discrete* variable because it can take on only integer values.
        b   "Weight" is a *continuous* variable, taking on any values associated with an interval on the real line.
        c   "Time" is a *continuous* variable.
        d   "Number of consumers" is integer-valued and hence *discrete*.

1.5   **a**   The experimental unit, the item or object on which variables are measured, is the vehicle.
        b   Type (qualitative); make (qualitative); carpool or not? (qualitative); one-way commute distance (quantitative continuous); age of vehicle (quantitative continuous)
        c   Since five variables have been measured, the data is *multivariate*.

1.7   The population of interest consists of voter opinions (for or against the candidate) <u>at the time of the election</u> for all persons voting in the election. Note that when a sample is taken (at some time prior or the election), we are not actually sampling from the population of interest. As time passes, voter opinions change. Hence, the population of voter opinions changes with time, and the sample may not be representative of the population of interest.

1.9   **a**   The variable "reading score" is a quantitative variable, which is usually integer-valued and hence discrete.
        b   The individual on which the variable is measured is the student.
        c   The population is hypothetical – it does not exist in fact – but consists of the reading scores for all students who could possibly be taught by this method.

1.11   **a-b**   The experimental unit is the pair of jeans, on which the qualitative variable "province" is measured.
        c-d   Construct a statistical table to summarize the data. The pie and bar charts are shown in the figures below and on the next page.

Province	Frequency	Fraction of Total	Sector Angle
ON	9	.36	129.6
QC	8	.32	115.2
MB	8	.32	115.2

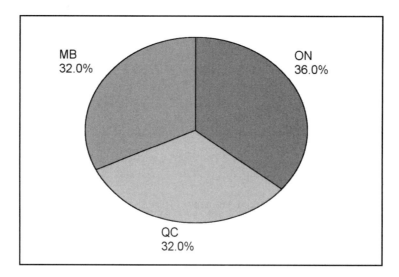

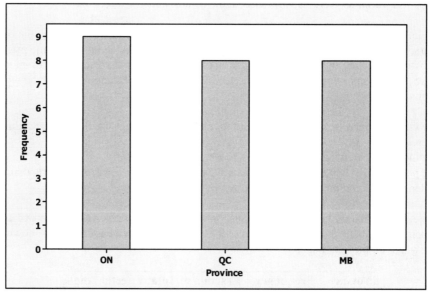

  e From the table or the chart, Quebec produced $8/25 = 0.32$ of the jeans.

  f The highest bar represents Ontario, which produced the most pairs of jeans.

  g Since the bars and the sectors are almost equal in size, the three provinces produced roughly the same number of pairs of jeans.

1.13 **a** No, a few more Islamic countries (Iraq, Pakistan, Afghanistan, Syria, etc.) can be added in the table.

  b A bar chart is appropriate.

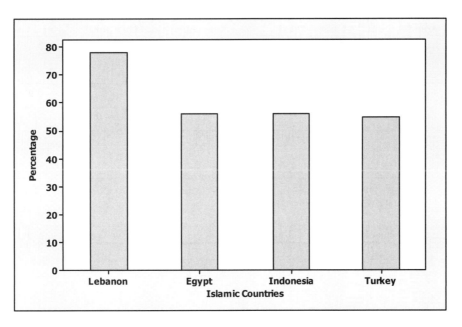

    d   Answers will vary.

1.15   **a**   Yes. The total percentage of education level in each bar graph is 100.
    b   Yes. There is a significant increase (from 39% to 46%) in the post secondary education attainment over the years.
    c   The pie chart is shown below. The bar chart is probably more interesting to look at.

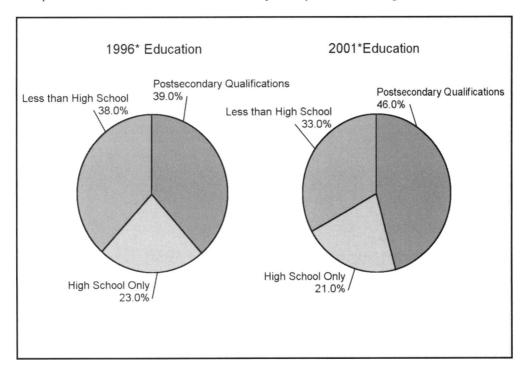

1.17   Refer to the table in Exercise 1.16. Answers will vary, depending on the student's choice of class width. If the class width is substantially larger than the minimum, the number of classes needed may be slightly fewer than specified in the table. Here is one possible solution.

Introduction to Probability and Statistics, 2ce

Number of measurements	Smallest and largest values	Convenient starting point	First Two Classes
75	0.5 to 1.0	0.5	0.5 to < 0.58
			0.58 to < 0.66
25	0 to 100	0	0 to < 20
			20 to < 40
200	1200 to 1500	1200	1200 to < 1235
			1235 to < 1270

1.19 **a** For $n = 5$, use between 8 and 10 classes.

b

Class i	Class Boundaries	Tally	f_i	Relative frequency, f_i/n
1	1.6 to < 2.1	11	2	.04
2	2.1 to < 2.6	11111	5	.10
3	2.6 to < 3.1	11111	5	.10
4	3.1 to < 3.6	11111	5	.10
5	3.6 to < 4.1	11111 11111 1111	14	.28
6	4.1 to < 4.6	11111 11	7	.14
7	4.6 to < 5.1	11111	5	.10
8	5.1 to < 5.6	11	2	.04
9	5.6 to < 6.1	111	3	.06
10	6.1 to < 6.6	11	2	.04

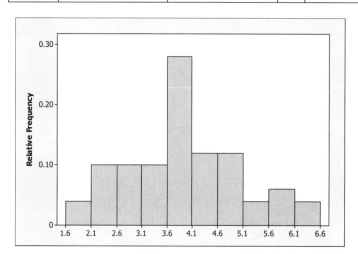

c From **b**, the fraction less than 5.1 is that fraction lying in classes 1-7, or
$$(2 + 5 + \cdots + 7 + 5)/50 = 43/50 = 0.86$$

d From **b**, the fraction larger than 3.6 lies in classes 5-10, or.
$$(14 + 7 + \cdots + 3 + 2)/50 = 33/50 = 0.66$$

e The stem and leaf display has a more peaked mound-shaped distribution than the relative frequency histogram because of the smaller number of groups.

1.21 **a** Since the variable of interest can only take the values 0, 1, or 2, the classes can be chosen as the integer values 0, 1, and 2. The table below shows the classes, their corresponding frequencies and their relative frequencies. The relative frequency histogram is shown on the next page.

Value	Frequency	Relative Frequency
0	5	.25
1	9	.45
2	6	.30

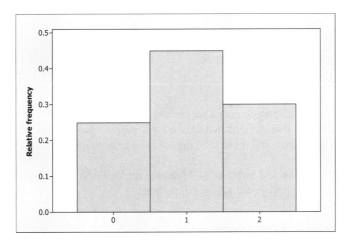

b Using the table in part **a**, the proportion of measurements greater then 1 is the same as the proportion of "2"s, or 0.30.

c The proportion of measurements less than 2 is the same as the proportion of "0"s and "1"s, or $0.25 + 0.45 = .70$.

d The probability of selecting a "2" in a random selection from these twenty measurements is $6/20 = 30$.

e There are no outliers in this relatively symmetric, mound-shaped distribution.

1.23 The line chart plots "day" on the horizontal axis and "time" on the vertical axis. The line chart shown below reveals that learning is taking place, since the time decreases each successive day.

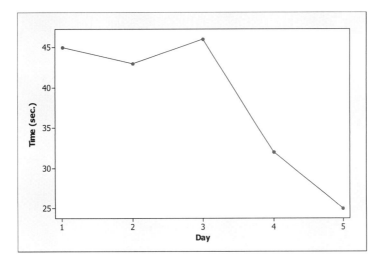

1.25 a The test scores are graphed using a stem and leaf plot generated by *Minitab*.

Stem-and-Leaf Display: Scores
```
Stem-and-leaf of Scores   N = 20
Leaf Unit = 1.0

    2    5   57
    5    6   123
    8    6   578
    9    7   2
   (2)   7   56
    9    8   24
    7    8   6679
    3    9   134
```

      b-c   The distribution is not mound-shaped, but is rather bimodal with two peaks centred around the scores 65 and 85. This might indicate that the students are divided into two groups – those who understand the material and do well on exams, and those who do not have a thorough command of the material.

1.27   **a**   The data represent the average annual incomes of Albertans divided into five educational categories. A bar chart would be the most appropriate graphical method.

       b   The bar chart is shown below.

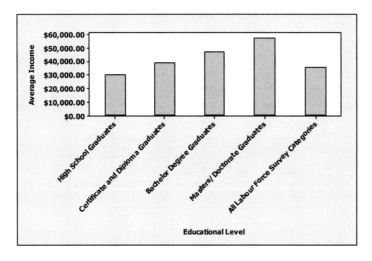

       c   The average salary for Albertans residents increases substantially as the person's educational level increases.

1.29   **a**   Similar to previous exercises. The pie chart is shown below.

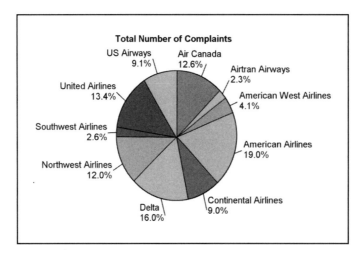

b The Pareto chart is a bar chart with the heights of the bars ordered from large to small. This display is more effective than the pie chart.

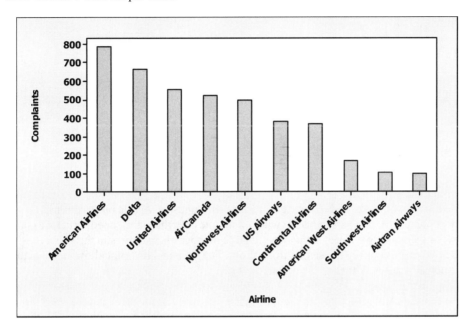

c The larger the airline, the more difficult it may be to serve passengers without any complaints. Other variable might be "size of airline" or "number of passengers served" or "airfare".

1.31 a The data ranges from .2 to 5.2, or 5.0 units. Since the number of class intervals should be between five and twenty, we choose to use eleven class intervals, with each class interval having length 0.50 ($5.0/11 = .45$, which, rounded to the nearest convenient fraction, is .50). We must now select interval boundaries such that no measurement can fall on a boundary point. The subintervals .1 to < .6, .6 to < 1.1, and so on, are convenient and a tally is constructed.

Class i	Class Boundaries	Tally	f_i	Relative frequency, f_i/n
1	0.1 to < 0.6	11111 11111	10	.167
2	0.6 to < 1.1	11111 11111 11111	15	.250
3	1.1 to < 1.6	11111 11111 11111	15	.250
4	1.6 to < 2.1	11111 11111	10	.167
5	2.1 to < 2.6	1111	4	.067
6	2.6 to < 3.1	1	1	.017
7	3.1 to < 3.6	11	2	.033
8	3.6 to < 4.1	1	1	.017
9	4.1 to < 4.6	1	1	.017
10	4.6 to < 5.1		0	.000
11	5.1 to < 5.6	1	1	.017

The relative frequency histogram is shown on the next page.

Introduction to Probability and Statistics, 2ce

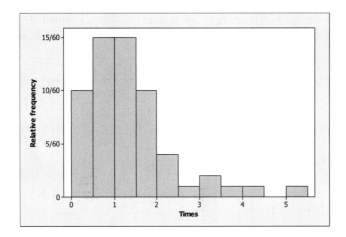

- **a** The distribution is skewed to the right, with several unusually large observations.
- **b** For some reason, one person had to wait 5.2 minutes. Perhaps the supermarket was understaffed that day, or there may have been an unusually large number of customers in the store.
- **c** The two graphs convey the same information. The stem and leaf plot allows us to actually recreate the actual data set, while the histogram does not.

1.33
- **b** The stem and leaf plot is constructed using the tens place as the stem and the ones place as the leaf. *Minitab* divides each stem into two parts to create a better descriptive picture. Notice that the distribution is roughly mound-shaped.

    Stem-and-Leaf Display: Age
    ```
    Stem-and-leaf of Age   N  = 22
    Leaf Unit = 1.0

       1    3   9
       1    4
       8    4   5666778
      11    5   124
      11    5   579
       8    6   01
       6    6   5669
       2    7   04
    ```

- **c** All five youngest prime ministers – Clark, Mulroney, Harper, Campbell and Meighen – are conservative. Although this does not tell us much, we can guess that the trend of transferring the leadership to younger generations in Conservative party is faster.

1.35
- **a** Histograms will vary from student to student. A typical histogram, generated by *Minitab* is shown on the next page.

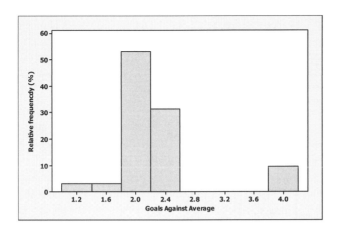

     b   Since all 20 players have averages above 0.400, the chance is 20 out of 20 or $20/20 = 1 = 100\%$.

1.37   **a**   The variable being measured is a discrete variable – the number of contaminated waste sites in each province or territory in Canada.

     b   **Stem-and-Leaf Display: Number of Contaminated Waste Sites**

```
Stem-and-leaf of Sites   N  = 13
Leaf Unit = 1.0

   1    1   7
   2    2   1
   3    3   4
   5    4   44
  (3)   5   567
   5    6
   5    7   88

HI 205, 235, 300
```

         The distribution is skewed to the right, with three unusually high numbers of contaminated waste sites (Ontario, British Columbia and Quebec).

     c   In comparison to other provinces, the three largest provinces – Quebec, Ontario and British Columbia have very large numbers of contaminated sites. However, despite their large land sizes, Nunavut and Northwest territory have relatively fewer contaminated sites. The pattern is not very clear. Some other variables like – population size, number of industries, etc. can help to explain the relationship better.

1.39   To determine whether a distribution is likely to be skewed, look for the likelihood of observing extremely large or extremely small values of the variable of interest.

     a   The distribution of non-secured loan sizes might be skewed (a few extremely large loans are possible).

     b   The distribution of secured loan sizes is not likely to contain unusually large or small values.

     c   Not likely to be skewed.

     d   Not likely to be skewed.

     e   If a package is dropped, it is likely that all the shells will be broken. Hence, a few large number of broken shells is possible. The distribution will be skewed.

     f   If an animal has one tick, he is likely to have more than one. There will be some "0"s with uninfected rabbits, and then a larger number of large values. The distribution will not be symmetric.

1.41   **a**   Weight is continuous, taking any positive real value.

     b   Body temperature is continuous, taking any real value.

     c   Number of people is discrete, taking the values 0, 1, 2, …

     d   Number of properties is discrete.

     e   Number of claims is discrete.

1.43 a Stem and leaf displays may vary from student to student. The most obvious choice is to use the tens digit as the stem and the ones digit as the leaf.

```
 7 | 8 9
 8 | 0 1 7
 9 | 0 1 2 4 4 5 6 6 6 8 8
10 | 1 7 9
11 | 2
```

The display is fairly mound-shaped, with a large peak in the middle.

1.45 a-b Answers will vary from student to student. The students should notice that the distribution is skewed to the right with a few pennies being unusually old. A typical histogram is shown below.

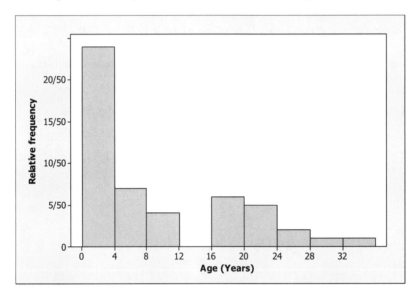

1.47 Answers will vary from student to student. The students should notice that the distribution is mound-shaped with the number of seats won in most of the elections being between 75 and 125. There is an unusually high number of seats won in two of the elections (208 and 211 seats).

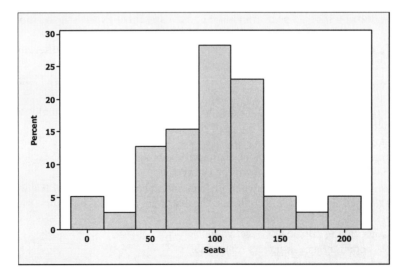

1.49 **a** The line chart is shown below. The year skated does not appear to have an effect on his winning time.

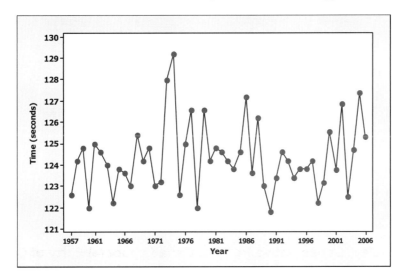

b Since the year of the race is not important in describing the data set, the distribution can be described using a relative frequency histogram. The distribution shown below is roughly mound-shaped with a few unusually slow ($x = 129.2, x = 128.0$) race times.

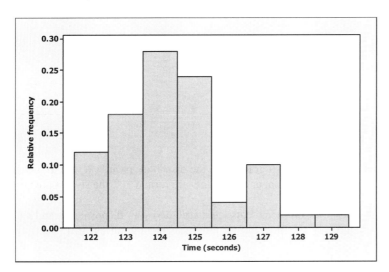

1.51 **a** Most of the provinces/territories have very few conservative seats (nine out of 13 have 10 or fewer seats won), the distribution should be skewed to the right.

b-c Histograms will vary from student to student, but should resemble the histogram generated by **Minitab** in the figure on the next page. The distribution is indeed skewed to the right, with one outlier – Ontario ($x = 40$).

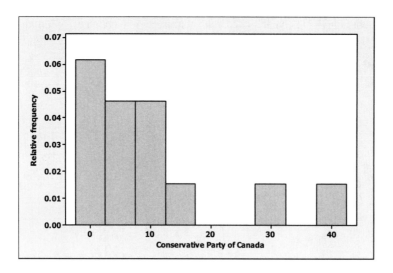

1.53 **Stem-and-Leaf Display: Conservative Party of Canada, Liberal Party of Canada**

```
Stem-and-leaf of                          Stem-and-leaf of
Conservative Party of Canada   N = 13     Liberal Party of Canada   N = 13
Leaf Unit = 1.0                           Leaf Unit = 1.0

  (7)   0   0000333                         (8)   0   00112344
   6    0   8                                5    0   669
   5    1   02                               2    1   3
   3    1   7                                1    1
   2    2                                    1    2
   2    2   8                                1    2
   1    3                                    1    3
   1    3                                    1    3
   1    4   0                                1    4
                                             1    4
  HI   40                                    1    5   4

                                            HI   54
```

a-b As in the case of relative frequency histogram, both the distributions are skewed to the right with one outlier on each (Ontario). When the stem and leaf plots are turned 90°, the shapes are very similar to the histograms.

c Since the total of 308 House of Commons seats are distributed very disproportionately among different provinces and territories, with only four provinces having more than 15 seats, these graphs will be skewed right.

1.55 **a-b-c** Answers will vary. The line chart should look similar to the one shown below.

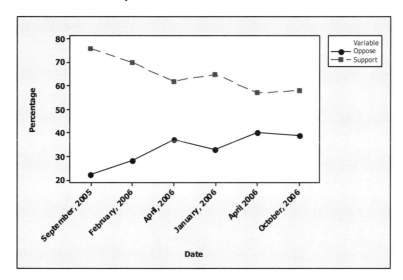

d The horizontal axis on the EKOS chart is not an actual time line, so that the time frame in which these changes occur may be distorted.

1.57 **a-b** Similar to previous exercises. The percentages add up to 100%, and the pie chart is shown below.

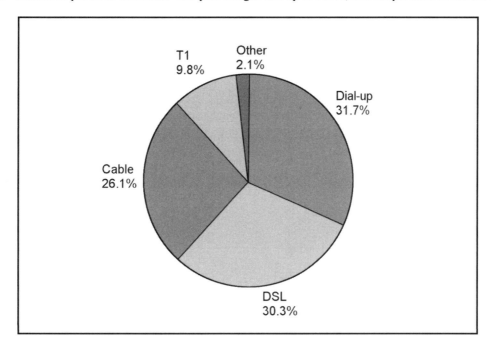

1.59 **a-b** Answers will vary. A typical histogram is shown below. Notice the gaps and the bimodal nature of the histogram, probably due to the fact that the samples were collected at different locations.

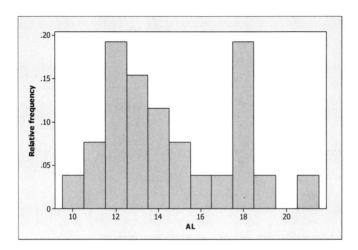

c The dotplot is shown below. The locations are indeed responsible for the unusual gaps and peaks in the relative frequency histogram.

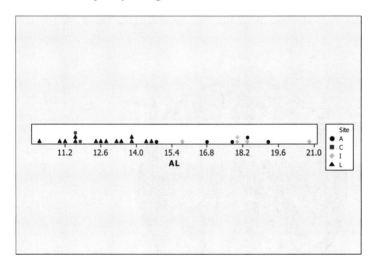

1.61 Answers will vary from student to student. Students should notice that the first distribution (12:00 – 1:30) is mound-shaped and the distribution (4:30 – 6:00) is slightly skewed.

1.63 **a-b** The *Minitab* stem and leaf plot is shown below. The distribution is slightly skewed to the left.

Stem-and-Leaf Display: Total Tax Component
```
Stem-and-leaf of Total Tax Component   N  = 15
Leaf Unit = 1.0

    1    2  2
    2    2  4
    4    2  66
    4    2
    7    3  000
   (1)   3  3
```

```
                    7   3
                    7   3   667
                    4   3   8899
```

    c   There are no unusually high or low gasoline taxes in the data.

1.65   The data should be displayed with either a bar chart or a pie chart. Because of the large number of categories, the bar chart is probably more effective.

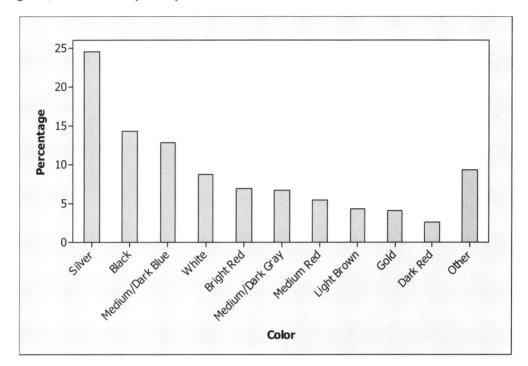

1.67

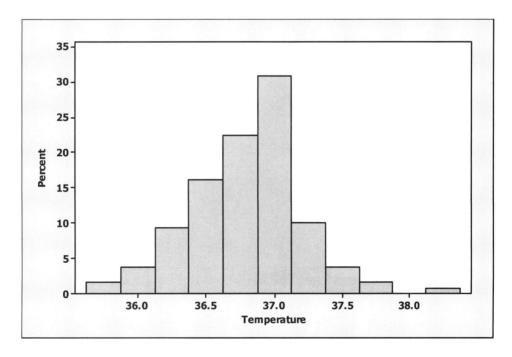

Introduction to Probability and Statistics, 2ce

a-b The distribution is approximately mound-shaped, with one unusual measurement, in the class with midpoint at 38.5° ($x = 38.22$). Perhaps the person whose temperature was 38.22 has some sort of illness.

c The value 37° is slightly to the right of centre.

1.69

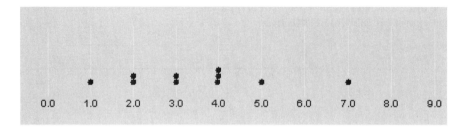

a The distribution is somewhat mound-shaped (as much as a small set can be); there are no outliers.

b $2/10 = 0.2$

1.71

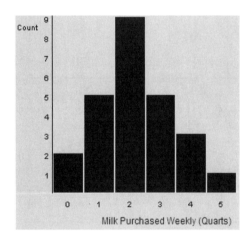

1.73 **a** There are a few extremely large numbers, indicating that the distribution is probably skewed to the right.

b-c The distribution is indeed skewed right with three possible outliers – Yahoo!, Time Warner and MSN-Microsoft.

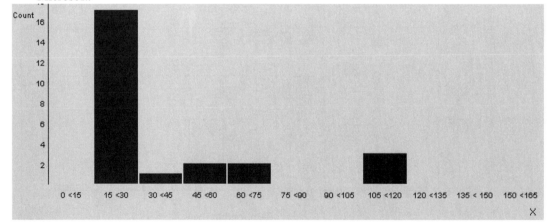

Case Study:
How is Your Blood Pressure?

1. The following variables have been measured on the participants in this study: sex (qualitative); age in years (quantitative discrete); diastolic blood pressure (quantitative continuous, but measured to an integer value) and systolic blood pressure (quantitative continuous, but measured to an integer value). For each person, both systolic and diastolic readings are taken, making the data bivariate.

2. The important variables in this study are diastolic and systolic blood pressure, which can be described singly with histograms in various categories (male vs. female or by age categories). Further, the relationship between systolic and diastolic blood pressure can be displayed together using a scatterplot or a bivariate histogram.

3. Answers will vary from student to student, depending on the choice of class boundaries or the software package which is used. The histograms should look fairly mound-shaped. A typical side by side histogram, generated by *Minitab* is shown below.

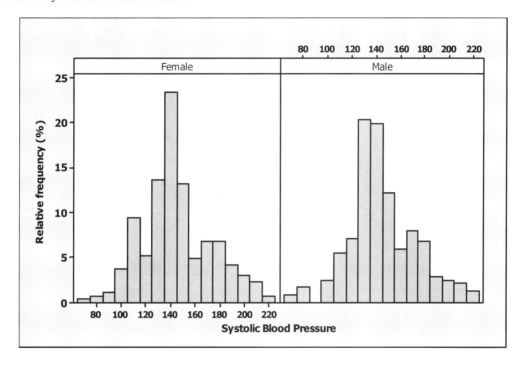

4. Answers will vary from student to student.

5. In determining how a student's blood pressure compares to those in a comparable sex and age group, female students (ages 15-20) must compare to the population of females, while male students (ages 15-20) must compare to the population of males. The student should use his or her blood pressure and compare it to the scatterplot generated in part 4.

Introduction to Probability and Statistics, 2ce

Project 1-A: Five Tips for Keeping Your Home Safe This Summer

 a The population is all households in that particular subdivision in the city of North York. The sample is the 300 households in that subdivision that were surveyed.

 b The collected data is based on population, in the sense that the sample taken is randomly selected from the population and meant to be representative of the population.

 c The experimental units are the households.

 d The variable being measured is the type of tip employed.

 e The variable is qualitative.

 f Neither. The variable is qualitative. The counts, however, are discrete.

 g The bar chart is shown below. The chart graphically portrays the counts for each type of tip.

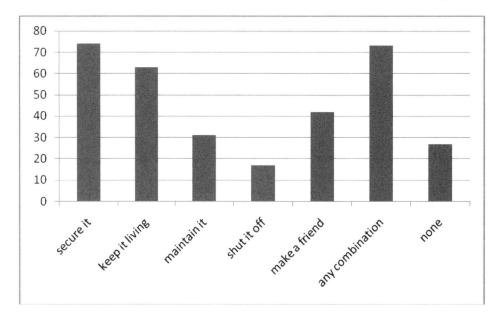

 h The relative frequencies are obtained by dividing the counts by the total sample size (300):

Type of Tips	Number of Households	Relative Frequency
secure it	74	0.246667
keep it living	63	0.210000
maintain it	31	0.103333
shut it off	17	0.056667
make a friend	42	0.140000
any combination	73	0.243333
none	27	0.090000

i A relative frequency bar chart is shown below.

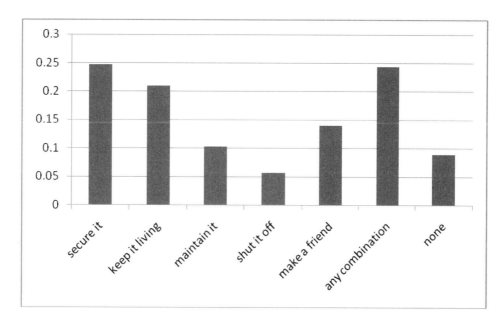

j A pie chart (by count) is shown below.

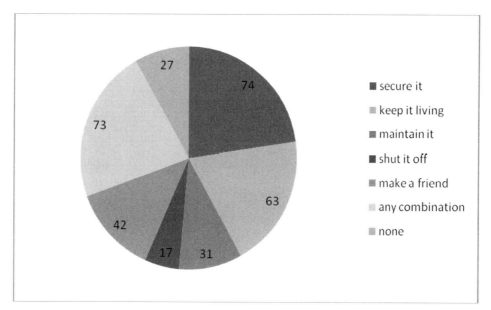

k The proportion of respondents who chose either "Secure it" or "Make a friend" is the sum of their respective relative frequencies: $0.246667 + 0.140000 = 0.386667$ or 38.67%.

Project 1-B: Handwashing Saves Lives: It's In Your Hands

a   The experimental units are the students.

b   The variable is the time (in seconds) students take to wash their hands.

c   The variable is quantitative.

d   The variable is discrete because it is only measured to the nearest second.

e   A dot plot is shown below. The value that occurs most often is 5. The range of the data is from 0 to 20. The data is distributed more to the lower values. There are a few gaps in the data as well.

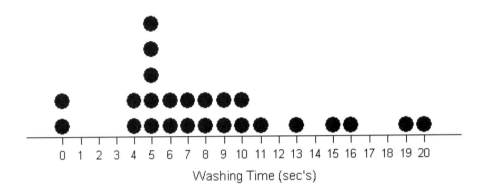

f   The distribution of data is skewed to the lower values, as most students took 10 seconds or less to wash their hands.

g The line chart was constructed by using the student's numerical order (student 1, 2,...25) as the *x*-variable and the amount of time they washed their hands as the *y*-variable.

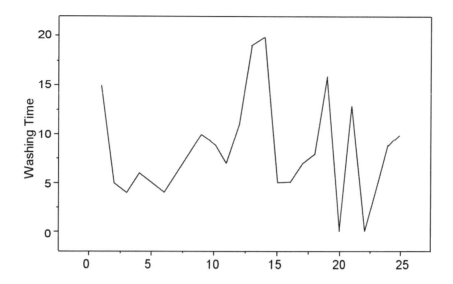

h The frequency histogram is:

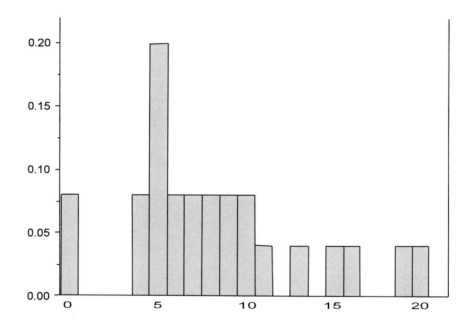

i We count 17 students that washed their hands for less than 10 seconds, or 17/25 = 68%.

j We count 21 students that washed their hands for at least 5 seconds, or 21/25 = 84%.

k No, we cannot comfortably state that most students wash their hands for 5 seconds or less. This is because only 9 students out of 25 (36%) wash their hands for 5 seconds or less. This leaves the majority of students washing their hands for more than 5 seconds.

l The stem and leaf plot is below. Note that the colon represents the decimal point in this case.

$$
\begin{aligned}
&0:00\\
&1:\\
&2:\\
&3:\\
&4:00\\
&5:00000\\
&6:00\\
&7:00\\
&8:00\\
&9:00\\
&10:00\\
&11:0\\
&12:\\
&13:0\\
&14:\\
&15:0\\
&16:0\\
&17:\\
&18:\\
&19:0\\
&20:0
\end{aligned}
$$

m The data is skewed right, as the tail of the distribution is on the right (i.e. the higher values).

n Points 0, 19 and 20 appear to be potential outliers.

Chapter 2: Describing Data with Numerical Measures

2.1   a   The dotplot shown below plots the five measurements along the horizontal axis. Since there are two "1"s, the corresponding dots are placed one above the other. The approximate centre of the data appears to be around 1.

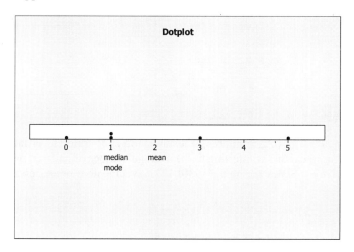

     b   The mean is the sum of the measurements divided by the number of measurements, or
$$\bar{x} = \frac{\sum x_i}{n} = \frac{0+5+1+1+3}{5} = \frac{10}{5} = 2$$
To calculate the median, the observations are first ranked from smallest to largest: 0, 1, 1, 3, 5. Then since $n = 5$, the position of the median is $0.5(n+1) = 3$, and the median is the 3rd ranked measurement, or $m = 1$. The mode is the measurement occurring most frequently, or mode = 1.

     c   The three measures in part **b** are located on the dotplot. Since the median and mode are to the left of the mean, we conclude that the measurements are skewed to the right.

2.3   a $\bar{x} = \frac{\sum x_i}{n} = \frac{58}{10} = 5.8$

     b   The ranked observations are: 2, 3, 4, 5, 5, 6, 6, 8, 9, 10. Since $n = 10$, the median is halfway between the 5th and 6th ordered observations, or $m = (5+6)/2 = 5.5$.

     c   There are two measurements, 5 and 6, which both occur twice. Since this is the highest frequency of occurrence for the data set, we say that the set is *bimodal* with modes at 5 and 6.

2.5   a   Although there may be a few households who own more than one DVD player, the majority should own either 0 or 1. The distribution should be slightly skewed to the right.

     b   Since most households will have only one DVD player, we guess that the mode is 1.

     c   The mean is
$$\bar{x} = \frac{\sum x_i}{n} = \frac{1+0+\cdots+1}{25} = \frac{27}{25} = 1.08$$
To calculate the median, the observations are first ranked from smallest to largest: There are six 0s, thirteen 1s, four 2s, and two 3s. Then since $n = 25$, the position of the median is $0.5(n+1) = 13$, which is the 13th ranked measurement, or $m = 1$. The mode is the measurement occurring most frequently, or mode = 1.

     d   The relative frequency histogram is shown on the next page, with the three measures superimposed. Notice that the mean falls slightly to the right of the median and mode, indicating that the measurements are slightly skewed to the right.

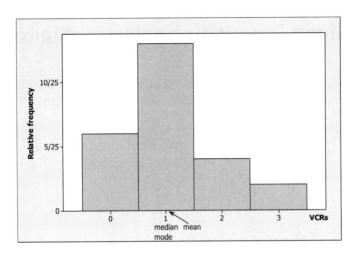

2.7 It is obvious that any one family cannot have 2.5 children, since the number of children per family is a quantitative discrete variable. The researcher is referring to the average number of children per family calculated for all families in the United States during the 1930s. The average does not necessarily have to be integer-valued.

2.9 The distribution of sports salaries will be skewed to the right, because of the very high salaries of some sports figures. Hence, the median salary would be a better measure of centre than the mean.

2.11 **a** Similar to previous exercises.
$$\bar{x} = \frac{\sum x_i}{n} = \frac{417}{18} = 23.17$$
The ranked observations are:

| 4 | 6 | 7 | 10 | 12 | 16 | 19 | 19 | 20 |
| 20 | 21 | 22 | 23 | 34 | 39 | 40 | 40 | 65 |

The median is the average of the 9th and 10th observations or $m = (20 + 20)/2 = 20$ and the mode is the most frequently occurring observation—mode = 19, 20, 40.

b Since the mean is larger than the median, the data are skewed to the right.

c The dotplot is shown below. Yes, the distribution is skewed to the right.

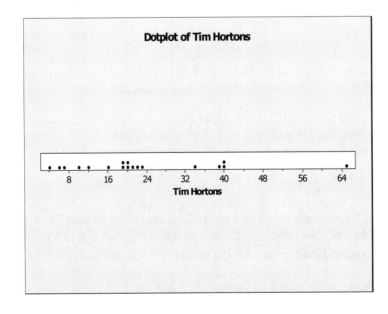

2.13 a   The sample mean is $\bar{x} = \frac{8+6+3+\cdots+8+10}{17} = 6.412$.

b   We arrange the data in increasing order: {0,1,2,2,3,4,5,5,6,6,7,8,8,8,10,11,23}. The median is the number in the (n+1)/2 = 18/2 = 9th position, which is 6. The mode is the number which occurs most often, which in this case is 8.

c   Since the mean and median are only slightly off from each other, there is only slight skewness.

d   The use of the median is probably better than the mean in this case as the point with value 23 is likely an outlier (and the median is less sensitive to outliers). The mode is rarely employed in practice.

2.15 a   The range is $R = 4 - 1 = 3$.

b $\bar{x} = \frac{\sum x_i}{n} = \frac{17}{8} = 2.125$

c   Calculate $\sum x_i^2 = 4^2 + 1^2 + \cdots + 2^2 = 45$. Then

$$s^2 = \frac{\sum x_i^2 - \frac{(\sum x_i)^2}{n}}{n-1} = \frac{45 - \frac{(17)^2}{8}}{7} = \frac{8.875}{7} = 1.2679 \text{ and } s = \sqrt{s^2} = \sqrt{1.2679} = 1.126.$$

2.17 a   The range is $R = 2.39 - 1.28 = 1.11$.

b   Calculate $\sum x_i^2 = 1.28^2 + 2.39^2 + \cdots + 1.51^2 = 15.415$. Then

$$s^2 = \frac{\sum x_i^2 - \frac{(\sum x_i)^2}{n}}{n-1} = \frac{15.451 - \frac{(8.56)^2}{5}}{4} = \frac{.76028}{4} = .19007 \text{ and } s = \sqrt{s^2} = \sqrt{.19007} = .436$$

c   The range, $R = 1.11$, is $1.11/.436 = 2.5$ standard deviations.

2.19 a   The range of the data is $R = 6 - 1 = 5$ and the range approximation with $n = 10$ is $s \approx \frac{R}{3} = 1.67$

b   The standard deviation of the sample is

$$s = \sqrt{s^2} = \sqrt{\frac{\sum x_i^2 - \frac{(\sum x_i)^2}{n}}{n-1}} = \sqrt{\frac{130 - \frac{(32)^2}{10}}{9}} = \sqrt{3.0667} = 1.751$$

which is very close to the estimate for part **a**.

c-e   From the dotplot on the next page, you can see that the data set is not mound-shaped. Hence you can use Tchebysheff's Theorem, but not the Empirical Rule to describe the data.

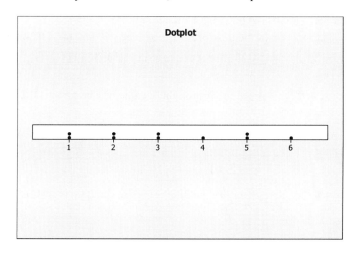

2.21 **a**   The interval from 40 to 60 represents $\mu \pm \sigma = 50 \pm 10$. Since the distribution is relatively mound-shaped, the proportion of measurements between 40 and 60 is 68% according to the Empirical Rule and is shown below.

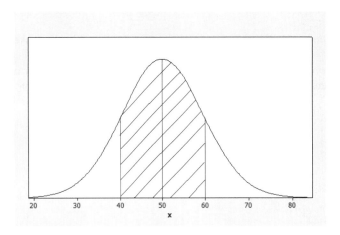

b   Again, using the Empirical Rule, the interval $\mu \pm 2\sigma = 50 \pm 2(10)$ or between 30 and 70 contains approximately 95% of the measurements.

c   Refer to the figure below.

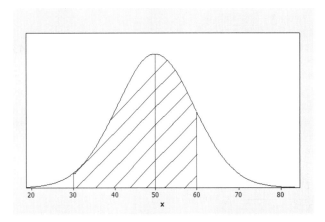

Since approximately 68% of the measurements are between 40 and 60, the symmetry of the distribution implies that 34% of the measurements are between 50 and 60. Similarly, since 95% of the measurements are between 30 and 70, approximately 47.5% are between 30 and 50. Thus, the proportion of measurements between 30 and 60 is $0.34 + 0.475 = 0.815$

d   From the figure in part **a**, the proportion of the measurements between 50 and 60 is 0.34 and the proportion of the measurements which are greater than 50 is 0.50. Therefore, the proportion that are greater than 60 must be $0.5 - 0.34 = 0.16$

2.23 **a**   The range of the data is $R = 1.1 - 0.5 = 0.6$ and the approximate value of s is $s \approx \dfrac{R}{3} = 0.2$

b   Calculate $\sum x_i = 7.6$ and $\sum x_i^2 = 6.02$, the sample mean is $\bar{x} = \dfrac{\sum x_i}{n} = \dfrac{7.6}{10} = .76$ and the standard deviation of the sample is $s = \sqrt{s^2} = \sqrt{\dfrac{\sum x_i^2 - \dfrac{(\sum x_i)^2}{n}}{n-1}} = \sqrt{\dfrac{6.02 - \dfrac{(7.6)^2}{10}}{9}} = \sqrt{\dfrac{0.244}{9}} = 0.165$

which is very close to the estimate from part **a**.

2.25 According to the Empirical Rule, if a distribution of measurements is approximately mound-shaped,

    a approximately 68% or 0.68 of the measurements fall in the interval $\mu \pm \sigma = 12 \pm 2.3$ or 9.7 to 14.3

    b approximately 95% or 0.95 of the measurements fall in the interval $\mu \pm 2\sigma = 12 \pm 4.6$ or 7.4 to 16.6

    c approximately 99.7% or 0.997 of the measurements fall in the interval $\mu \pm 3\sigma = 12 \pm 6.9$ or 5.1 to 18.9
Therefore, approximately 0.3% or 0.003 will fall outside this interval.

2.27
    a The centre of the distribution should be approximately halfway between 0 and 9 or $(0+9)/2 = 4.5$.

    b The range of the data is $R = 9 - 0 = 9$. Using the range approximation, $s \approx R/4 = 9/4 = 2.25$.

    c Using the data entry method the students should find $\bar{x} = 4.586$ and $s = 2.892$, which are fairly close to our approximations.

2.29
    a Although most of the animals will die at around 32 days, there may be a few animals that survive a very long time, even with the infection. The distribution will probably be skewed right.

    b Using Tchebysheff's Theorem, at least 3/4 of the measurements should be in the interval $\mu \pm \sigma \Rightarrow 32 \pm 72$ or 0 to 104 days.

2.31
    a We choose to use 12 classes of length 1.0. The tally and the relative frequency histogram follow.

Class i	Class Boundaries	Tally	f_i	Relative frequency, f_i/n
1	2 to < 3	1	1	1/70
2	3 to < 4	1	1	1/70
3	4 to < 5	111	3	3/70
4	5 to < 6	11111	5	5/70
5	6 to < 7	11111	5	5/70
6	7 to < 8	11111 11111 11	12	12/70
7	8 to < 9	11111 11111 11111 111	18	18/70
8	9 to < 10	11111 11111 11111	15	15/70
9	10 to < 11	11111 1	6	6/70
10	11 to < 12	111	3	3/70
11	12 to < 13		0	0
12	13 to < 14	1	1	1/70

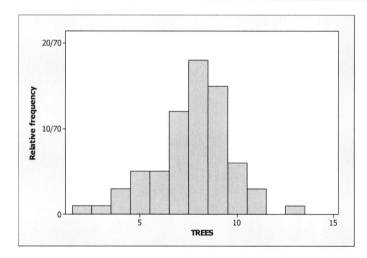

    b Calculate $n = 70$, $\sum x_i = 541$ and $\sum x_i^2 = 4453$. Then $\bar{x} = \dfrac{\sum x_i}{n} = \dfrac{541}{70} = 7.729$ is an estimate of μ.

Introduction to Probability and Statistics, 2ce

c The sample standard deviation is

$$s = \sqrt{\frac{\sum x_i^2 - \frac{(\sum x_i)^2}{n}}{n-1}} = \sqrt{\frac{4453 - \frac{(541)^2}{70}}{69}} = \sqrt{3.9398} = 1.985$$

The three intervals, $\bar{x} \pm ks$ for $k = 1, 2, 3$ are calculated below. The table shows the actual percentage of measurements falling in a particular interval as well as the percentage predicted by Tchebysheff's Theorem and the Empirical Rule. Note that the Empirical Rule should be fairly accurate, as indicated by the mound-shape of the histogram in part **a**.

k	$\bar{x} \pm ks$	Interval	Fraction in Interval	Tchebysheff	Empirical Rule
1	7.729 ± 1.985	5.744 to 9.714	50/70 = 0.71	at least 0	≈ 0.68
2	7.729 ± 3.970	3.759 to 11.699	67/70 = 0.96	at least 0.75	≈ 0.95
3	7.729 ± 5.955	1.774 to 13.684	70/70 = 1.00	at least 0.89	≈ 0.997

2.33 **a-b** Calculate $R = 93 - 51 = 42$ so that $s \approx R/4 = 42/4 = 10.5$.

c Calculate $n = 30$, $\sum x_i = 2145$ and $\sum x_i^2 = 158,345$. Then

$$s^2 = \frac{\sum x_i^2 - \frac{(\sum x_i)^2}{n}}{n-1} = \frac{158,345 - \frac{(2145)^2}{30}}{29} = 171.6379 \text{ and } s = \sqrt{171.6379} = 13.101$$

which is fairly close to the approximate value of s from part **b**.

d The two intervals are calculated below. The proportions agree with Tchebysheff's Theorem, but are not to close to the percentages given by the Empirical Rule. (This is because the distribution is not quite mound-shaped.)

k	$\bar{x} \pm ks$	Interval	Fraction in Interval	Tchebysheff	Empirical Rule
2	71.5 ± 26.20	45.3 to 97.7	30/30 = 1.00	at least 0.75	≈ 0.95
3	71.5 ± 39.30	32.2 to 110.80	30/30 = 1.00	at least 0.89	≈ 0.997

2.35 **a** Answers will vary. A typical stem and leaf plot is generated by *Minitab*.
Stem-and-Leaf Display: Goals

```
Stem-and-leaf of Goals   N  = 21
Leaf Unit = 1.0

    1   0   9
    3   1   16
    6   2   335
    8   3   18
   (4)  4   0016
    9   5   1245
    5   6   2
    4   7   13
    2   8   7
    1   9   2
```

b Calculate $n = 21$, $\sum x_i = 940$ and $\sum x_i^2 = 53036$. Then $\bar{x} = \frac{\sum x_i}{n} = \frac{940}{21} = 44.76$,

$$s^2 = \frac{\sum x_i^2 - \frac{(\sum x_i)^2}{n}}{n-1} = \frac{53036 - \frac{(940)^2}{21}}{20} = 547.99 \text{ and } s = \sqrt{s^2} = \sqrt{547.99} = 23.41.$$

c Calculate $\bar{x} \pm 2s \Rightarrow 44.76 \pm 46.82$ or -2.06 to 91.58. From the original data set, 20 of the measurements, or 95.24% fall in this interval.

2.37 **a** Calculate $n=15$, $\sum x_i = 21$ and $\sum x_i^2 = 49$. Then $\bar{x} = \dfrac{\sum x_i}{n} = \dfrac{21}{15} = 1.4$ and

$$s^2 = \dfrac{\sum x_i^2 - \dfrac{(\sum x_i)^2}{n}}{n-1} = \dfrac{49 - \dfrac{(21)^2}{15}}{14} = 1.4$$

b Using the frequency table and the grouped formulas, calculate
$$\sum x_i f_i = 0(4) + 1(5) + 2(2) + 3(4) = 21$$
$$\sum x_i^2 f_i = 0^2(4) + 1^2(5) + 2^2(2) + 3^2(4) = 49$$

Then, as in part **a**,
$$\bar{x} = \dfrac{\sum x_i f_i}{n} = \dfrac{21}{15} = 1.4$$

$$s^2 = \dfrac{\sum x_i^2 f_i - \dfrac{(\sum x_i f_i)^2}{n}}{n-1} = \dfrac{49 - \dfrac{(21)^2}{15}}{14} = 1.4$$

2.39 **a** The data in this exercise have been arranged in a frequency table.

x_i	0	1	2	3	4	5	6	7	8	9	10
f_i	10	5	3	2	1	0	1	0	0	1	1

Using the frequency table and the grouped formulas, calculate
$$\sum x_i f_i = 0(10) + 1(5) + \cdots + 10(1) = 46$$
$$\sum x_i^2 f_i = 0^2(10) + 1^2(5) + \cdots + 10^2(1) = 268$$

Then
$$\bar{x} = \dfrac{\sum x_i f_i}{n} = \dfrac{46}{24} = 1.917$$

$$s^2 = \dfrac{\sum x_i^2 f_i - \dfrac{(\sum x_i f_i)^2}{n}}{n-1} = \dfrac{268 - \dfrac{(46)^2}{24}}{23} = 7.819 \text{ and } s = \sqrt{7.819} = 2.796.$$

b-c The three intervals $\bar{x} \pm ks$ for $k = 1, 2, 3$ are calculated in the table along with the actual proportion of measurements falling in the intervals. Tchebysheff's Theorem is satisfied and the approximation given by the Empirical Rule are fairly close for $k = 2$ and $k = 3$.

k	$\bar{x} \pm ks$	Interval	Fraction in Interval	Tchebysheff	Empirical Rule
1	1.917 ± 2.796	−0.879 to 4.713	21/24 = 0.875	at least 0	≈ 0.68
2	1.917 ± 5.592	−3.675 to 7.509	22/24 = 0.917	at least 0.75	≈ 0.95
3	1.917 ± 8.388	−6.471 to 10.305	24/24 = 1.00	at least 0.89	≈ 0.997

2.41 The data have already been sorted. Find the positions of the quartiles, and the measurements that are just above and below those positions. Then find the quartiles by interpolation.

Sorted Data Set	Position of Q_1	Above and below	Q_1	Position of Q_3	Above and below	Q_3
1, 1.5, 2, 2, 2.2	.25(6) = 1.5	1 and 1.5	1.25	.75(6) = 4.5	2 and 2.2	2.1
0, 1.7, 1.8, 3.1, 3.2, 7, 8, 8.8, 8.9, 9, 10	.25(12) = 3	None	1.8	.75(12) = 9	None	8.9
.23, .30, .35, .41, .56, .58, .76, .80	.25(9) = 2.25	.30 and .35	.30 + .25(.05) = .3125	.75(9) = 6.75	.58 and .76	.58 + .75(.18) = .7150

2.43 The ordered data are: 0, 1, 5, 6, 7, 8, 9, 10, 12, 12, 13, 14, 16, 19, 19
With $n = 15$, the median is in position $0.5(n+1) = 8$, so that $m = 10$. The lower quartile is in position $0.25(n+1) = 4$ so that $Q_1 = 6$ and the upper quartile is in position $0.75(n+1) = 12$ so that $Q_3 = 14$. Then the five-number summary is

Min	Q_1	Median	Q_3	Max
0	6	10	14	19

and $IQR = Q_3 - Q_1 = 14 - 6 = 8$.

2.45 The ordered data are: 2, 3, 4, 5, 6, 6, 6, 7, 8, 9, 9, 10, 22
For $n = 13$, the position of the median is $0.5(n+1) = 0.5(13+1) = 7$ and $m = 6$. The positions of the quartiles are $0.25(n+1) = 3.5$ and $0.75(n+1) = 10.5$, so that $Q_1 = 4.5$, $Q_3 = 9$, and $IQR = 9 - 4.5 = 4.5$. The *lower and upper fences* are:
$$Q_1 - 1.5IQR = 4.5 - 6.75 = -2.25$$
$$Q_3 + 1.5IQR = 9 + 6.75 = 15.75$$
The value $x = 22$ lies outside the upper fence and is an outlier. The box plot is shown below. The lower whisker connects the box to the smallest value that is not an outlier, which happens to be the minimum value, $x = 2$. The upper whisker connects the box to the largest value that is not an outlier or $x = 10$.

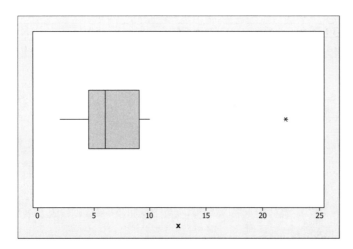

2.47   a   The ordered data are shown below:

1.70	101.00	209.00	264.00	316.00	445.00
1.72	118.00	218.00	278.00	318.00	481.00
5.90	168.00	221.00	286.00	329.00	485.00
8.80	180.00	241.00	314.00	397.00	
85.40	183.00	252.00	315.00	406.00	

For $n = 28$, the position of the median is $0.5(n+1) = 14.5$ and the positions of the quartiles are $0.25(n+1) = 7.25$ and $0.75(n+1) = 21.75$. The lower quartile is ¼ the way between the 7th and 8th measurements or $Q_1 = 118 + 0.25(168 - 118) = 130.5$ and the upper quartile is ¾ the way between the 21st and 22nd measurements or $Q_3 = 316 + 0.75(318 - 316) = 317.5$. Then the five-number summary is

Min	Q_1	Median	Q_3	Max
1.70	130.5	246.5	317.5	485

b   Calculate $IQR = Q_3 - Q_1 = 317.5 - 130.5 = 187$. Then the *lower and upper fences* are:
$$Q_1 - 1.5IQR = 130.5 - 280.5 = -150$$
$$Q_3 + 1.5IQR = 317.5 + 280.5 = 598$$

The box plot is shown below. Since there are no outliers, the whiskers connect the box to the minimum and maximum values in the ordered set.

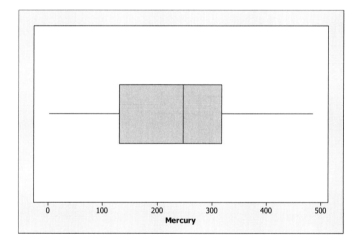

c-d   The boxplot does not identify any of the measurements as outliers, mainly because the large variation in the measurements cause the IQR to be large. However, the student should notice the extreme difference in the magnitude of the first four observations taken on young dolphins. These animals have not been alive long enough to accumulate a large amount of mercury in their bodies.

2.49 **a** For $n = 15$, the position of the median is $0.5(n+1) = 8$ and the positions of the quartiles are $0.25(n+1) = 4$ and $0.75(n+1) = 12$. The sorted measurements are shown below.

Lemieux: 1, 6, 7, 17, 19, 28, 35, 44, 45, 50, 54, 69, 69, 70, 85
Hull: 0, 1, 25, 29, 30, 32, 37, 39, 41, 42, 54, 57, 70, 72, 86

For Mario Lemieux, $m = 44$, $Q_1 = 17$ and $Q_3 = 69$.
For Brett Hull, $m = 39$, $Q_1 = 29$ and $Q_3 = 57$.

Then the five-number summaries are

	Min	Q_1	Median	Q_3	Max
Lemieux	1	17	44	69	85
Hull	0	29	39	57	86

b For Mario Lemieux, calculate $IQR = Q_3 - Q_1 = 69 - 17 = 52$. Then the *lower and upper fences* are:
$$Q_1 - 1.5 IQR = 17 - 78 = -61$$
$$Q_3 + 1.5 IQR = 69 + 78 = 147$$
For Brett Hull $IQR = Q_3 - Q_1 = 57 - 29 = 28$ Then the *lower and upper fences* are:
$$Q_1 - 1.5 IQR = 29 - 42 = -13$$
$$Q_3 + 1.5 IQR = 57 + 42 = 99$$

There are no outliers, and the box plots are shown below.

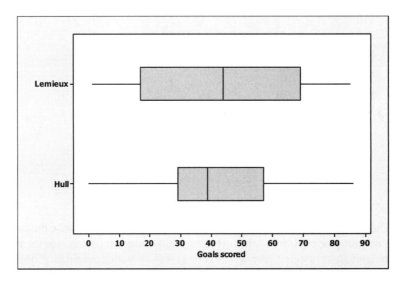

c Answers will vary. The Lemieux distribution is roughly symmetric, while the Hull distribution seems little skewed. The Lemieux distribution is slightly more variable; it has a higher IQR and a higher median number of goals scored.

2.51 **a** Just by scanning through the 25 measurements, it seems that there are a few unusually large measurements, which would indicate a distribution that is skewed to the right.

b The position of the median is $0.5(n+1) = 0.5(25+1) = 13$ and $m = 24.4$. The mean is

$$\bar{x} = \frac{\sum x_i}{n} = \frac{960}{25} = 38.4$$

which is larger than the median, indicate a distribution skewed to the right.

c The positions of the quartiles are $0.25(n+1) = 6.5$ and $0.75(n+1) = 19.5$, so that $Q_1 = 18.7$, $Q_3 = 48.9$, and $IQR = 48.9 - 18.7 = 30.2$. The *lower and upper fences* are:

$$Q_1 - 1.5 IQR = 18.7 - 45.3 = -26.6$$
$$Q_3 + 1.5 IQR = 48.9 + 45.3 = 94.2$$

The box plot is shown below. There are three outliers in the upper tail of the distribution, so the upper whisker is connected to the point $x = 69.2$. The long right whisker and the median line located to the left of the centre of the box indicates that the distribution that is skewed to the right.

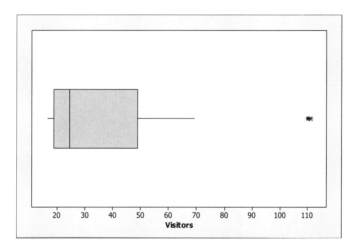

2.53 Answers will vary. The student should notice the outliers in the female group, that the median female temperature is higher than the median male temperature.

2.55 **a** Calculate $n = 14$, $\sum x_i = 367$ and $\sum x_i^2 = 9641$. Then $\bar{x} = \frac{\sum x_i}{n} = \frac{367}{14} = 26.214$ and

$$s = \sqrt{\frac{\sum x_i^2 - \frac{(\sum x_i)^2}{n}}{n-1}} = \sqrt{\frac{9641 - \frac{(367)^2}{14}}{13}} = 1.251$$

b Calculate $n = 14$, $\sum x_i = 366$ and $\sum x_i^2 = 9644$. Then $\bar{x} = \frac{\sum x_i}{n} = \frac{366}{14} = 26.143$ and

$$s = \sqrt{\frac{\sum x_i^2 - \frac{(\sum x_i)^2}{n}}{n-1}} = \sqrt{\frac{9644 - \frac{(366)^2}{14}}{13}} = 2.413$$

c The centres are roughly the same; the Sunmaid raisins appear slightly more variable.

2.57 **a** The largest observation found in the data from Exercise 1.26 is 32.3, while the smallest is 0.2. Therefore the range is $R = 32.3 - 0.2 = 32.1$.

b Using the range, the approximate value for s is: $s \approx R/4 = 32.1/4 = 8.025$.

c Calculate $n = 50$, $\sum x_i = 418.4$ and $\sum x_i^2 = 6384.34$. Then

$$s = \sqrt{\frac{\sum x_i^2 - \frac{(\sum x_i)^2}{n}}{n-1}} = \sqrt{\frac{6384.34 - \frac{(418.4)^2}{50}}{49}} = 7.671$$

2.59 The ordered data are shown below.

0.2	2.0	4.3	8.2	14.7
0.2	2.1	4.4	8.3	16.7
0.3	2.4	5.6	8.7	18.0
0.4	2.4	5.8	9.0	18.0
1.0	2.7	6.1	9.6	18.4
1.2	3.3	6.6	9.9	19.2
1.3	3.5	6.9	11.4	23.1
1.4	3.7	7.4	12.6	24.0
1.6	3.9	7.4	13.5	26.7
1.6	4.1	8.2	14.1	32.3

Since $n = 50$, the position of the median is $0.5(n+1) = 25.5$ and the positions of the lower and upper quartiles are $0.25(n+1) = 12.75$ and $0.75(n+1) = 38.25$.

Then $m = (6.1 + 6.6)/2 = 6.35$, $Q_1 = 2.1 + 0.75(2.4 - 2.1) = 2.325$ and $Q_3 = 12.6 + 0.25(13.5 - 12.6) = 12.825$. Then $IQR = 12.825 - 2.325 = 10.5$.

The *lower and upper fences* are:
$$Q_1 - 1.5 IQR = 2.325 - 15.75 = -13.425$$
$$Q_3 + 1.5 IQR = 12.825 + 15.75 = 28.575$$

and the box plot is shown below. There is one outlier, $x = 32.3$. The distribution is skewed to the right.

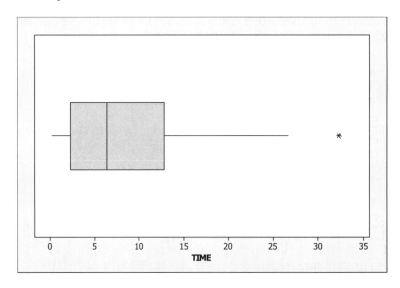

2.61 First calculate the intervals:

$\bar{x} \pm s = 0.17 \pm 0.01$ or   0.16 to 0.18
$\bar{x} \pm 2s = 0.17 \pm 0.02$ or   0.15 to 0.19
$\bar{x} \pm 3s = 0.17 \pm 0.03$ or   0.14 to 0.20

a If no prior information as to the shape of the distribution is available, we use Tchebysheff's Theorem. We would expect at least $(1 - 1/1^2) = 0$ of the measurements to fall in the interval 0.16 to 0.18; at least $(1 - 1/2^2) = 3/4$ of the measurements to fall in the interval 0.15 to 0.19; at least $(1 - 1/3^2) = 8/9$ of the measurements to fall in the interval 0.14 to 0.20.

b According to the Empirical Rule, approximately 68% of the measurements will fall in the interval 0.16 to 0.18; approximately 95% of the measurements will fall between 0.15 to 0.19; approximately 99.7% of the measurements will fall between 0.14 and 0.20. Since mound-shaped distributions are so frequent, if we do have a sample size of 30 or greater, we expect the sample distribution to be mound-shaped. Therefore, in this exercise, we would expect the Empirical Rule to be suitable for describing the set of data.

c If the chemist had used a sample size of four for this experiment, the distribution would not be mound-shaped. Any possible histogram we could construct would be non-mound-shaped. We can use at most 4 classes, each with frequency 1, and we will not obtain a histogram that is even close to mound-shaped. Therefore, the Empirical Rule would not be suitable for describing $n = 4$ measurements.

2.63 The following information is available:

$$n = 400, \bar{x} = 600, s^2 = 4900$$

The standard deviation of these scores is then 70, and the results of Tchebysheff's Theorem follow:

k	$\bar{x} \pm ks$	Interval	Tchebysheff
1	600 ± 70	530 to 670	at least 0
2	600 ± 140	460 to 740	at least 0.75
3	600 ± 210	390 to 810	at least 0.89

If the distribution of scores is mound-shaped, we use the Empirical Rule, and conclude that approximately 68% of the scores would lie in the interval 530 to 670 (which is $\bar{x} \pm s$). Approximately 95% of the scores would lie in the interval 460 to 740.

2.65 a Answers will vary. A typical histogram is shown below. The distribution is slightly skewed to the left.

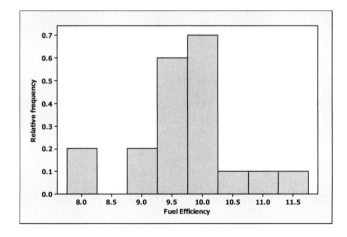

b Calculate $n = 20$, $\sum x_i = 193.1$, $\sum x_i^2 = 1876.65$. Then

$$\bar{x} = \frac{\sum x_i}{n} = 9.655$$

$$s = \sqrt{\frac{\sum x_i^2 - \frac{(\sum x_i)^2}{n}}{n-1}} = \sqrt{\frac{1876.65 - \frac{(193.1)^2}{20}}{19}} = \sqrt{.646} = 0.804$$

c The sorted data is shown below:

```
7.9    8.1    8.9    8.9    9.4    9.4    9.5    9.5    9.6    9.7
9.8    9.8    9.9    9.9    10.0   10.1   10.2   10.3   10.9   11.3
```

The z-scores for $x = 7.9$ and $x = 11.3$ are

$$z = \frac{x - \bar{x}}{s} = \frac{7.9 - 9.655}{.804} = -2.18 \text{ and } z = \frac{x - \bar{x}}{s} = \frac{11.3 - 9.655}{.804} = 2.05$$

Since neither of the z-scores are greater than 3 in absolute value, the measurements are not judged to be outliers.

d The position of the median is $0.5(n+1) = 10.5$ and the median is $m = (9.7 + 9.8)/2 = 9.75$

e The positions of the quartiles are $0.25(n+1) = 5.25$ and $0.75(n+1) = 15.75$. Then
$Q_1 = 9.4 + 0.25(9.4 - 9.4) = 9.4$ and $Q_3 = 10.0 + 0.75(10.1 - 10.0) = 10.075$.

2.67 a The range is $R = 71 - 40 = 31$ and the range approximation is $s \approx R/4 = 31/4 = 7.75$

b Calculate $n = 10$, $\sum x_i = 592$, $\sum x_i^2 = 36014$. Then

$$\bar{x} = \frac{\sum x_i}{n} = \frac{592}{10} = 59.2$$

$$s = \sqrt{\frac{\sum x_i^2 - \frac{(\sum x_i)^2}{n}}{n-1}} = \sqrt{\frac{36014 - \frac{(592)^2}{10}}{9}} = \sqrt{107.5111} = 10.369$$

The sample standard deviation calculated above is of the same order as the approximated value found in part **a**.

c The ordered set is: 40, 49, 52, 54, 59, 61, 67, 69, 70, 71
Since $n = 10$, the positions of m, Q_1, and Q_3 are 5.5, 2.75 and 8.25 respectively, and
$m = (59 + 61)/2 = 60$, $Q_1 = 49 + 0.75(52 - 49) = 51.25$, $Q_3 = 69.25$ and $IQR = 69.25 - 51.25 = 18.0$.
The *lower and upper fences* are:

$$Q_1 - 1.5 IQR = 51.25 - 27.00 = 24.25$$
$$Q_3 + 1.5 IQR = 69.25 + 27.00 = 96.25$$

and the box plot is shown below. There are no outliers and the data set is slightly skewed left.

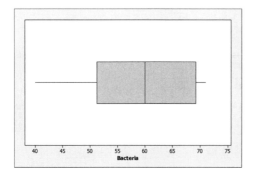

2.69 If the distribution is mound-shaped with mean μ, then almost all of the measurements will fall in the interval $\mu \pm 3\sigma$, which is an interval 6σ in length. That is, the range of the measurements should be approximately 6σ. In this case, the range is $800 - 200 = 600$, so that $\sigma \approx 600/6 = 100$.

2.71 a The range is $R = 172 - 108 = 64$ and the range approximation is $s \approx R/4 = 64/4 = 16$

   b Calculate $n = 15$, $\sum x_i = 2041, \sum x_i^2 = 281,807$. Then

$$\bar{x} = \frac{\sum x_i}{n} = \frac{2041}{15} = 136.07$$

$$s = \sqrt{\frac{\sum x_i^2 - \frac{(\sum x_i)^2}{n}}{n-1}} = \sqrt{\frac{281,807 - \frac{(2041)^2}{15}}{14}} = \sqrt{292.495238} = 17.102$$

   c According to Tchebysheff's Theorem, with $k = 2$, at least 3/4 or 75% of the measurements will lie within $k = 2$ standard deviations of the mean. For this data, the two values, a and b, are calculated as

$$\bar{x} \pm 2s \Rightarrow 136.07 \pm 2(17.10) \Rightarrow 136.07 \pm 34.20 \text{ or } a = 101.87 \text{ and } b = 170.27.$$

2.73 a The range is $R = 19 - 4 = 15$ and the range approximation is $s \approx R/4 = 15/4 = 3.75$

   b Calculate $n = 15$, $\sum x_i = 175, \sum x_i^2 = 2237$. Then

$$\bar{x} = \frac{\sum x_i}{n} = \frac{175}{15} = 11.67$$

$$s = \sqrt{\frac{\sum x_i^2 - \frac{(\sum x_i)^2}{n}}{n-1}} = \sqrt{\frac{2237 - \frac{(175)^2}{15}}{14}} = \sqrt{13.95238} = 3.735$$

   c Calculate the interval $\bar{x} \pm 2s \Rightarrow 11.67 \pm 2(3.735) \Rightarrow 11.67 \pm 7.47$ or 4.20 to 19.14. Referring to the original data set, the fraction of measurements in this interval is $14/15 = .93$.

2.75 a The relative frequency histogram for these data is shown below.

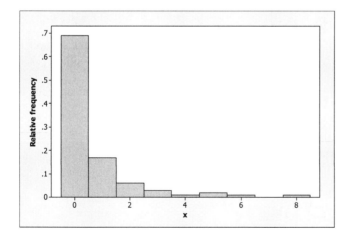

b Refer to the formulas given in Exercise 2.37. Using the frequency table and the grouped formulas, calculate $n = 100$, $\sum x_i f_i = 66$, $\sum x_i^2 f_i = 234$. Then

$$\bar{x} = \frac{\sum x_i f_i}{n} = \frac{66}{100} = 0.66$$

$$s^2 = \frac{\sum x_i^2 f_i - \frac{(\sum x_i f_i)^2}{n}}{n-1} = \frac{234 - \frac{(66)^2}{100}}{99} = 1.9236 \text{ and } s = \sqrt{1.9236} = 1.39.$$

c The three intervals, $\bar{x} \pm ks$ for $k = 2, 3$ are calculated in the table along with the actual proportion of measurements falling in the intervals. Tchebysheff's Theorem is satisfied and the approximation given by the Empirical Rule are fairly close for $k = 2$ and $k = 3$.

k	$\bar{x} \pm ks$	Interval	Fraction in Interval	Tchebysheff	Empirical Rule
2	0.66 ± 2.78	-2.12 to 3.44	$95/100 = 0.95$	at least 0.75	≈ 0.95
3	0.66 ± 4.17	-3.51 to 4.83	$96/100 = 0.96$	at least 0.89	≈ 0.997

2.77 We must estimate s and compare with the student's value of 0.263. In this case, $n = 20$ and the range is $R = 17.4 - 16.9 = 0.5$. The estimated value for s is then $s \approx R/4 = 0.5/4 = 0.125$

which is less than 0.263. It is important to consider the magnitude of the difference between the "rule of thumb" and the calculated value. For example, if we were working with a standard deviation of 100, a difference of 0.142 would not be great. However, the student's calculation is twice as large as the estimated value. Moreover, two standard deviations, or $2(0.263) = 0.526$, already exceeds the range. Thus, the value $s = 0.263$ is probably incorrect. The correct value of s is

$$s = \sqrt{\frac{\sum x_i^2 - \frac{(\sum x_i)^2}{n}}{n-1}} = \sqrt{\frac{5851.95 - \frac{117032.41}{20}}{19}} = \sqrt{0.0173} = 0.132$$

2.79 a Use the information in the exercise. For 1957 – 80, $IQR = 21$, and the upper fence is

$$Q_3 + 1.5 IQR = 52 + 1.5(21) = 83.5$$

For 1957 – 75, $IQR = 16$, and the upper fence is

$$Q_3 + 1.5 IQR = 52.25 + 1.5(16) = 76.25$$

b Although the maximum number of goals in both distribution is the same (77 goals), the upper fence is different in 1957 – 80, so that the record number of goals, $x = 77$ is no longer an outlier.

2.81 The variable of interest is the environmental factor in terms of the threat it poses to Canada. Each bulleted statement produces a percentile.
- x = toxic chemicals is the 61st percentile.
- x = air pollution and smog is the 55th percentile.
- x = global warming is the 52nd percentile.

2.83 a Calculate $n = 25$, $\sum x_i = 104.9$, $\sum x_i^2 = 454.810$. Then

$$\bar{x} = \frac{\sum x_i}{n} = \frac{104.9}{25} = 4.196$$

$$s = \sqrt{\frac{\sum x_i^2 - \frac{(\sum x_i)^2}{n}}{n-1}} = \sqrt{\frac{454.810 - \frac{(104.9)^2}{25}}{24}} = \sqrt{.610} = .781$$

b The ordered data set is shown below:

```
2.5   3.0   3.1   3.3   3.6
3.7   3.8   3.8   3.9   3.9
4.1   4.2   4.2   4.2   4.3
4.3   4.4   4.7   4.7   4.8
4.8   5.2   5.3   5.4   5.7
```

c The z-scores for $x = 2.5$ and $x = 5.7$ are
$$z = \frac{x - \bar{x}}{s} = \frac{2.5 - 4.196}{.781} = -2.17 \text{ and } z = \frac{x - \bar{x}}{s} = \frac{5.7 - 4.196}{.781} = 1.93$$
Since neither of the z-scores are greater than 3 in absolute value, the measurements are not judged to be unusually large or small.

2.85
- **a** When the applet loads, the mean and median are shown in the upper left-hand corner:
$\bar{x} = 6.6$ and $m = 6.0$
- **b** When the largest value is changed to $x = 13$, $\bar{x} = 7.0$ and $m = 6.0$.
- **c** When the largest value is changed to $x = 33$, $\bar{x} = 11.0$ and $m = 6.0$. The mean is larger when there is one unusually large measurement.
- **d** Extremely large values cause the mean to increase, but not the median.

2.87
- **a** When the applet loads, the mean and median are shown in the upper left-hand corner:
$\bar{x} = 31.6$ and $m = 32.0$
- **b** When the smallest value is changed to $x = 25$, $\bar{x} = 31.2$ and $m = 32.0$.
- **c** When the smallest value is changed to $x = 5$, $\bar{x} = 27.2$ and $m = 32.0$. The mean is smaller when there is one unusually small measurement.
- **d** $x = 29.0$
- **e** The largest and smallest possible values for the median are $32 \leq m \leq 34$.
- **f** Extremely small values cause the mean to decrease, but not the median.

2.89
- **a** Answers will vary from student to student. Students should notice that, when the estimators are compared in the long run, the standard deviation when dividing by $n-1$ is closer to $\sigma = 29.2$. When dividing by n, the estimate is closer to 27.7.
- **b** When the sample size is larger, the estimate is not as far from the true value $\sigma = 29.2$. The difference between the two estimators is less noticeable.

2.91 The box plot shows a distribution that is slightly skewed to the right, with no outliers. The student should estimate values for m, Q_1, and Q_3 that are close to the true values: $m = 12$, $Q_1 = 8.75$, and $Q_3 = 18.5$.

Case Study:
The Boys of Winter

1 The *Minitab* computer package was used to analyze the data. In the printout below, various descriptive statistics as well as histograms and box plots are shown.

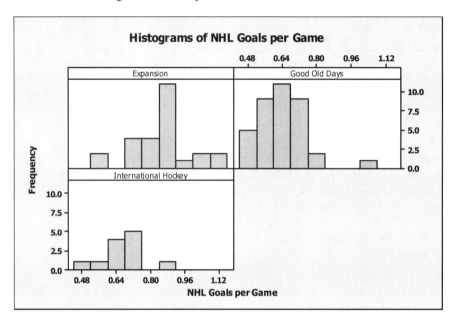

Descriptive Statistics: Average

```
                Total
Variable   Era  Count     Mean   SE Mean    StDev   Minimum        Q1    Median        Q3
Average     1      37   0.6345    0.0180   0.1095    0.4667    0.5423    0.6286    0.7113
            2      26   0.8505    0.0280   0.1426    0.5658    0.7594    0.8562    0.9067
            3      12   0.6633    0.0249   0.0864    0.5000    0.6159    0.6585    0.7128

Variable   Era  Maximum    Range      IQR
Average     1    1.0000   0.5333   0.1690
            2    1.1500   0.5842   0.1473
            3    0.8415   0.3415   0.0969
```

2 Notice that the average goals per game is the least in the Good Old Days Era (1 = 1931 – 1967) and the highest in the Expansion Era (2 = 1968 – 1993). The Expansion Era is the most variable. Although the International Hockey Era (3 = 1994 – 2006) has slightly higher average goals per game than Good Old Days Era, it is noticeably less variable.

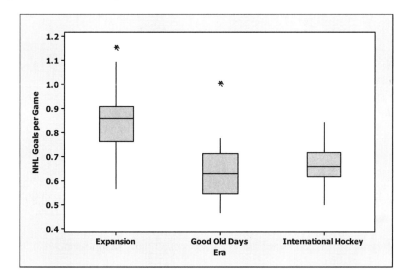

3 The box plot shows that each of the Expansion and Good Old Days Eras has one outlier. There is no outlier in the International Hockey Era, the least variable Era.

2 In summary, the Expansion Era is quite different than the other two Eras; it has higher mean and median number of goals per game. The outlier in the Expansion Era indicates the season with the record high goals per game. Notice that there is very little difference between the Good Old Days and International Hockey Era.

Project 2: Ignorance is Not Bliss (Project 1-B continued)

a The sample mean is $\bar{x} = \frac{\sum x_i}{n}$. For this example, we have,

$$\bar{x} = \frac{22+19+21+\cdots+27+33}{25} = \frac{544}{25} = 21.76.$$

The sample mode is the value that occurs most often. For our data set, there are two: at 19 and at 22. Thus, there are two modes.

To find the median, first put the data into increasing order, as follows:

0	10	14	15	16	17	17	18	19	19	19	20
20	21	21	22	22	22	23	27	29	30	33	35
55											

The median is the value that is found in the middle position. For 25 data points, the middle value is the 13th largest, which in this case is 20.

From the histogram below, it can be concluded that the data is essentially mound-shaped.

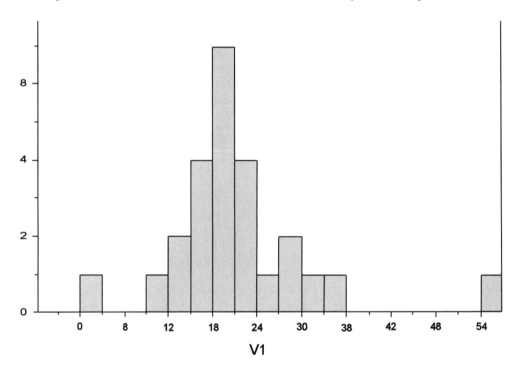

b Since we have a small data set with a large outlier, the median would be the best choice in this situation. The sample median is less sensitive to outliers and thus gives a more accurate representation of the centre of this distribution. The mean of small samples, such as this one, is heavily influenced by outliers, such as the point $x = 55$ here, and therefore does not give an accurate representation of centre.

c   The sample standard deviation can be calculated as follows:

$$s = \sqrt{\frac{\sum(x_i^2) - \frac{(\sum x_i)^2}{n}}{n-1}} = \sqrt{\frac{14234 - \frac{(544)^2}{25}}{25-1}} = \sqrt{\frac{2396.56}{24}} \Rightarrow s = 9.992831$$

The range (R) is simply the difference between the maximum value and the minimum value. For our data set, $R = 55 - 0 = 55$. The approximation for s based on R (as espoused in Section 2.5 of the text) is $s \approx \frac{R}{4}$ or $55/4 = 13.75$, which is a decent approximation, but certainly not very good.

d   If all data points were increased by 4%, the mean would also increase by 4%, or be multiplied by 1.04. This can be proven as follows. Assume that c is a constant. Then, we obtain:

$$\bar{x} = \frac{\sum x_i}{n} \Rightarrow \frac{\sum cx_i}{n} = \frac{c\sum x_i}{n} = c\bar{x}$$

Thus, the result follows if we let c equal 1.04.

e   If all data points were raised by 5%, the standard deviation would also be raised by 5%. Recall the formula for the standard deviation and let c be a constant. Then,

$$s \equiv \sqrt{\frac{\sum(x_i^2) - \frac{(\sum x_i)^2}{n}}{n-1}} \Rightarrow \sqrt{\frac{\sum([cx_i]^2) - \frac{(\sum cx_i)^2}{n}}{n-1}} = \sqrt{\frac{\sum(c^2 x_i^2) - \frac{(c\sum x_i)^2}{n}}{n-1}}$$

$$= \sqrt{\frac{c^2 \sum(x_i^2) - \frac{c^2(\sum x_i)^2}{n}}{n-1}}$$

$$\Rightarrow \sqrt{\frac{c^2\left\{\sum(x_i^2) - \frac{(\sum x_i)^2}{n}\right\}}{n-1}} = \sqrt{c^2}\sqrt{\frac{\sum(x_i^2) - \frac{(\sum x_i)^2}{n}}{n-1}} = c \cdot s$$

Thus, we see that the standard deviation will be multiplied by the constant multiple c, in this case, 1.05.

f   From (a), it was calculated that $\bar{x} = 21.76$ and $s = 9.992831$. Then, the interval $\bar{x} \pm s$ becomes $x \in [11.767169, 31.752831]$. Counting the number of x values between 12 and 31 (inclusive), there are 20 entries. Thus, $\frac{20}{25} = 80\%$ of the entries are within the interval $\bar{x} \pm s$. Now, computing the domain of the interval $\bar{x} \pm 2s$, it can be seen that $x \in [1.774338, 41.745662]$. There are 23 values of x between 2 and 41 (inclusive), thus $\frac{23}{25} = 92\%$. Comparing to the Empirical Rule, normal distributions should have approximately 68% of the total values of x within $\bar{x} \pm s$ and 95% of the total values within $\bar{x} \pm 2s$. Thus, it can be seen that there are more measurements than predicted for the first interval and slightly less than predicted for the second interval. This discrepancy can be accounted for by the non-normal behaviour in the distribution and the small amount of data points in the sample. Finally, Tchebysheff's Theorem predicts that at least 0% of the measurements are within $\bar{x} \pm s$ and that at least 75% of the measurements are within $\bar{x} \pm 2s$. As such, this sample follows Tchebysheff's Theorem.

g   Yes, Tchebysheff's Theorem can be used to describe this data set, since it can be used for *any* distribution.

h The Empirical Rule has a limited use in describing this sample. The data is relatively mound-shaped, and so the Empirical Rule is somewhat appropriate. However, due to the outliers in the data set, the Empirical Rule fails to accurately predict the percentage of measurements within the interval $\bar{x} \pm s$. It provides a better approximation for the interval $\bar{x} \pm 2s$.

i Referring to plot made in part **a**, it seems as though points 0 and 55 could be outliers, as these are far from the bulk of the data in the center.

j Given that $n = 25$, the 25th percentile is found at $x = 0.25(n+1) = 0.25(26) = 6.5$. Thus, the 25th percentile is the average of the 6th and 7th measurements when they are arranged in increasing order, $Q_1 = \frac{17+17}{2} = 17$. Likewise, $x = 0.50(26) = 13$ and thus $Q_2 = 20$. Lastly, $x = 0.75(26) = 19.5$, thus $Q_3 = \frac{23+27}{2} = 25$. The interquartile range is therefore $IQR = Q_3 - Q_1 = 25 - 17 = 8$.

k The range is 55 as calculated in part **c**, whereas the $IQR = 8$. Thus, about 50% of the data can be found in a very narrow middle area of the data range, suggesting that a mound-shaped distribution is likely.

l A box plot is shown below. The box plot shows 4 outliers, as indicated by the points beyond the whiskers.

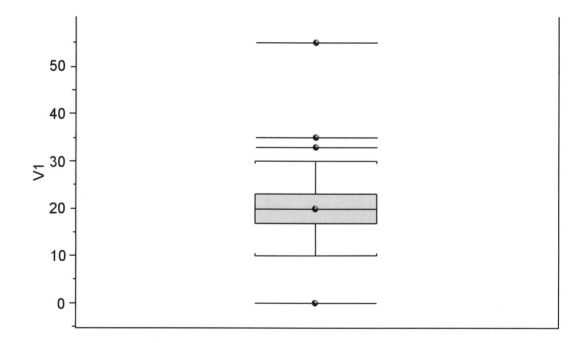

m The side-by-side box plots for the Hand Washing Time after (box 1) and before (box 2) the training session are given below:

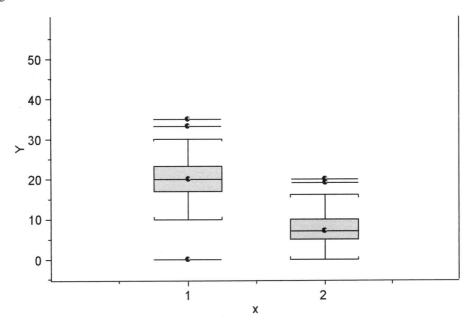

n It is clear from the box plots that the average time spent washing hands improved after the training session.

o Yes, we can conclude that the session was useful, and that the conjecture was true. A difference in median times of 20 seconds and 7 seconds is substantial.

p The histogram for the both data sets taken together is shown below. There is clear evidence of 2 distinct peaks, suggesting that the data comprises of 2 distinct populations (which we know is true).

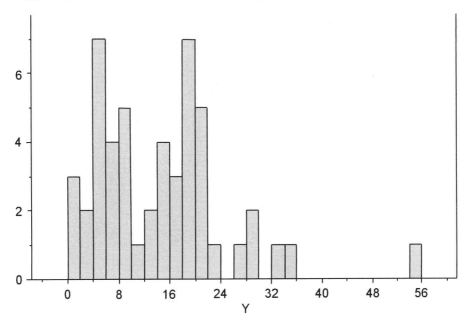

Introduction to Probability and Statistics, 2ce

Chapter 3: Describing Bivariate Data

3.1 **a** The side-by-side pie charts are constructed as in Chapter 1 for each of the two groups (men and women) and are displayed below using the percentages shown in the table below.

	Group 1	Group 2	Group 3	Total
Men	23%	31%	46%	100%
Women	8%	57%	35%	100%

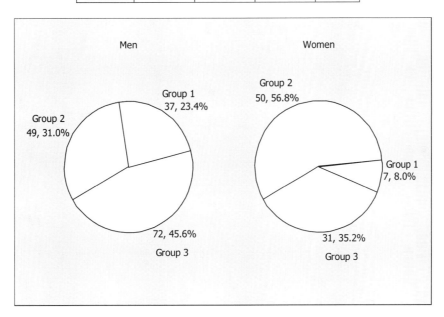

b-c The side-by-side and stacked bar charts in the next two figures measure the frequency of occurrence for each of the three groups. A separate bar (or portion of a bar) is used for men and women.

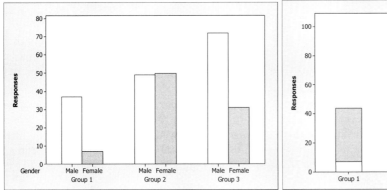
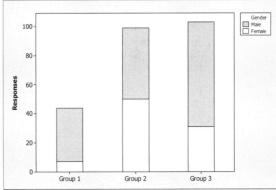

d The differences in the proportions of men and women in the three groups is most graphically portrayed by the pie charts, since the unequal number of men and women tend to confuse the interpretation of the bar charts. However, the bar charts are useful in retaining the actual frequencies of occurrence in each group, which is lost in the pie chart.

3.3 **a** Similar to Exercises 3.1 and 3.2. Any of the comparative charts (side-by-side pie charts, stacked or side-by-side bar charts) can be used.

b-c The two types of comparative bar charts are shown below. The amounts spent in each of the four categories seem to be quite different for men and women, except in category C. In category C which involves the largest dollar amount of purchase, there is little difference between the genders.

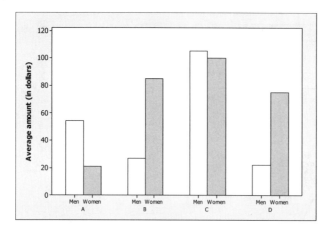

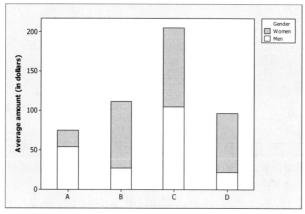

d Although it is really a matter of preference, the only advantage to the stacked chart is that the reader can easily see the total dollar amount for each category. For comparison purposes, the side-by-side chart may be better.

3.5 **a** The population of interest is the population of responses to the question about free time for all parents and children in the United States. The sample is the set of responses generated for the 198 parents and 200 children in the survey.

b The data can be considered bivariate if, for each person interviewed, we record the person's relationship (Parent or Child) and their response to the question (just the right amount, not enough, too much, don't know). Since the measurements are not numerical in nature, the variables are qualitative.

c The entry in a cell represents the number of people who fell into that relationship-opinion category.

d A pie chart is created for both the "parents" and the "children" categories. The size of each sector angle is proportional to the fraction of measurements falling into that category.

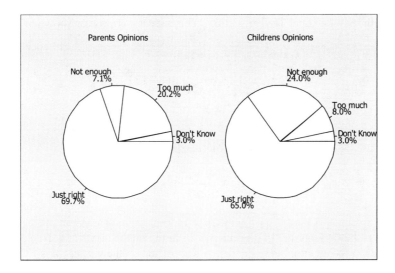

e   Either stacked or comparative bar charts could be used, but since the height of the bar represents the frequency of occurrence (and hence is tied to the sample size), this type of chart would be misleading. The comparative pie charts are the best choice.

3.7 a   The side-by-side bar chart is shown below.

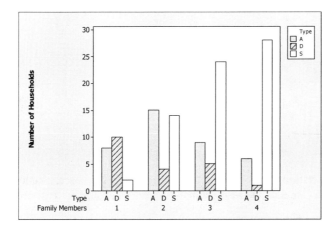

b-c   The stacked bar chart is shown below. Both charts indicate that the more family members there are, the more likely it is that the family lives in a duplex or a single residence. The fewer the number of family members, the more likely it is that the family lives in an apartment.

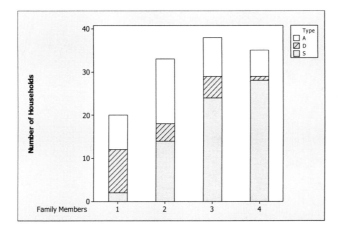

Introduction to Probability and Statistics, 2ce   49

3.9 Follow the instructions in the My Personal Trainer section. The correct answers are shown in the table.

x	y	xy	Calculate:	Covariance
1	6	6	$n = 3$	$s_{xy} = \dfrac{20 - \dfrac{(6)(12)}{3}}{2} = -2$
3	2	6		
			$s_x = 1$	
2	4	8	$s_y = 2$	**Correlation Coefficient**
$\sum x = 6$	$\sum y = 12$	$\sum xy = 20$		$r = \dfrac{-2}{1(2)} = -1$

3.11 a The first variable (x) is the first number in the pair and is plotted on the horizontal axis, while the second variable (y) is the second number in the pair and is plotted on the vertical axis. The scatterplot is shown in the figure below.

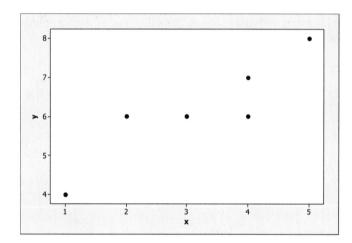

b There appears to be a positive relationship between x and y; that is, as x increases, so does y.

c Use your scientific calculator to calculate the sums, sums of square sand sum of cross products for the pairs (x_i, y_i).

$\sum x_i = 19;\ \sum y_i = 37;\ \sum x_i^2 = 71;\ \sum y_i^2 = 237;\ \sum x_i y_i = 126$

Then the covariance is

$$s_{xy} = \dfrac{\sum x_i y_i - \dfrac{(\sum x_i)(\sum y_i)}{n}}{n-1} = \dfrac{126 - \dfrac{(19)37}{6}}{5} = 1.76667$$

and the sample standard deviations are

$$s_x = \sqrt{\dfrac{\sum x_i^2 - \dfrac{(\sum x_i)^2}{n}}{n-1}} = \sqrt{\dfrac{71 - \dfrac{(19)^2}{6}}{5}} = 1.472 \text{ and } s_y = \sqrt{\dfrac{\sum y_i^2 - \dfrac{(\sum y_i)^2}{n}}{n-1}} = \sqrt{\dfrac{237 - \dfrac{(37)^2}{6}}{5}} = 1.329$$

The correlation coefficient is $r = \dfrac{s_{xy}}{s_x s_y} = \dfrac{1.76667}{(1.472)(1.329)} = 0.902986 \approx 0.903$

d The slope and y-intercept of the regression line are

$$b = r\frac{s_y}{s_x} = 0.902986\left(\frac{1.329}{1.472}\right) = 0.81526 \text{ and } a = \bar{y} - b\bar{x} = \frac{37}{6} - 0.81526\left(\frac{19}{6}\right) = 3.58$$

and the equation of the regression line is $y = 3.58 + 0.815x$.

The graph of the data points and the best fitting line is shown below. The line fits through the data points.

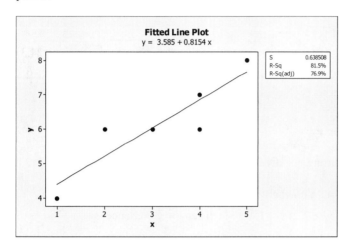

3.13 a Similar to Exercise 3.11. The scatterplot is shown below.

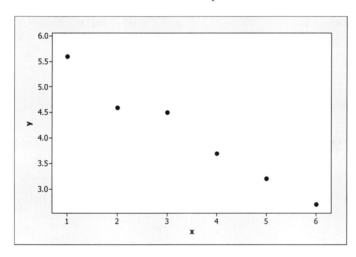

b There appears to be a negative relationship between x and y; that is, as x increase, y decreases.

c Use your scientific calculator to calculate the sums, sums of squares and sum of cross products for the pairs (x_i, y_i).

$$\sum x_i = 21;\ \sum y_i = 24.3;\ \sum x_i^2 = 91;\ \sum y_i^2 = 103.99;\ \sum x_i y_i = 75.3$$

Then the covariance is

$$s_{xy} = \frac{\sum x_i y_i - \frac{(\sum x_i)(\sum y_i)}{n}}{n-1} = \frac{75.3 - \frac{(21)24.3}{6}}{5} = -1.95$$

and the sample standard deviations are

$$s_x = \sqrt{\frac{\sum x_i^2 - \frac{(\sum x_i)^2}{n}}{n-1}} = \sqrt{\frac{91 - \frac{(21)^2}{6}}{5}} = 1.8708 \text{ and } s_y = \sqrt{\frac{\sum y_i^2 - \frac{(\sum y_i)^2}{n}}{n-1}} = \sqrt{\frac{103.99 - \frac{(24.3)^2}{6}}{5}} = 1.0559$$

The correlation coefficient is $r = \frac{s_{xy}}{s_x s_y} = \frac{-1.95}{(1.8708)(1.0559)} = -0.987$

This value of r indicates a strong negative relationship between x and y.

3.15 a Use your scientific calculator. You can verify that $\sum x_i = 17;\ \sum y_i = 441.67;$
$\sum x_i^2 = 59;\ \sum y_i^2 = 38{,}917.2579;\ \sum x_i y_i = 1509.51$. Then the covariance is

$$s_{xy} = \frac{\sum x_i y_i - \frac{(\sum x_i)(\sum y_i)}{n}}{n-1} = \frac{1509.51 - \frac{17(441.67)}{6}}{5} = 51.62233$$

The sample standard deviations are $s_x = 1.47196$ and $s_y = 35.79160$ so that $r = 0.9799$. Then

$$b = r\frac{s_y}{s_x} = 23.8257 \text{ and } a = \bar{y} - b\bar{x} = 73.611667 - 23.8257(2.8333) = 6.106$$

and the equation of the regression line is $y = 6.106 + 23.826x$.

b The graph of the data points and the best fitting line is shown below.

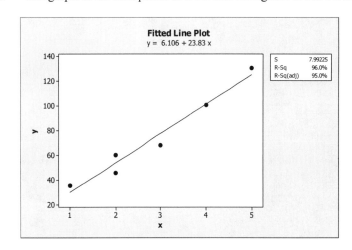

c When $x = 6$, the estimated value of y is $y = 6.106 + 23.826(6) = 149.06$. However, it is risky to try to estimate the value of y for a value of x outside of the experimental region – that is, the range of x values for which you have collected data.

3.17 **a-b** The scatterplot is shown below. There is a slight positive trend between pre- and post-test scores, but the trend is not too pronounced.

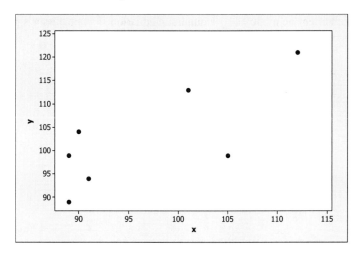

c Calculate $n = 7$; $\sum x_i = 677$; $\sum y_i = 719$; $\sum x_i^2 = 65,993$; $\sum y_i^2 = 74,585$; $\sum x_i y_i = 70,006$. Then the covariance is

$$s_{xy} = \frac{\sum x_i y_i - \frac{(\sum x_i)(\sum y_i)}{n}}{n-1} = 78.071429$$

The sample standard deviations are $s_x = 9.286447$ and $s_y = 11.056134$ so that $r = 0.760$. This is a relatively strong positive correlation, confirming the interpretation of the scatterplot.

3.19 **a** Since we would be interested in predicting the price of a TV based on its size, the price is the dependent variable (y) and size is the independent variable (x).

b The scatterplot is shown below. The relationship is somewhat linear, but has a bit of a curve to it. The relationship may in fact be slightly curvilinear.

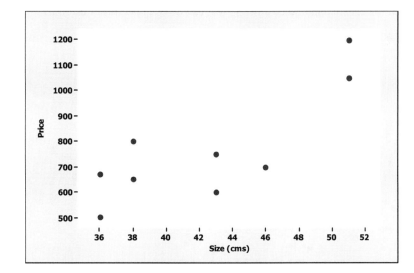

3.21 **a-b** The data is graphed as a scatterplot in the figure below, with the time in months plotted on the horizontal axis and the number of books on the vertical axis. The data points are then connected to form a line graph. There is a very distinct pattern in this data, with the number of books increasing with time, a response which might be modeled by a quadratic equation. The professor's productivity appears to increase, with less time required to write later books.

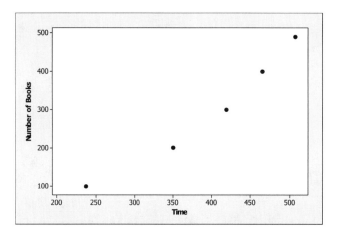

3.23 **a** The following variables are recorded: automobiles sold or produced year (quantitative), the number of sales or production (quantitative), and the automobile brand or type (foreign vs. Japanese, passenger cars vs. trucks/buses, qualitative).

b The populations of interest are the populations of all import sales of automobiles and domestic automobile production in Japan. Although the source of this data is not given, it is probably based on census information, in which case the data represents the entire population.

c A comparative (side-by-side) bar chart has been used. An alternative presentation can be obtained by using comparative pie charts.

d Since there are not much change in both sales and production over the period 1998 – 2005, there seems to be very little pattern between total sales and total production.

3.25 The scatterplot for these two quantitative variables is shown below. Notice the almost perfect positive correlation. The correlation coefficient and best fitting line can be calculated for descriptive purposes.

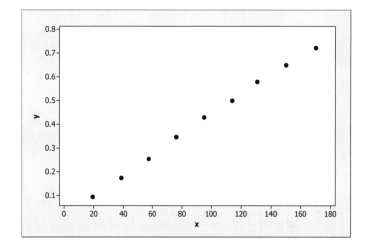

Calculate $n = 9$; $\sum x_i = 850.8$; $\sum y_i = 3.755$; $\sum x_i^2 = 101{,}495.78$; $\sum y_i^2 = 1.941467$; $\sum x_i y_i = 443.7727$.

Then the covariance is

$$s_{xy} = \frac{\sum x_i y_i - \frac{(\sum x_i)(\sum y_i)}{n}}{n-1} = 11.10000417$$

The sample standard deviations are $s_x = 51.31620$ and $s_y = 0.216448$ so that $r = 0.999$. Then

$$b = r\frac{s_y}{s_x} = 0.004215 \text{ and } a = \bar{y} - b\bar{x} = 0.417222 - 0.004215(94.5333) = 0.0187$$

and the equation of the regression line is $y = 0.0187 + 0.0042x$.

3.27   **a**   Using the *Minitab* output, calculate

$$r = \frac{s_{xy}}{s_x s_y} = \frac{1232.231}{\sqrt{(412.528)(4437.109)}} = 0.9108$$

       b   Since the first weekend's gross would tend to explain the total gross, we could consider $x =$ first weekend's gross to be the independent variable, and $y =$ total gross to be the dependent variable.

       c   Using $r = 0.9108$, $s_x = \sqrt{412.528}$ and $s_y = \sqrt{4437.109}$, calculate

$$b = r\frac{s_y}{s_x} = 2.987 \text{ and } a = \bar{y} - b\bar{x} = 86.71 - 2.987(25.66) = 10.06$$

and the equation of the regression line is $y = 10.06 + 2.987x$

       d   Answers will vary slightly depending on rounding. For the regression line $y = 10.06 + 2.987x$, the predicted value of y when $x = 30$ will be $y = 10.06 + 2.987(30) = 99.67$ or 99.67 million dollars.

3.29   **a-c**   There does not seem a very strong relationship between the variables. Some data points seem unusual; for example, we the 9th observation (Ontario; $y = 2449$, $x = 662$) seems to be an unusual observation. However, most of the data points show a decreasing trend from left to right indicating a negative relationship.

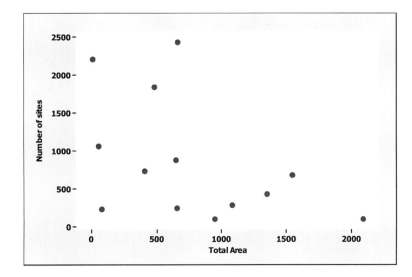

Introduction to Probability and Statistics, 2ce

b The pattern described in parts **a** and **c** would indicate a weak correlation:

$$s_{xy} = \frac{\sum x_i y_i - \frac{(\sum x_i)(\sum y_i)}{n}}{n-1} = -231750.5833$$

$$r = \frac{s_{xy}}{s_x s_y} = \frac{-231750.5833}{(622)(812)} = -0.459$$

3.31 **a** The variables measured are amount of aluminum oxide (quantitative), and archeological site (qualitative).

b Looking at the box plots, you can see that there are higher levels of aluminum oxide at the Ashley Rails and Island Thorns sites. The variability is about the same at the three sites.

3.33 **a** The number of High Speed Internet users (per 100 members of the population; quantitative discrete) have been measured, along with the year (quantitative) and the type of high speed internet (qualitative).

b-c Answers will vary. We choose to use a line chart for the two types of broadband connections. As the number of DSL increases, the number of Cable Modem also increases. Every year a noticeable increase in the users of both types of connections can be observed.

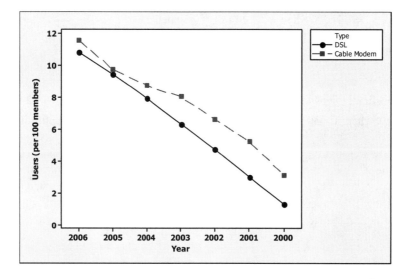

3.35 **a** The scatterplot is shown below. There is a positive linear relationship between armspan and height.

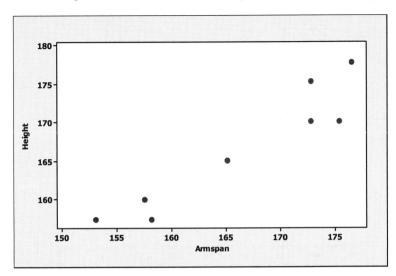

b Calculate
$n = 8$; $\sum x_i = 1330.90$; $\sum y_i = 1333.60$; $\sum x_i^2 = 222001.79$; $\sum y_i^2 = 222749.52$; $\sum x_i y_i = 222342$.

Then the covariance is $s_{xy} = \dfrac{\sum x_i y_i - \dfrac{(\sum x_i)(\sum y_i)}{n}}{n-1} = 68.71$

The sample standard deviations are $s_x = 9.18$ and $s_y = 7.91$ so that $r = 0.946$.

c Since DaVinci indicated that a person's armspan is roughly equal to his height, the slope of the line should be approximately equal to 1.

d Calculate (using full accuracy)

$b = r\dfrac{s_y}{s_x} = .815$ and $a = \bar{y} - b\bar{x} = 166.70 - .815(166.36) = 31.1$

and the equation of the regression line is $y = 31.1 + .815x$.

e For $x = 157.5$, $y = 31.1 + .815(157.5) = 159.46$

3.37 **a-b** The scatterplot is shown below. There is a strong positive linear relationship between x and y.

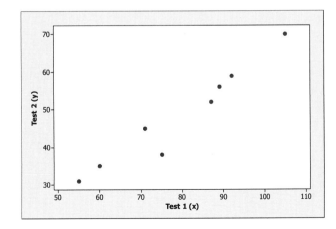

3.39 **a-b** The scatterplot is shown below. There is a strong positive relationship between the total number of complaints and number of passengers served.

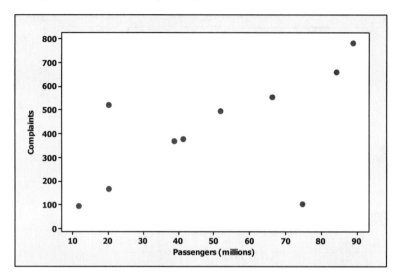

c-d There is one outlier in the scatterplot—corresponding to Southwest Airlines. The airline appears to serve a large number of passengers with fewer than the typical number of complaints. Southwest is doing better than the other airlines with respect to customer satisfaction. Air Canada also seems somewhat extreme and it is doing worse than the other airlines.

3.40-3.41 Students should use the **Building a Scatterplot** applets to create scatterplots, and check their accuracy using Figures 3.4 and 3.5 in the text.

3.43 **a** Calculate $n = 8$; $\sum x_i = 451$; $\sum y_i = 555$; $\sum x_i^2 = 29{,}619$; $\sum y_i^2 = 43{,}205$; $\sum x_i y_i = 35{,}082$.

Then the covariance is $s_{xy} = \dfrac{\sum x_i y_i - \dfrac{(\sum x_i)(\sum y_i)}{n}}{n-1} = 541.9821$

The sample standard deviations are $s_x = 24.4770$ and $s_y = 25.9171$ so that $r = 0.8544$.

b-c The scatterplot should look like the one shown below. The correlation coefficient should be close to $r = 0.85$. There is a strong positive trend.

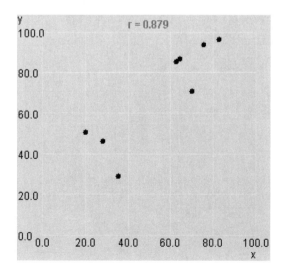

3.45

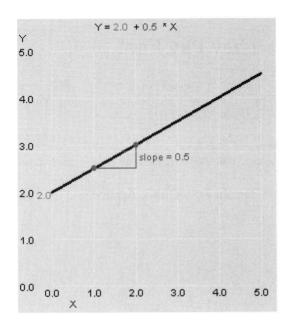

a   The slope is $b = 0.5$, the change in y for a one-unit change in x.
b   As x increases, so does y, as indicated by the positive slope.
c   When $x = 0$, $y = 2.0$. This is the y-intercept $= a$.
d   When $x = 2.5$, $y = 2.0 + 0.5(2.5) = 3.25$. When $x = 4$, $y = 2.0 + 0.5(4) = 4.0$.

Case Study:
Do Hockey Teams Get What They Pay For?

1. Almost all the variables are approximately symmetric. Payroll and Points maybe slightly skewed to the left. On the other hand, overtime loss is slightly skewed right. There are two unusually low Wins ($x = 21$, $x = 22$); and one of the teams has extremely high "Goals for" ($x = 314$).

2. Although the correlations are not very strong, as you can expect, payroll is positively correlated with Points, Wins, Goals for and is negatively correlated with Losses, Overtime Losses and Goals Against.

Correlations: Payroll, Points, Wins, Losses, Goals For, Goals Against, OT Losses

```
               Payroll    Points    Wins      Losses    Goals For  Goals A.
Points         0.568
Wins           0.554      0.987
Losses        -0.566     -0.983    -0.940
Goals For      0.390      0.767     0.760    -0.749
Goals Against -0.451     -0.806    -0.774     0.818    -0.387
OT Losses     -0.176     -0.375    -0.519     0.197    -0.307      0.176
```

3. Answers will vary. Yes, the price of the NHL team does convey something about its quality – higher priced teams are in general performing slightly better than poorly priced teams. One way to find this would be to compare the relation between the variables payroll and points or payroll and wins.

Project 3: Child Safety Seat Survey

a   Some of the variables measured in this survey are Type of Restraint (i.e., Rear-facing Infant Seat, Forward-facing Infant Seat, Booster Seat, and Seat Belt Only) and Age Group (Infants, Toddlers, School Aged: 4-9 years, and Older than 9 years). The Restraint Type category consists of qualitative variables, whereas the Age Group category consists of quantitative variables.

b   The side-by-side bar chart is shown below.

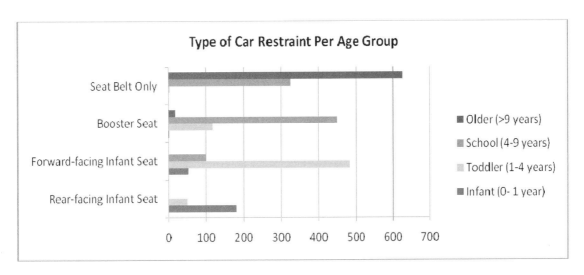

The stacked bar chart is shown below.

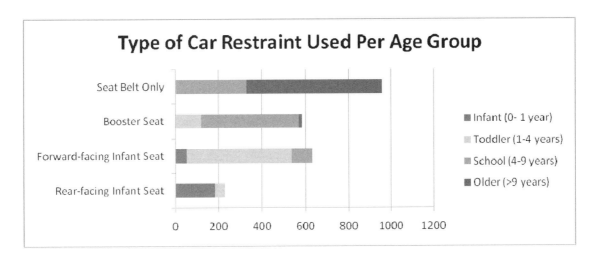

Introduction to Probability and Statistics, 2ce

c   Pie charts for each age group are shown below.

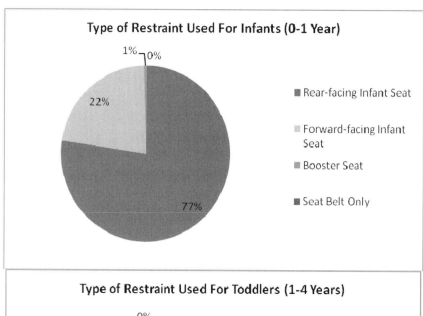

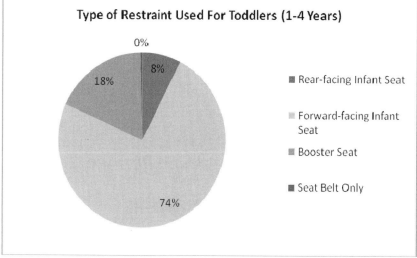

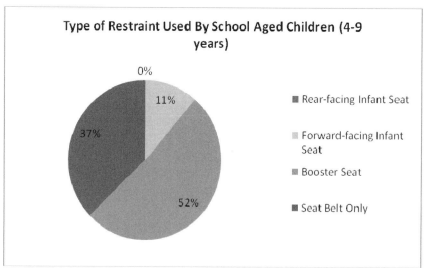

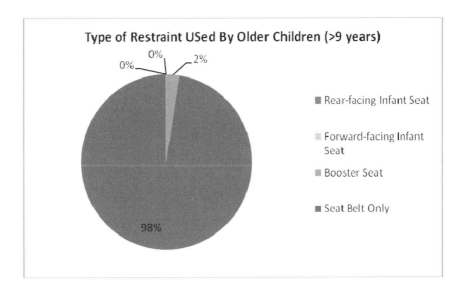

The pie chart titled "**Type of Restraint Used For Toddlers (1-4 Years)**" best depicts the differences and similarities of use for toddler restraint types. It can be clearly seen that approximately three-quarters of the seat restraints used for this age category are forward-facing infant seats, while the other quarter consists of mainly of booster seats, and under half as many rear-facing infant seats. Although we judge the pie chart to be visually more informative, the bar chart does an adequate job as well.

d By looking at these graphs, one can ascertain information such as which restraints were used the most, least, and so on, as well as which age groups tended to prefer which types of restraints, and differences and similarities in use of the restraints between age groups. The bar graph is more effective in this case than the pie graph, because it can graphically portray all of the variables in one chart, while many charts are needed if pie graphs are used. By combining all of these variables, bar graphs make it easy to compare them. The stacked bar chart is the more effective of the two bar graphs. It combines the best features of the other two graph types, by containing all of the information that the side by side bar graph does, while portraying it in a more compact manner. However, if we take "effective" to refer to visual impact, we might say that the pie charts, although more cumbersome, are somewhat more revealing than the bar charts.

e Other graphical techniques which could be used to display this data include histograms, dot diagrams, box plots, and stem and leaf plots. Each type of graphical representation has its own advantages and disadvantages. It is difficult to say which would be most effective.

Child Safety Seat Survey, continued

a A scatterplot of female weight against height is shown below.

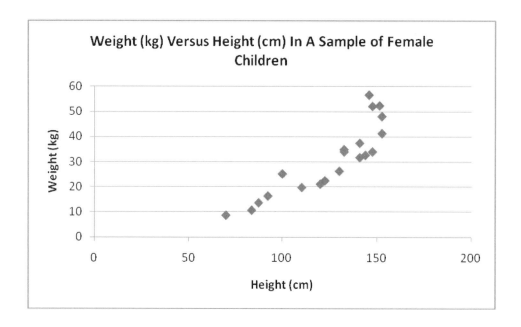

There is evidence of a positive relationship between the variables weight and height. The relationship, however, is not perfectly linear, but somewhat curvilinear.

b Based on the scatterplot from part (a), there do not seem to be any outliers.

c A scatterplot of male weight against height is shown below.

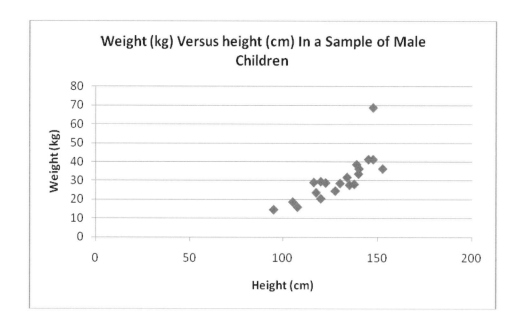

Once again, there is evidence of a positive relationship, this time, more linear. Due to the fact that the points are clustered closer to a straight line (apart from the one outlier), there is a much stronger linear relationship with the males than with the females.

d In the scatterplot for males, there is a strong linear correlation, with one very clear outlier. The data is also clustered tightly within a smaller range (95 to 152.5 cm for height, and 14.51 to 41.28 kg), with the exception of the outlier. The scatterplot for female children, however, is more spread out. It has a larger range for its data values (70 to 152.5 cm for height, and 8.62 to 52.39 kg for weight), and is less linear in its distribution, although the values for weight tend to still increase with the values for height. Aside from the females having a larger variance in their weight, many of them tend to have a larger weight than the males of the same height.

e Side-by-side box plots for height are shown below.

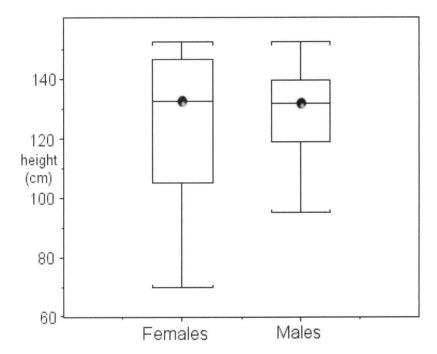

Clearly as you can see from the box plots, the variability of height in females is higher than that of males. Also, the median height for females is slightly larger than for males.

f The correlation coefficient is $r = \frac{s_{xy}}{s_x s_y} = 0.881546$. The value is relatively high, indicating a rather strong linear association between female height and weight.

g This correlation coefficient reinforces the earlier assertions made based on the scatterplot. It is a positive value (0.881546), indicating that the relationship is positive. As well, since it is close to 1, it indicates a relatively strong correlation, as was evident in the scatterplot (close clustering around a straight line relationship). The slope of the regression line can be found by multiplying the correlation coefficient, r, by s_y, and then dividing by s_x.

h The regression line can be found using any statistical software package. Alternately, one could use the formulas from the textbook. In either case, the fitted line is $y = -29.648 + 0.4842x$, where x represents female height and y represents female weight.

Introduction to Probability and Statistics, 2ce

i When the female has a height of 160 cm, we can predict her weight by plugging in $x=160$ into the regression equation and finding y. That is, $y = -29.648 + 0.4842(160) = 47.824$ kg as the predicted weight.

j No. A new regression line for males should be fitted because the weight and height relationship is different for the two sexes (as was seen in the preceding parts of this question).

k No. The two sexes exhibit different weight-height relationships. Although ignoring the sex category might not produce very unreliable projections, there does not seem to be any compelling reason to do so. More formally, we would be ignoring a lurking variable (sex) for no good reason.

Chapter 4: Probability and Probability Distributions

4.1 **a** This experiment involves tossing a single die and observing the outcome. The sample space for this experiment consists of the following simple events:
E_1: Observe a 1 E_4: Observe a 4
E_2: Observe a 2 E_5: Observe a 5
E_3: Observe a 3 E_6: Observe a 6

b Events A through F are composed of the simple events in the following manner:
A: (E_2) D: (E_2)
B: (E_2, E_4, E_6) E: (E_2, E_4, E_6)
C: (E_3, E_4, E_5, E_6) F: contains no simple events

c Since the simple events E_i, $i = 1, 2, 3, \ldots, 6$ are equally likely, $P(E_i) = 1/6$.

d To find the probability of an event, we sum the probabilities assigned to the simple events in that event. For example,
$$P(A) = P(E_2) = \frac{1}{6}$$
Similarly, $P(D) = 1/6$; $P(B) = P(E) = P(E_2) + P(E_4) + P(E_6) = \frac{3}{6} = \frac{1}{2}$; and $P(C) = \frac{4}{6} = \frac{2}{3}$. Since event F contains no simple events, $P(F) = 0$.

4.3 It is given that $P(E_1) = .45$ and that $3P(E_2) = .45$, so that $P(E_2) = .15$. Since $\sum_S P(E_i) = 1$, the remaining 8 simple events must have probabilities whose sum is $P(E_3) + P(E_4) + \ldots + P(E_{10}) = 1 - .45 - .15 = .4$. Since it is given that they are equiprobable, $P(E_i) = \frac{.4}{8} = .05$ for $i = 3, 4, \ldots, 10$

4.5 **a** The experiment consists of choosing three coins at random from four. The order in which the coins are drawn is unimportant. Hence, each simple event consists of a triplet, indicating the three coins drawn. Using the letters N, D, Q, and L to represent the nickel, dime, quarter, and loonie, respectively, the four possible simple events are listed below.
E_1: (NDQ) E_2: (NDL) E_3: (NQL) E_4: (DQL)

b The event that a loonie is chosen is associated with the simple events E_2, E_3, and E_4. Hence,
$P[\text{choose a loonie}] = P(E_2) + P(E_3) + P(E_4) = \frac{1}{4} + \frac{1}{4} + \frac{1}{4} = \frac{3}{4}$ since each simple event is equally likely.

c The simple events along with their monetary values follow:
E_1 NDQ $0.40
E_2 NDL $1.15
E_3 NQL $1.30
E_4 DQL $1.35
Hence, $P[\text{total amount is \$1.10 or more}] = P(E_2) + P(E_3) + P(E_4) = 3/4$.

4.7 Label the five balls as R_1, R_2, R_3, Y_1 and Y_2. The selection of two balls is accomplished in two stages to produce the simple events in the tree diagram below.

First Ball	Second Ball	Simple Events		First Ball	Second Ball	Simple Events
R_1	R_2	R_1R_2		Y_1	R_1	Y_1R_1
	R_3	R_1R_3			R_2	Y_1R_2
	Y_1	R_1Y_1			R_3	Y_1R_3
	Y_2	R_1Y_2			Y_2	Y_1Y_2
R_2	R_1	R_2R_1		Y_2	R_1	Y_2R_1
	R_3	R_2R_3			R_2	Y_2R_2
	Y_1	R_2Y_1			R_3	Y_2R_3
	Y_2	R_2Y_2			Y_1	Y_2Y_1
R_3	R_1	R_3R_1				
	R_2	R_3R_2				
	Y_1	R_3Y_1				
	Y_2	R_3Y_2				

4.9 The four possible outcomes of the experiment, or simple events, are represented as the cells of a 2×2 table, and have probabilities as given in the table.
a P[adult judged to need glasses] $= .44 + .14 = .58$
b P[adult needs glasses but does not use them] $= .14$
c P[adult uses glasses] $= .44 + .02 = .46$

4.11 **a** *Experiment*: Select three people and record their gender (M or F).
b Extend the tree diagram in Figure 4.3 of the text to include one more coin toss (a total of $n = 3$). Then replace the H and T by M and F to obtain the 8 possible simple events shown below:
FFF FMM MFM MMF
MFF FMF FFM MMM
c Since there are $N = 8$ equally likely simple events, each is assigned probability, $P(E_i) = 1/N = 1/8$.
d-e Sum the probabilities of the appropriate simple events:

$$P(\text{only one man}) = P(MFF) + P(FMF) + P(FFM) = 3\left(\frac{1}{8}\right) = \frac{3}{8}$$

$$P(\text{all three are women}) = P(FFF) = \frac{1}{8}$$

4.13 **a** *Experiment*: A taster tastes and ranks three varieties of tea A, B, and C, according to preference.
b Simple events in S are in triplet form.
$E_1 : (1,2,3)$ $E_4 : (2,3,1)$
$E_2 : (1,3,2)$ $E_5 : (3,2,1)$
$E_3 : (2,1,3)$ $E_6 : (3,1,2)$
Here 1 is assigned to the most desirable, 2 to the next most desirable, and 3 to the least desirable.
c Define the events   D: variety A is ranked first
                           F: variety A is ranked third
Then $P(D) = P(E_1) + P(E_2) = 1/6 + 1/6 = 1/3$
The probability that A is least desirable is $P(F) = P(E_5) + P(E_6) = 1/6 + 1/6 = 1/3$

4.15 Similar to Exercise 4.9. The four possible outcomes of the experiment, or simple events, are represented as the cells of a 2×2 table, and have probabilities (when divided by 300) as given in the table.
a $P(\text{normal eyes and normal wing size}) = 140/300 = .467$
b $P(\text{vermillion eyes}) = (3 + 151)/300 = 154/300 = .513$
c $P(\text{either vermillion eyes or miniature wings or both}) = (3 + 151 + 6)/300 = 160/300 = .533$

4.17 Use the *mn* Rule. There are $10(8) = 80$ possible pairs.

4.19 a $P_3^5 = \dfrac{5!}{2!} = 5(4)(3) = 60$

  b $P_9^{10} = \dfrac{10!}{1!} = 3,628,800$

  c $P_6^6 = \dfrac{6!}{0!} = 6! = 720$

  d $P_1^{20} = \dfrac{20!}{19!} = 20$

4.21 Since order is important, you use *permutations* and $P_5^8 = \dfrac{8!}{3!} = 8(7)(6)(5)(4) = 6720$.

4.23 Use the extended *mn* Rule. The first die can fall in one of 6 ways, *and* the second and third die can each fall in one of 6 ways. The total number of simple events is $6(6)(6) = 216$.

4.25 Since order is unimportant, you use *combinations* and $C_3^{10} = \dfrac{10!}{3!7!} = \dfrac{10(9)(8)}{3(2)(1)} = 120$.

4.27 This exercise involves the arrangement of 6 different cities in all possible orders. Each city will be visited once and only once. Hence, order is important and elements are being chosen from a single set. Permutations are used and the number of arrangements is $P_6^6 = \dfrac{6!}{0!} = 6! = 6(5)(4)(3)(2)(1) = 720$

4.29 a Each student has a choice of 52 cards, since the cards are replaced between selections. The *mn* Rule allows you to find the total number of configurations for three students as $52(52)(52) = 140,608$.

  b Now each student must pick a different card. That is, the first student has 52 choices, but the second and third students have only 51 and 50 choices, respectively. The total number of configurations is found using the *mn* Rule or the rule for permutations:
  $$mnt = 52(51)(50) = 132,600 \quad \text{or} \quad P_3^{52} = \dfrac{52!}{49!} = 132,600.$$

  c Let A be the event of interest. Since there are 52 different cards in the deck, there are 52 configurations in which all three students pick the same card (one for each card). That is, there are $n_A = 52$ ways for the event A to occur, out of a total of $N = 140,608$ possible configurations from part **a**. The probability of interest is $P(A) = \dfrac{n_A}{N} = \dfrac{52}{140,608} = .00037$

  d Again, let A be the event of interest. There are $n_A = 132,600$ ways (from part **b**) for the event A to occur, out of a total of $N = 140,608$ possible configurations from part **a**, and the probability of interest is $P(A) = \dfrac{n_A}{N} = \dfrac{132,600}{140,608} = .943$

4.31 a Since the order of selection for the five-card hand is unimportant, use *combinations* to find the number of possible hands as $N = C_5^{52} = \dfrac{52!}{5!47!} = \dfrac{52(51)(50)(49)(48)}{5(4)(3)(2)(1)} = 2,598,960$.

  b Since there are only four different suits, there are $n_A = 4$ ways to get a royal flush.

  c From parts **a** and **b**,
  $$P(\text{royal flush}) = \dfrac{n_A}{N} = \dfrac{4}{2,598,960} = .000001539$$

4.33 Notice that a sample of 10 nurses will be the same no matter in which order they were selected. Hence, order is unimportant and combinations are used. The number of samples of 10 selected from a total of 90 is

$$C_{10}^{90} = \frac{90!}{10!80!} = \frac{2.0759076(10^{19})}{3.6288(10^6)} = 5.720645(10^{12})$$

4.35 **a** Use the *mn* Rule. The Group I team can be chosen in one of $m = 10$ ways, while there are 5 ways to choose the Group II team, for a total of $N = mn = 10(5) = 50$ possible pairings.

b You must choose Calgary from the first group and Ottawa from the second group, so that $n_A = (1)(1) = 1$ and the probability is $n_A/N = 1/50$.

c Since there are two New York teams in the Group I, there are two choices for the first team and five choices for the second team. Hence $n_A = (2)(5) = 10$ and the probability is $n_A/N = 10/50 = 1/5$.

4.37 The situation presented here is analogous to drawing 5 items from a jar (the five members voting in favor of the plaintiff). If the jar contains 5 red and 3 white items (5 women and 3 men), what is the probability that all five items are red? That is, if there is no sex bias, five of the eight members are randomly chosen to be those voting for the plaintiff. What is the probability that all five are women? There are

$$N = C_5^8 = \frac{8!}{5!3!} = 56$$

simple events in the experiment, only one of which results in choosing 5 women. Hence,

$$P(\text{five women}) = \frac{1}{56}.$$

4.39 The monkey can place the twelve blocks in any order. Each arrangement will yield a simple event and hence the total number of simple events (arrangements) is $P_{12}^{12} = 12!$ It is necessary to determine the number of simple events in the event of interest (that he draws three of each kind, in order). First, he may draw the four different <u>types</u> of blocks in any order. Thus we need the number of ways of arranging these four items, which is $P_4^4 = 4!$ Once this order has been chosen, the three squares can be arranged in $P_3^3 = 3!$ ways, the three triangles can be arranged in $P_3^3 = 3!$ ways, and so on. Thus the total number of simple events in the event of interest is $P_4^4(P_3^3)^4$ and the associated probability is

$$\frac{P_4^4(P_3^3)^4}{P_{12}^{12}} = \frac{4!(3!)^4}{12!}$$

4.41 Follow the instructions given in the MyPersonal Trainer section. The answers are given in the table.

P(A)	P(B)	Conditions for events A and B	P(A ∩ B)	P(A ∪ B)	P(A|B)
.3	.4	Mutually exclusive	0	.3 + .4 = .7	0
.3	.4	Independent	.3(.4) = .12	.3 + .4 − (.3)(.4) = .58	.3
.1	.5	Independent	.1(.5) = .05	.1 + .5 − (.1)(.5) = .55	.1
.2	.5	Mutually exclusive	0	.2 + .5 = .7	0

4.43 **a** $P(A^C) = 1 - P(A) = 1 - \frac{2}{5} = \frac{3}{5}$

b $P(A \cap B)^C = 1 - P(A \cup B) = 1 - \frac{1}{5} = \frac{4}{5}$

4.45 **a** $P(A \cup B) = P(A) + P(B) - P(A \cap B) = 2/5 + 4/5 - 1/5 = 5/5 = 1$

b $P(A \cap B) = P(A|B)P(B) = (1/4)(4/5) = 1/5$

c $P(B \cap C) = P(B|C)P(C) = (1/2)(2/5) = 1/5$

4.47 Refer to the solution to Exercise 4.1 where the six simple events in the experiment are given, with $P(E_i) = 1/6$.

a $S = \{E_1, E_2, E_3, E_4, E_5, E_6\}$ and $P(S) = 6/6 = 1$

b $P(A|B) = \dfrac{P(A \cap B)}{P(B)} = \dfrac{1/3}{1/3} = 1$

c $B = \{E_1, E_2\}$ and $P(B) = 2/6 = 1/3$

d $A \cap B \cap C$ contains no simple events, and $P(A \cap B \cap C) = 0$

e $P(A \cap B) = P(A|B)P(B) = 1(1/3) = 1/3$

f $A \cap C$ contains no simple events, and $P(A \cap C) = 0$

g $B \cap C$ contains no simple events, and $P(B \cap C) = 0$

h $A \cup C = S$ and $P(A \cup C) = 1$

i $B \cup C = \{E_1, E_2, E_4, E_5, E_6\}$ and $P(B \cup C) = 5/6$

4.49 **a** Since A and B are independent, $P(A \cap B) = P(A)P(B) = .4(.2) = .08$.

b $P(A \cup B) = P(A) + P(B) - P(A \cap B) = .4 + .2 - (.4)(.2) = .52$

4.51 **a** Use the definition of conditional probability to find

$$P(B|A) = \dfrac{P(A \cap B)}{P(A)} = \dfrac{.12}{.4} = .3$$

b Since $P(A \cap B) \neq 0$, A and B are not mutually exclusive.

c If $P(B) = .3$, then $P(B) = P(B|A)$ which means that A and B are independent.

4.53 **a** From Exercise 4.52, since $P(A \cap B) = .34$, the two events are not mutually exclusive.

b From Exercise 4.52, $P(A|B) = .425$ and $P(A) = .49$. The two events are not independent.

4.55 Define the following events:
  A: project is approved for funding
  D: project is disapproved for funding
For the first group, $P(A_1) = .2$ and $P(D_1) = .8$. For the second group, $P[\text{same decision as first group}] = .7$ and $P[\text{reversal}] = .3$. That is, $P(A_2|A_1) = P(D_2|D_1) = .7$ and $P(A_2|D_1) = P(D_2|A_1) = .3$.

a $P(A_1 \cap A_2) = P(A_1)P(A_2|A_1) = .2(.7) = .14$

b $P(D_1 \cap D_2) = P(D_1)P(D_2|D_1) = .8(.7) = .56$

c $P(D_1 \cap A_2) + P(A_1 \cap D_2) = P(D_1)P(A_2|D_1) + P(A_1)P(D_2|A_1) = .8(.3) + .2(.3) = .30$

4.57 Refer to Exercise 4.56.

a From the table, $P(A \cap B) = .1$ while $P(A)P(B|A) = (.4)(.25) = .1$

b From the table, $P(A \cap B) = .1$ while $P(B)P(A|B) = (.37)(.1/.37) = .1$

c From the table, $P(A \cup B) = .1 + .27 + .30 = .67$ while $P(A) + P(B) - P(A \cap B) = .4 + .37 - .10 = .67$.

4.59 Fix the birth date of the first person entering the room. Then define the following events:

A_2: second person's birthday differs from the first
A_3: third person's birthday differs from the first and second
A_4: fourth person's birthday differs from all preceding
⋮
A_n: n^{th} person's birthday differs from all preceding

Then $P(A) = P(A_2)P(A_3)\cdots P(A_n) = \left(\dfrac{364}{365}\right)\left(\dfrac{363}{365}\right)\cdots\left(\dfrac{365-n+1}{365}\right)$

since at each step, one less birth date is available for selection. Since event B is the complement of event A,
$$P(B) = 1 - P(A)$$

a For $n = 3$, $P(A) = \dfrac{(364)(363)}{(365)^2} = .9918$ and $P(B) = 1 - .9918 = .0082$

b For $n = 4$, $P(A) = \dfrac{(364)(363)(362)}{(365)^3} = .9836$ and $P(B) = 1 - .9836 = .0164$

4.61 Let events A and B be defined as follows: A: article gets by the first inspector
B: article gets by the second inspector
The event of interest is then the event $A \cap B$, that the article gets by both inspectors. It is given that $P(A) = .1$, and also that $P(B \mid A) = .5$. Applying the Multiplication Rule,
$$P(A \cap B) = P(A)P(B \mid A) = (.1)(.5) = .05$$

4.63 Define   A: smoke is detected by device A
B: smoke is detected by device B
If it is given that $P(A) = .95$, $P(B) = .98$, and $P(A \cap B) = .94$.

a $P(A \cup B) = P(A) + P(B) - P(A \cap B) = .95 + .98 - .94 = .99$

b $P(A^C \cap B^C) = 1 - P(A \cup B) = 1 - .99 = .01$

4.65 Similar to Exercise 4.56.

a $P(A) = \dfrac{54}{1029}$

b $P(F) = \dfrac{517}{1029}$

c $P(A \cap F) = \dfrac{37}{1029}$

d $P(F \mid A) = \dfrac{P(F \cap A)}{P(A)} = \dfrac{37/1029}{54/1029} = \dfrac{37}{54}$

e $P(F \mid B) = \dfrac{P(F \cap B)}{P(B)} = \dfrac{64/1029}{99/1029} = \dfrac{64}{99}$

f $P(F \mid C) = \dfrac{P(F \cap C)}{P(C)} = \dfrac{138/1029}{241/1029} = \dfrac{138}{241}$

g $P(C \mid M) = \dfrac{P(C \cap M)}{P(M)} = \dfrac{103/1029}{512/1029} = \dfrac{103}{512}$

h $P(B^C) = 1 - P(B) = 1 - \dfrac{99}{1029} = \dfrac{930}{1029} = \dfrac{310}{343}$

4.67 Define the following events: A: player makes first shot
B: player makes second shot
The probabilities of events A and B will depend on which player is shooting.

a For Alex, $P(A) = P(B) = 0.228$. Then
$P(A \cap B) = P(A)P(B) = .228(.228) = .052$, since the shots are independent.

b For Jordan, $P(A) = P(B) = 0.26$. The event that Jordan makes exactly one of the two shots will occur if he makes the first and misses the second, or vice versa. Then
$$P(\text{Jordan makes exactly one}) = P(A \cap B^C) + P(A^C \cap B)$$
$$= .26(.74) + .74(.26) = .3848$$

c This probability is the intersection of the individual probabilities for both Jordan and Alex.
$P(\text{Jordan makes both and Alex makes neither}) = [.26(.26)][.772(.772)] = .0403$

4.69 **a** Use the Law of Total Probability, writing
$$P(A) = P(S_1)P(A \mid S_1) + P(S_2)P(A \mid S_2) = .7(.2) + .3(.3) = .23$$

b Use the results of part a in the form of Bayes' Rule:
$$P(S_i \mid A) = \frac{P(S_i)P(A \mid S_i)}{P(S_1)P(A \mid S_1) + P(S_2)P(A \mid S_2)}$$

For $i = 1$, $P(S_1 \mid A) = \frac{.7(.2)}{.7(.2) + .3(.3)} = \frac{.14}{.23} = .6087$

For $i = 2$, $P(S_2 \mid A) = \frac{.3(.3)}{.7(.2) + .3(.3)} = \frac{.09}{.23} = .3913$

4.71 Use the Law of Total Probability, writing $P(A) = P(S_1)P(A \mid S_1) + P(S_2)P(A \mid S_2) = .6(.3) + .4(.5) = .38$.

4.73 Define A: machine produces a defective item
B: worker follows instructions

Then $P(A \mid B) = .01$, $P(B) = .90$, $P(A \mid B^C) = .03$, $P(B^C) = .10$. The probability of interest is
$$P(A) = P(A \cap B) + P(A \cap B^C)$$
$$= P(A \mid B)P(B) + P(A \mid B^C)P(B^C)$$
$$= .01(.90) + .03(.10) = .012$$

4.75 Define L: play goes to the left
R: play goes to the right
S: right guard shifts his stance

a It is given that $P(L) = .3$, $P(R) = .7$, $P(S \mid R) = .8$, $P(S^C \mid L) = .9$, $P(S \mid L) = .1$, and $P(S^C \mid R) = .2$.
Using Bayes' Rule,
$$P(L \mid S^C) = \frac{P(L)P(S^C \mid L)}{P(L)P(S^C \mid L) + P(R)P(S^C \mid R)} = \frac{.3(.9)}{.3(.9) + .7(.2)} = \frac{.27}{.41} = .6585$$

b From part a, $P(R \mid S^C) = 1 - P(L \mid S^C) = 1 - .6585 = .3415$.

c Given that the guard takes a balanced stance, it is more likely (.6585 versus .3415) that the play will go to the left.

Introduction to Probability and Statistics, 2ce

4.77 The probability of interest is $P(A|H)$ which can be calculated using Bayes' Rule and the probabilities given in the exercise.

$$P(A|H) = \frac{P(A)P(H|A)}{P(A)P(H|A) + P(B)P(H|B) + P(C)P(H|C)}$$

$$= \frac{.01(.90)}{.01(.90) + .005(.95) + .02(.75)} = \frac{.009}{.02875} = .3130$$

4.79 **a** Using the probability table,

$P(D) = .08 + .02 = .10$ $\qquad$ $P(D^C) = 1 - P(D) = 1 - .10 = .90$

$P(N|D^C) = \frac{P(N \cap D^C)}{P(D^C)} = \frac{.85}{.90} = .94$ $\qquad$ $P(N|D) = \frac{P(N \cap D)}{P(D)} = \frac{.02}{.10} = .20$

b Using Bayes' Rule, $P(D|N) = \frac{P(D)P(N|D)}{P(D)P(N|D) + P(D^C)P(N|D^C)} = \frac{.10(.20)}{.10(.20) + .90(.94)} = .023$

c Using the definition of conditional probability,

$$P(D|N) = \frac{P(N \cap D)}{P(N)} = \frac{.02}{.87} = .023$$

d $P(\text{false positive}) = P(P|D^C) = \frac{P(P \cap D^C)}{P(D^C)} = \frac{.05}{.90} = .056$

e $P(\text{false negative}) = P(N|D) = \frac{P(N \cap D)}{P(D)} = \frac{.02}{.10} = .20$

f The probability of a false negative is quite high, and would cause concern about the reliability of the screening method.

4.81 **a** The increase in length of life achieved by a cancer patient as a result of surgery is a continuous random variable, since an increase in life (measured in units of time) can take on any of an infinite number of values in a particular interval.

b The tensile strength, in pounds per square inch, of one-inch diameter steel wire cable is a continuous random variable.

c The number of deer killed per year in a state wildlife preserve is a discrete random variable taking the values 0, 1, 2, …

d The number of overdue accounts in a department store at a particular point in time is a discrete random variable, taking the values 0, 1, 2, ….

e Blood pressure is a continuous random variable.

4.83 **a** Since one of the requirements of a probability distribution is that $\sum_x p(x) = 1$, we need

$p(3) = 1 - (.1 + .3 + .3 + .1) = 1 - .8 = .2$

b The probability histogram is shown below.

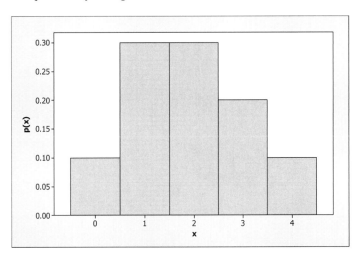

c For the random variable x given here, $\mu = E(x) = \sum xp(x) = 0(.1) + 1(.3) + \cdots + 4(.1) = 1.9$
The variance of x is defined as
$$\sigma^2 = E\left[(x-\mu)^2\right] = \sum (x-\mu)^2 p(x) = (0-1.9)^2(.1) + (1-1.9)^2(.3) + \cdots + (4-1.9)^2(.1) = 1.29$$
and $\sigma = \sqrt{1.29} = 1.136$.

d Using the table form of the probability distribution given in the exercise, $P(x > 2) = .2 + .1 = .3$.

e $P(x \le 3) = 1 - P(x = 4) = 1 - .1 = .9$.

4.85 For the probability distribution given in this exercise,
$$\mu = E(x) = \sum xp(x) = 0(.1) + 1(.4) + 2(.4) + 3(.1) = 1.5.$$

4.87 **a-b** On the first try, the probability of selecting the proper key is 1/4. If the key is not found on the first try, the probability changes on the second try. Let F denote a failure to find the key and S denote a success. The random variable is x, the number of keys tried before the correct key is found. The four associated simple events are shown below.

E_1: S $(x = 1)$ $\quad$ E_3: FFS $(x = 3)$
E_2: FS $(x = 2)$ $\quad$ E_4: FFFS $(x = 4)$

c-d Then
$$p(1) = P(x = 1) = P(S) = 1/4$$
$$p(2) = P(x = 2) = P(FS) = P(F)P(S) = (3/4)(1/3) = 1/4$$
$$p(3) = P(x = 3) = P(FFS) = P(F)P(F)P(S) = (3/4)(2/3)(1/2) = 1/4$$
$$p(4) = P(x = 4) = P(FFFS) = P(F)P(F)P(F)P(S) = (3/4)(2/3)(1/2)(1) = 1/4$$

The probability distribution and probability histogram follow.

x	1	2	3	4
$p(x)$	1/4	1/4	1/4	1/4

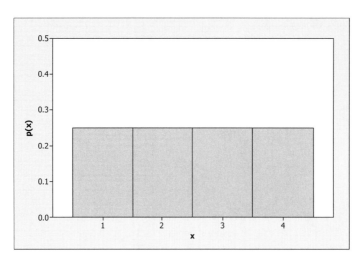

4.89 a-b Let W_1 and W_2 be the two women while M_1, M_2 and M_3 are the three men. There are 10 ways to choose the two people to fill the positions. Let x be the number of women chosen. The 10 equally likely simple events are:

E_1: W_1W_2 $(x=2)$ E_6: W_2M_2 $(x=1)$

E_2: W_1M_1 $(x=1)$ E_7: W_2M_3 $(x=1)$

E_3: W_1M_2 $(x=1)$ E_8: M_1M_2 $(x=0)$

E_4: W_1M_3 $(x=1)$ E_9: M_1M_3 $(x=0)$

E_5: W_2M_1 $(x=1)$ E_{10}: M_2M_3 $(x=0)$

The probability distribution for x is then $p(0) = 3/10$, $p(1) = 6/10$, $p(2) = 1/10$. The probability histogram is shown below.

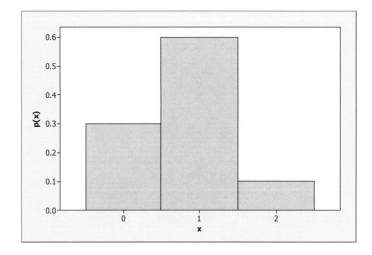

4.91 Let x be the number of drillings until the first success (oil is struck). It is given that the probability of striking oil is $P(O) = .1$, so that the probability of no oil is $P(N) = .9$

a $p(1) = P[\text{oil struck on first drilling}] = P(O) = .1$

$p(2) = P[\text{oil struck on second drilling}]$. This is the probability that oil is not found on the first drilling, but is found on the second drilling. Using the Multiplication Law,
$$p(2) = P(NO) = (.9)(.1) = .09.$$
Finally, $p(3) = P(NNO) = (.9)(.9)(.1) = .081$.

b-c For the first success to occur on trial x, $(x-1)$ failures must occur before the first success. Thus, $p(x) = P(NNN\ldots NNO) = (.9)^{x-1}(.1)$ since there are $(x-1)$ N's in the sequence. The probability histogram is shown below.

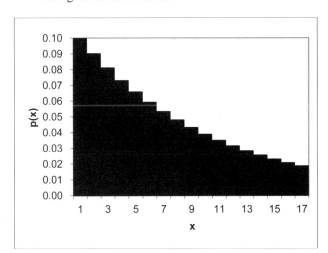

4.93 Refer to Exercise 4.92. For a general value $P(A) = p$ and $P(B) = 1 - p$, we showed that the probability distribution for x is

$$p(3) = p^3 + (1-p)^3$$
$$p(4) = 3p^3(1-p) + 3p(1-p)^3$$
$$p(5) = 6p^3(1-p)^2 + 6p^2(1-p)^3$$

For the three values of p given in this exercise, the probability distributions and values of $E(x)$ are given in the table below.

$P(A) = .6$		$P(A) = .5$		$P(A) = .9$	
x	$p(x)$	x	$p(x)$	x	$p(x)$
3	.2800	3	.25	3	.7300
4	.3744	4	.375	4	.2214
5	.3456	5	.375	5	.0486

a $E(x) = 4.0656$ **b** $E(x) = 4.125$ **c** $E(x) = 3.3186$

4.95 The random variable G, total gain to the insurance company, will be D if there is no theft, but D – 50,000 if there is a theft during a given year. These two events will occur with probability .99 and .01, respectively. Hence, the probability distribution for G is given below.

G	$p(G)$
D	.99
D – 50,000	.01

The expected gain is
$$E(G) = \sum Gp(G) = .99D + .01(D - 50,000)$$
$$= D - 50,000$$

In order that $E(G) = 1000$, it is necessary to have $1000 = D - 500$ or $D = \$1500$.

Introduction to Probability and Statistics, 2ce

4.97 **a** Similar to Exercise 4.91. For the first person who does not find time to eat vegetables on call x, $(x-1)$ people who do find time to eat vegetable must be called. Since 41% Canadians have no time for enough vegetables, we take the percentage of Canadians, who do find time to eat vegetables to be 51%. Thus, $p(x) = P(NNN\ldots NNY) = (.59)^{x-1}(.41)$

b As with other phone surveys, there is always a problem of non-response – people who do not answer the telephone or decline to participate in the survey. Also, there is a problem of truthfulness of the response for a question such as this which may be a sensitive subject for some people.

4.99 We are asked to find the premium that the insurance company should charge in order to break even. Let c be the unknown value of the premium and x be the gain to the insurance company caused by marketing the new product. There are three possible values for x. If the product is a failure or moderately successful, x will be negative; if the product is a success, the insurance company will gain the amount of the premium and x will be positive. The probability distribution for x follows:

x	$p(x)$
c	.94
$-80,000 + c$	.01
$-25,000 + c$	.05

In order to break even, $E(x) = \sum xp(x) = 0$

$$.94(c) + .01(-80,000 + c) + (.05)(-25,000 + c) = 0$$

Therefore, $-800 - 1250 + (.01 + .05 + .94)c = 0$

$c = 2050$

Hence, the insurance company should charge a premium of $2,050.

4.101 Define the following events: A: worker fails to report fraud
B: worker suffers reprisal

It is given that $P(B \mid A^C) = .23$ and $P(A) = .69$. The probability of interest is

$$P(A^C \cap B) = P(B \mid A^C)P(A^C) = .23(.31) = .0713$$

4.103 Refer to Exercise 4.102. There are 12 simple events, each with equal probability 1/12. By summing the probabilities of simple events in the events of interest we have

$P(A) = 6/12 = 1/2$ $P(A \cup B) = 10/12 = 5/6$

$P(B) = 8/12 = 2/3$ $P(C) = 2/12 = 1/6$

$P(A \cap B) = 4/12 = 1/3$ $P(A \cap C) = 0$

$P(A \cup C) = 8/12 = 2/3$

4.105 Two systems are selected from seven, three of which are defective. Denote the seven systems as G_1, G_2, G_3, G_4, D_1, D_2, D_3 according to whether they are good or defective. Each simple event will represent a particular pair of systems chosen for testing, and the sample space, consisting of 21 pairs, is shown below.

G_1G_2 G_1D_1 G_2D_3 G_4D_2 G_1G_3 G_1G_2 G_3D_1 G_4D_3
G_1G_4 G_1D_3 G_3D_2 D_1D_2 G_2G_3 G_2D_1 G_3D_3 D_1D_3
G_2G_4 G_2D_2 G_4D_1 D_2D_3 G_3G_4

Note that the two systems are drawn simultaneously and that order is unimportant in identifying a simple event. Hence, the pairs G_1G_2 and G_2G_1 are not considered to represent two different simple events. The event A, "no defectives are selected", consists of the simple events G_1G_2, G_1G_3, G_1G_4, G_2G_3, G_2G_4, G_3G_4. Since the systems are selected at random, any pair has an equal probability of being selected. Hence, the probability assigned to each simple event is 1/21 and $P(A) = 6/21 = 2/7$.

4.107 The random variable x, defined as the number of householders insured against fire, can assume the values 0, 1, 2, 3 or 4. The probability that, on any of the four draws, an insured person is found is .6; hence, the probability of finding an uninsured person is .4. Note that each numerical event represents the intersection of the results of four independent draws.

a $P[x=0] = (.4)(.4)(.4)(.4) = .0256$, since all four people must be uninsured.

b $P[x=1] = 4(.6)(.4)(.4)(.4) = .1536$ (Note: the 4 appears in this expression because $x=1$ is the union of four mutually exclusive events. These represent the 4 ways to choose the single insured person from the fours.)

c $P[x=2] = 6(.6)(.6)(.4)(.4) = .3456$, since the two insured people can be chosen in any of 6 ways.

d $P[x=3] = 4(.6)^3(.4) = .3456$ and $P[x=4] = (.6)^4 = .1296$.

Then $P[\text{at least three insured}] = p(3) + p(4) = .3456 + .1296 = .4752$

4.109 a Similar to Exercise 4.15. $P(\text{cold}) = \dfrac{49+43+34}{276} = \dfrac{126}{276} = .4565$

b Define: F: person has four or five relationships
S: person has six or more relationships
Then for the two people chosen from the total 276,

$P(\text{one F and one S}) = P(F \cap S) + P(S \cap F)$

$= \left(\dfrac{100}{276}\right)\left(\dfrac{96}{275}\right) + \left(\dfrac{96}{276}\right)\left(\dfrac{100}{275}\right) = .2530$

c $P(\text{Three or fewer} \mid \text{cold}) = \dfrac{P(\text{three or fewer} \cap \text{cold})}{P(\text{cold})} = \dfrac{49/276}{126/276} = \dfrac{49}{126} = .3889$

4.111 Similar to Exercise 4.7. An investor can invest in three of five recommended stocks. Unknown to him, only 2 out of 5 will show substantial profit. Let P_1 and P_2 be the two profitable stocks. A typical simple event might be ($P_1P_2N_3$), which represents the selection of two profitable and one nonprofitable stock. The ten simple events are listed below:

E_1: ($P_1P_2N_1$)　　E_2: ($P_1P_2N_2$)　　E_3: ($P_1P_2N_3$)　　E_4: ($P_2N_1N_2$)
E_5: ($P_2N_1N_3$)　　E_6: ($P_2N_2N_3$)　　E_7: ($N_1N_2N_3$)　　E_8: ($P_1N_1N_2$)
E_9: ($P_1N_1N_3$)　　E_{10}: ($P_1N_2N_3$)

Then $P[\text{investor selects two profitable stocks}] = P(E_1) + P(E_2) + P(E_3) = 3/10$ since the simple events are equally likely, with $P(E_i) = 1/10$. Similarly,

$P[\text{investor selects only one of the profitable stocks}] = P(E_4) + P(E_5) + P(E_6) + P(E_8) + P(E_9) + P(E_{10}) = 6/10$

4.113 a Define the following events: B_1: client buys on first contact
B_2: client buys on second contact
Since the client may buy on either the first of the second contact, the desired probability is

$P[\text{client will buy}] = P[\text{client buys on first contact}] + P[\text{client doesn't buy on first, but buys on second}]$

$= P(B_1) + (1 - P(B_1))P(B_2) = .4 + (1-.4)(.55)$

$= .73$

b The probability that the client will not buy is one minus the probability that the client will buy, or $1 - .73 = .27$.

4.115 Define the following events: A: first system fails
B: second system fails

A and B are independent and $P(A) = P(B) = .001$. To determine the probability that the combined missile system does not fail, we use the complement of this event; that is,

$$P[\text{system does not fail}] = 1 - P[\text{system fails}] = 1 - P(A \cap B)$$
$$= 1 - P(A)P(B) = 1 - (.001)^2$$
$$= .999999$$

4.117 a. If the numbers are drawn without replacement, 6 numbers can be picked from 49 in
$C_6^{49} = \dfrac{49!}{6!43!} = 13,983,816$ ways.

b. The probability of being a 6/49 winner with a single ticket purchase $= \dfrac{1}{C_6^{49}} = \dfrac{1}{13,983,816}$.

c. There are $6! = 6 \times 5 \times 4 \times 3 \times 2 \times 1 = 720$ different orders of the numbers 5, 11, 20, 30, 37 and 43.

4.119 a Consider a single trial which consists of tossing two coins. A match occurs when either HH or TT is observed. Hence, the probability of a match on a single trial is $P(HH) + P(TT) = 1/4 + 1/4 = 1/2$. Let MMM denote the event "match on trials 1, 2, and 3". Then
$$P(MMM) = P(M)P(M)P(M) = (1/2)^3 = 1/8.$$

b On a single trial the event A, "two trails are observed" has probability $P(A) = P(TT) = 1/4$. Hence, in three trials
$$P(AAA) = P(A)P(A)P(A) = (1/4)^3 = 1/64$$

c This low probability would not suggest collusion, since the probability of three matches is low only if we assume that each student is merely guessing at each answer. If the students have studied together or if they both know the correct answer, the probability of a match on a single trial is no longer 1/2, but is substantially higher. Hence, the occurrence of three matches is not unusual.

4.121 Define R: the employee remains 10 years or more

a The probability that the man will stay less than 10 years is $P(R^C) = 1 - P(R) = 1 - 1/6 = 5/6$

b The probability that the man and the woman, acting independently, will both work less than 10 years is $P(R^C R^C) = P(R^C)P(R^C) = (5/6)^2 = 25/36$

c The probability that either or both people work 10 years or more is
$$1 - P(R^C R^C) = 1 - (5/6)^2 = 1 - 25/36 = 11/36$$

4.123 Define the events: A: the man waits five minutes or longer
B: the woman waits five minutes or longer

The two events are independent, and $P(A) = P(B) = .2$.

a $P(A^C) = 1 - P(A) = .8$

b $P(A^C B^C) = P(A^C)P(B^C) = (.8)(.8) = .64$

c $P[\text{at least one waits five minutes or longer}]$
$= 1 - P[\text{neither waits five minutes or longer}] = 1 - P(A^C B^C) = 1 - .64 = .36$

4.125 It is given that 40% of all people in a community favor the development of a mass transit system. Thus, given a person selected at random, the probability that the person will favor the system is .4. Since the pollings are independent events, when four people are selected at random,

$P[\text{all 4 favor the system}] = (.4)^4 = .0256$

Similarly, $P[\text{none favor the system}] = (1-.4)^4 = .1296$

4.127 Since the first pooled test is positive, we are interested in the probability of requiring five single tests to detect the disease in the single affected person. There are $(5)(4)(3)(2)(1)$ ways of ordering the five tests, and there are $4(3)(2)(1)$ ways of ordering the tests so that the diseased person is given the final test. Hence, the desired probability is $\dfrac{4!}{5!} = \dfrac{1}{5}$.

If two people are diseased, six tests are needed if the last two tests are given to the diseased people. There are $3(2)(1)$ ways of ordering the tests of the other three people and $2(1)$ ways of ordering the tests of the two diseased people. Hence, the probability that six tests will be needed is $\dfrac{2!3!}{5!} = \dfrac{1}{10}$.

4.129 The necessary probabilities can be found by summing the necessary cells in the probability table and dividing by 220, the total number of firms. Define the following events:
   A: the company is located in City A
   B: the company is located in city B
   F: the company has flextime schedules

a $P(A) = \dfrac{114}{220} = .5182$

b $P(B \cap F) = \dfrac{25}{220} = .1136$

c $P(F^C) = \dfrac{156}{220} = .7091$

d $P(B \mid F) = \dfrac{P(B \cap F)}{P(F)} = \dfrac{25/220}{64/220} = \dfrac{25}{64} = .3906$

4.131 a Define P: shopper prefers Pepsi and C: shopper prefers Coke. Then if there is actually no difference in the taste, $P(P) = P(C) = 1/2$ and

$P(\text{all four prefer Pepsi}) = P(PPPP) = [P(P)]^4 = \left(\dfrac{1}{2}\right)^4 = \dfrac{1}{16} = .0625$

b

$P(\text{exactly one prefers Pepsi}) = P(PCCC) + P(CPCC) + P(CCPC) + P(CCCP)$

$= 4P(P)[P(C)]^3 = 4\left(\dfrac{1}{2}\right)\left(\dfrac{1}{2}\right)^3 = \dfrac{4}{16} = .25$

4.133 **a**   There are six volunteers, from whom we must choose two people for the committee. The number of choices is then $C_2^6 = \dfrac{6(5)}{2(1)} = 15$

If the number of women chosen from the two women is x, and the number of men chosen from the four men must be $2-x$. Then

$$P(x=0) = \dfrac{C_0^2 C_2^4}{15} = \dfrac{6}{15}$$

$$P(x=1) = \dfrac{C_1^2 C_1^4}{15} = \dfrac{8}{15}$$

$$P(x=2) = \dfrac{C_2^2 C_0^4}{15} = \dfrac{1}{15}$$

and the probability distribution for x is shown in the table.

x	0	1	2
$p(x)$	6/15	8/15	1/15

b $P(x=2) = \dfrac{1}{15}$

$\sigma^2 = \sum (x-\mu)^2 p(x) = (0-\tfrac{2}{3})^2 \left(\dfrac{6}{15}\right) + (1-\tfrac{2}{3})^2 \left(\dfrac{8}{15}\right) + (2-\tfrac{2}{3})^2 \left(\dfrac{1}{15}\right) = \dfrac{48}{135} = \dfrac{16}{45}$

c $P(x=2) = \dfrac{1}{15}$

4.135   Refer to the **Tossing Dice** applet, in which the simple events for this experiment are displayed. Each simple event has a particular value of T associated with it, and by summing the probabilities of all simple events producing a particular value of T, the following probability distribution is obtained. The distribution is mound-shaped.

a-b

T	2	3	4	5	6	7	8	9	10	11	12
p(T)	$\dfrac{1}{36}$	$\dfrac{2}{36}$	$\dfrac{3}{36}$	$\dfrac{4}{36}$	$\dfrac{5}{36}$	$\dfrac{6}{36}$	$\dfrac{5}{36}$	$\dfrac{4}{36}$	$\dfrac{3}{36}$	$\dfrac{2}{36}$	$\dfrac{1}{36}$

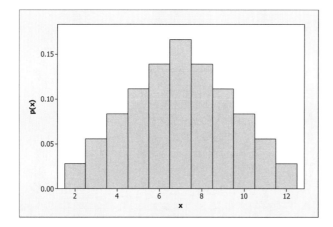

4.137 **a-b** When there is only one fair coin, $P(x=0) = P(T) = .5$ and $P(x=1) = P(T) = .5$. The probability histogram is shown below, along with the simulation using the **Flipping Fair Coins** applet.

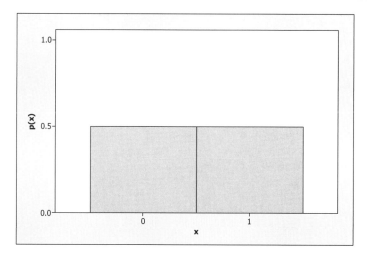

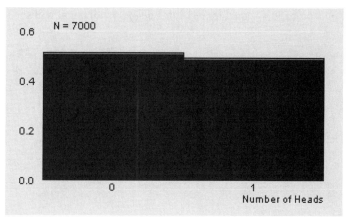

Case Study:
Probability and Decision Making in the Congo

Let p be the probability that a single person successfully completes the jump, and assume that the event that any one person successfully completes the jump is independent of the outcome of the other 11 individual jumps. Define the events:

    J: all members of the team successfully complete the jump
    A_1: Ross successfully completes the jump
    A_2: Elliot successfully completes the jump
    A_3: Munro successfully completes the jump
    $\vdots$
    A_{12}: Porter 8 successfully completes the jump

Then J is the intersection of the events $A_1, A_2, \ldots, A_{12}$ and

$$P(J) = P(A_1 \cap A_2 \cap A_3 \ldots \cap A_{12}) = P(A_1)P(A_2)\cdots P(A_{12}) = p(p)(p)\cdots p = p^{12}$$

Substituting $P(J) = .7980$ into this expression and solving for p, we find $p = .9814$. Thus Houston is basing its probability of a successful team jump on a probability equal to .9814 that a single individual will successfully land in the soft volcanic scree.

Project 4-A: Child Safety Seat Survey, Parts (Continued from Project 3-A)

a   The probability table is as follows:

Age Group	Rear-Facing Infant Seat	Forward-facing Infant Seat	Booster Seat	Seat Belt Only	Total
Infant	0.07535	0.02165	0.00042	0.00000	0.09742
Toddler	0.02040	0.20108	0.04871	0.00125	0.27144
School	0.00000	0.04080	0.18734	0.13530	0.36345
Older	0.00000	0.00000	0.00666	0.26103	0.26769
Total	0.09575	0.26353	0.24313	0.39759	1.00000

b   The probability that a random child is in a rear-facing infant seat is the total probability for that column, 0.09575.

c   The probability that a random child is an infant is the total probability in the first row, 0.09742.

d   Of the 652 toddlers, 117 use the booster seat, and so 117/652 = 0.17945 is the probability that a random child is in a booster seat, given that they are a toddler.

e   Of the 230 children who are in a rear-facing infant seat, 181 of them are infants, and so 181/230 = 0.78696 is the probability the child is an infant given that they are in a rear-facing infant seat.

f   The probability a random child is a toddler is 0.27144. The probability a random child is in a forward-facing infant seat is 0.26353. If we add up these two probabilities we obtain the correct answer, provided that we subtract their joint probability once (otherwise, we would be double-counting that specific joint probability). Thus, the answer is 0.27144 + 0.26353 - 0.20108 = 0.33389.

g   Of the 652 children who are toddlers, 483 are in a forward-facing infant seat, and so 483/652 = 0.74080 is the probability that the child is in a forward-facing infant seat given s(he) is a toddler.

h   No, the types of restraint the child uses and age are not mutually exclusive events. Both events can simultaneously occur with non-zero probability.

i   No, the types of restraint the child uses and age are not independent events. The type of restraint used clearly changes with age.

j   Given that 633 children were using a forward-facing infant seat and 483 of these were toddlers, the answer is 483/633 = 0.76303.

Project 4-B: False Results in Medical Testing

a The probability the test comes backs positive is equal to the probability that it is positive and the person truly has it .02(.9) = 0.018, plus the probability the test comes back positive but the person truly does not have the disease .98(.11) = 0.1078; the sum of both of these is 0.1258.

b No, this is not surprising, given that 11% of those who do not have the disease showed a false-positive result. We did not expect the number to be higher. An alternative way might be to set up the whole probability table.

c This conditional probability equals P(truly has the disease and the test was positive)/P(test was positive) or 0.018/0.1258 = 0.143084.

d This conditional probability equals P(truly has the disease and the test was negative)/P(test was negative) or 0.02(0.10)/[0.02(.10) + 0.98(0.89)] = 0.002288.

e A "false-positive" indicates that the person doesn't have the disease, but the test says they do. By contrast, a "false-negative" indicates that the person does truly have the disease, but the test says they don't.

Project 4-C: Selecting Condiments

a Let M denote mustard and N non-mustard. Then,

For $P(x=0)$, all 3 condiments must be non-mustard: $NNN = (7/10)(6/9)(5/8) = 0.291667$.

For $P(x=1)$, $MNN + NMN + NNM = (3/10)(7/9)(6/8) + (7/10)(3/9)(6/8) + (7/10)(6/9)(3/8) = 0.525$.

For $P(x=2)$, $MMN + MNM + NMM = (3/10)(2/9)(7/8) + (3/10)(7/9)(2/8) + (7/10)(3/9)(2/8) = 0.175$.

For $P(x=3)$, $MMM = (3/10)(2/9)(1/8) = 0.008333$.

In summary, $P(x=0) = 0.291667$; $P(x=1) = 0.525$; $P(x=2) = 0.175$; $P(x=3) = 0.008333$.

b $\mu = \text{Mean}(x) = 0(0.291667) + 1(0.525) + 2(0.175) + 3(0.008333) = 0.9$;

$\sigma^2 = \text{Var}(x) = \Sigma(x - \mu)^2 p(x) = (0-0.9)^2(0.291667) + (1-0.9)^2(0.525) + (2-0.9)^2(0.175) + (3-0.9)^2(0.008333) = 0.49$.

c The answer is $p(0) + p(1) = 0.291667 + 0.525 = 0.816667$.

d The answer is 1-P(no mustard was selected) = $1 - 0.291667 = 0.708333$.

e $P(x$ is within 1 standard deviation of the mean$) = P(0.9-0.7 < x < 0.9 + 0.7) = P(0.2 < x < 1.6) = P(x=1) = 0.525$.

f (i) The probability distribution is the same as in part (a), since there are the same number of ketchup and mustard. When ($x=0$), the total winnings, y, is -\$15(3) = -\$45; When ($x=1$), $y = \$25-\$15-\$15 = -\$5$; When ($x=2$), $y = \$25+\$25-\$15 = \$35$; and when ($x=3$), $y = \$25(3) = \$75$.

Thus, in summary, $P(y = -\$45) = .291667$; $P(y = -\$5) = 0.525$; $P(y = \$35) = 0.175$; $P(y = \$75) = 0.008333$.

(ii) Clearly, all of the probabilities are positive, and they all sum to 1.

(iii) The expected value of y is $-45(.291667) - 5(0.525) + 35(0.175) + 75(0.008333) = -\$9.00$ (i.e. the expected loss is \$9).

Chapter 5: Several Useful Discrete Distributions

5.1 Follow the instructions in the My Personal Trainer section. The answers are shown in the tables below.

k	0	1	2	3	4	5	6	7	8
$P(x \leq k)$	.000	.001	.011	.058	.194	.448	.745	.942	1.000

The Problem	List the Values of x	Write the probability	Rewrite the Probability	Find the probability
Three or less	0, 1, 2, 3	$P(x \leq 3)$		.058
Three or more	3, 4, 5, 6, 7, 8	$P(x \geq 3)$	1 - $P(x \leq 2)$	$1 - .011 = .989$
More than three	4, 5, 6, 7, 8	$P(x > 3)$	1 - $P(x \leq 3)$	$1 - .058 = .942$
Fewer than three	0, 1, 2	$P(x < 3)$	$P(x \leq 2)$	.011
Between 3 and 5 (inclusive)	3, 4, 5	$P(3 \leq x \leq 5)$	$P(x \leq 5) - P(x \leq 2)$	$.448 - .011 = .437$
Exactly three	3	$P(x = 3)$	$P(x \leq 3) - P(x \leq 2)$	$.058 - .011 = .047$

5.3 The random variable x is not a binomial random variable since the balls are selected without replacement. For this reason, the probability p of choosing a red ball changes from trial to trial.

5.5   a $C_2^8(.3)^2(.7)^6 = \dfrac{8(7)}{2(1)}(.09)(.117649) = .2965$

      b $C_0^4(.05)^0(.95)^4 = (.95)^4 = .8145$

      c $C_3^{10}(.5)^3(.5)^7 = \dfrac{10(9)(8)}{3(2)(1)}(.5)^{10} = .1172$

      d $C_1^7(.2)^1(.8)^6 = 7(.2)(.8)^6 = .3670$

5.7   a   For $n = 7$ and $p = .3$, $P(x = 4) = C_4^7(.3)^4(.7)^3 = .097$.

      b   These probabilities can be found individually using the binomial formula, or alternatively using the cumulative binomial tables in Appendix I.
$P(x \leq 1) = p(0) + p(1)$
$= C_0^7(.3)^0(.7)^7 + C_1^7(.3)^1(.7)^6$
$= (.7)^7 + 7(.3)(.7)^6 = .08235 + .24706 = .329$
or directly from the binomial tables in the row marked $a = 1$.

      c   Refer to part **b**. $P(x > 1) = 1 - P(x \leq 1) = 1 - .329 = .671$.

      d $\mu = np = 7(.3) = 2.1$

      e $\sigma = \sqrt{npq} = \sqrt{7(.3)(.7)} = \sqrt{1.47} = 1.212$

5.9 Notice that when $p = .8$, $p(x) = C_x^6(.8)^x(.2)^{6-x}$. In Exercise 5.8, with $p = .2$, $p(x) = C_x^6(.2)^x(.8)^{6-x}$. The probability that $x = k$ when $p = .8$ --- $C_k^6(.8)^k(.2)^{n-k}$ --- is the same as the probability that $x = n - k$ when $p = .2$ --- $C_{n-k}^6(.2)^{n-k}(.8)^k$. This follows because

$$C_k^n = \frac{n!}{k!(n-k)!} = C_{n-k}^n$$

Therefore, the probabilities $p(x)$ for a binomial random variable x when $n = 6$ and $p = .8$ will be the mirror images of those found in Exercise 5.8 as shown in the table. The probability histogram is shown on the next page.

x	0	1	2	3	4	5	6
$p(x)$	.000	.002	.015	.082	.246	.393	.262

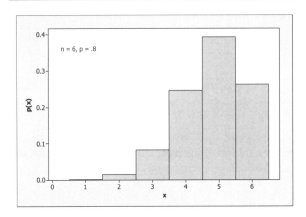

5.11 **a** For $n = 10$ and $p = .4$, $P(x = 4) = C_4^{10}(.4)^4(.6)^6 = .251$.

b To calculate $P(x \geq 4) = p(4) + p(5) + \cdots + p(10)$ it is easiest to write
$$P(x \geq 4) = 1 - P(x < 4) = 1 - P(x \leq 3).$$
These probabilities can be found individually using the binomial formula, or alternatively using the cumulative binomial tables in Appendix I.

$P(x = 0) = C_0^{10}(.4)^0(.6)^{10} = .006$ $P(x = 1) = C_1^{10}(.4)^1(.6)^9 = .040$

$P(x = 2) = C_2^{10}(.4)^2(.6)^8 = .121$ $P(x = 3) = C_3^{10}(.4)^3(.6)^7 = .215$

The sum of these probabilities gives $P(x \leq 3) = .382$ and $P(x \geq 4) = 1 - .382 = .618$.

c Use the results of parts **a** and **b**.
$$P(x > 4) = 1 - P(x \leq 4) = 1 - (.382 + .251) = .367$$

d From part **c**, $P(x \leq 4) = P(x \leq 3) + P(x = 4) = .382 + .251 = .633$.

e $\mu = np = 10(.4) = 4$

f $\sigma = \sqrt{npq} = \sqrt{10(.4)(.6)} = \sqrt{2.4} = 1.549$

5.13 **a** $P[x \geq 4] = 1 - P[x \leq 3] = 1 - .099 = .901$

b $P[x = 2] = P[x \leq 2] - P[x \leq 1] = .017 - .002 = .015$

c $P[x < 2] = P[x \leq 1] = .002$

d $P[x > 1] = 1 - P[x \leq 1] = 1 - .002 = .998$

5.15 **a** $P[x<12]=P[x\leq 11]=.748$

b $P[x\leq 6]=.610$

c $P[x>4]=1-P[x\leq 4]=1-.633=.367$

d $P[x\geq 6]=1-P[x\leq 5]=1-.034=.966$

e $P[3<x<7]=P[x\leq 6]-P[x\leq 3]=.828-.172=.656$

5.17 **a** $\mu=100(.01)=1;\ \sigma=\sqrt{100(.01)(.99)}=.99$

b $\mu=100(.9)=90;\ \sigma=\sqrt{100(.9)(.1)}=3$

c $\mu=100(.3)=30;\ \sigma=\sqrt{100(.3)(.7)}=4.58$

d $\mu=100(.7)=70;\ \sigma=\sqrt{100(.7)(.3)}=4.58$

e $\mu=100(.5)=50;\ \sigma=\sqrt{100(.5)(.5)}=5$

5.19 **a** $p(0)=C_0^{20}(.1)^0(.9)^{20}=.1215767$

$p(1)=C_1^{20}(.1)^1(.9)^{19}=.2701703$

$p(2)=C_2^{20}(.1)^2(.9)^{18}=.2851798$

$p(3)=C_3^{20}(.1)^3(.9)^{17}=.1901199$

$p(4)=C_4^{20}(.1)^4(.9)^{16}=.0897788$

so that $P[x\leq 4]=p(0)+p(1)+p(2)+p(3)+p(4)=.9568255$

b Using Table 1, Appendix I, $P[x\leq 4]$ is read directly as .957.

c Adding the entries for $x=0,1,2,3,4$, we have $P[x\leq 4]=.956826$.

d $\mu=np=20(.1)=2$ and $\sigma=\sqrt{npq}=\sqrt{1.8}=1.3416$

e For $k=1$, $\mu\pm\sigma=2\pm 1.342$ or .658 to 3.342 so that

$P[.658\leq x\leq 3.342]=P[1\leq x\leq 3]=.2702+.2852+.1901=.7455$

For $k=2$, $\mu\pm 2\sigma=2\pm 2.683$ or $-.683$ to 4.683 so that

$P[-.683\leq x\leq 4.683]=P[0\leq x\leq 4]=.9569$

For $k=3$, $\mu\pm 3\sigma=2\pm 4.025$ or -2.025 to 6.025 so that

$P[-2.025\leq x\leq 6.025]=P[0\leq x\leq 6]=.9977$

f The results are consistent with Tchebysheff's Theorem and the Empirical Rule.

5.21 Although there are trials (telephone calls) which result in either a person who will answer (S) or a person who will not (F), the number of trials, *n*, is not fixed in advance. Instead of recording *x*, the number of *successes* in *n* trials, you record *x*, the number of *trials* until the first success. This is *not* a binomial experiment.

5.23 Define *x* to be the number of alarm systems that are triggered. Then $p=P[\text{alarm is triggered}]=.99$ and $n=9$. Since there is a table available in Appendix I for $n=9$ and $p=.99$, you should use it rather than the binomial formula to calculate the necessary probabilities.

a $P[\text{at least one alarm is triggered}]=P(x\geq 1)=1-P(x=0)=1-.000=1.000$.

b $P[\text{more than seven}]=P(x>7)=1-P(x\leq 7)=1-.003=.997$

c $P[\text{eight or fewer}]=P(x\leq 8)=.086$

5.25 Define x to be the number of cars that are black. Then $p = P[\text{black}] = .1$ and $n = 25$. Use Table 1 in Appendix I.

a $P(x \geq 5) = 1 - P(x \leq 4) = 1 - .902 = .098$

b $P(x \leq 6) = .991$

c $P(x > 4) = 1 - P(x \leq 4) = 1 - .902 = .098$

d $P(x = 4) = P(x \leq 4) - P(x \leq 3) = .902 - .764 = .138$

e $P(3 \leq x \leq 5) = P(x \leq 5) - P(x \leq 2) = .967 - .537 = .430$

f $P(\text{more than 20 }not\text{ black}) = P(\text{less than 5 black}) = P(x \leq 4) = .902$

5.27 Define a success to be a patient who fails to pay his bill and is eventually forgiven. Assuming that the trials are independent and that p is constant from trial to trial, this problem satisfies the requirements for the binomial experiment with $n = 4$ and $p = .3$. You can use either the binomial formula or Table 1.

a $P[x = 4] = p(4) = C_4^4 (.3)^4 (.7)^0 = (.3)^4 = .0081$

b $P[x = 1] = p(1) = C_1^4 (.3)^1 (.7)^3 = 4(.3)(.7)^3 = .4116$

c $P[x = 0] = C_0^4 (.3)^0 (.7)^4 = (.7)^4 = .2401$

5.29 Define x to be the number of fields infested with whitefly. Then $p = P[\text{infected field}] = .1$ and $n = 100$.

a $\mu = np = 100(.1) = 10$

b Since n is large, this binomial distribution should be fairly mound-shaped, even though $p = .1$. Hence you would expect approximately 95% of the measurements to lie within two standard deviation of the mean with $\sigma = \sqrt{npq} = \sqrt{100(.1)(.9)} = 3$. The limits are calculated as
$$\mu \pm 2\sigma \Rightarrow 10 \pm 6 \text{ or from 4 to 16}$$

c From part **b**, a value of $x = 25$ would be very unlikely, assuming that the characteristics of the binomial experiment are met and that $p = .1$. If this value were actually observed, it might be possible that the trials (fields) are not independent. This could easily be the case, since an infestation in one field might quickly spread to a neighboring field. This is evidence of *contagion*.

5.31 Define x to be the number of travelers listing "traffic and other drivers" as their pet peeve. Then, $n = 8$ and $p = .33$.

a $P(x = 8) = C_8^8 (.33)^8 (.67)^0 = .0001406$

b From the Minitab **Probability Density Function**, read $P(x = 8) = .000141$.

c Use the Minitab **Cumulative Distribution Function** to find $P(x \leq 7) = .99986$.

5.33 Define x to be the number of Canadians who are "tasters". Then, $n = 20$ and $p = .7$. Using the binomial tables in Appendix I,

a $P(x \geq 17) = 1 - P(x \leq 16) = 1 - .893 = .107$

b $P(x \leq 15) = .762$

5.35 Follow the instructions in the My Personal Trainer section. The answers are shown in the table below.

Probability	Formula	Calculated value
$P(x=0)$	$\dfrac{2.5^0 e^{-2.5}}{0!}$	.0821
$P(x=1)$	$\dfrac{2.5^1 e^{-2.5}}{1!}$	.2052
$P(x=2)$	$\dfrac{2.5^2 e^{-2.5}}{2!}$	.2565
P(2 or fewer successes)	$P(x=0) + P(x=1) + P(x=2)$	.5438

5.37 Follow the instructions in the My Personal Trainer section. The answers are shown in the tables below.

k	0	1	2	3	4	5	6	7	8	9	10
$P(x \le k)$	.050	.199	.423	.647	.815	.916	.966	.988	.996	.999	1.000

5.39 Using $p(x) = \dfrac{\mu^x e^{-\mu}}{x!} = \dfrac{2^x e^{-2}}{x!}$,

a $\quad P[x=0] = \dfrac{2^0 e^{-2}}{0!} = .135335$

b $\quad P[x=1] = \dfrac{2^1 e^{-2}}{1!} = .27067$

c $\quad P[x>1] = 1 - P[x \le 1] = 1 - .135335 - .27067 = .593994$

d $\quad P[x=5] = \dfrac{2^5 e^{-2}}{5!} = .036089$

The Problem	List the Values of x	Write the probability	Rewrite the probability	Find the probability
Three or less	0, 1, 2, 3	$P(x \le 3)$		.647
Three or more	3, 4, …	$P(x \ge 3)$	1 - $P(x \le 2)$	1 − .423 = .577
More than three	4, 5, …	$P(x > 3)$	1 - $P(x \le 3)$	1 − .647 = .353
Fewer than three	0, 1, 2	$P(x < 3)$	$P(x \le 2)$	.423
Between 3 and 5 (inclusive)	3, 4, 5	$P(3 \le x \le 5)$	$P(x \le 5) - P(x \le 2)$	.916 − .423 = .493
Exactly three	3	$P(x = 3)$	$P(x \le 3) - P(x \le 2)$	.647 − .423 = .224

5.41 a $\quad$ Using Table 1, Appendix I, $P[x \le 2] = .677$

b $\quad$ With $\mu = np = 20(.1) = 2$, the approximation is $p(x) \approx \dfrac{2^x e^{-2}}{x!}$. Then

$$P[x \le 2] \approx \dfrac{2^0 e^{-0}}{0!} + \dfrac{2^1 e^{-1}}{1!} + \dfrac{2^2 e^{-2}}{2!}$$
$$= .135335 + .27067 + .27067 = .677$$

c $\quad$ The approximation is quite accurate.

5.43 Let x be the number of misses during a given month. Then x has a Poisson distribution with $\mu = 5$.

    a $p(0) = e^{-5} = .0067$

    b $p(5) = \dfrac{5^5 e^{-5}}{5!} = .1755$

    c $P[x \geq 5] = 1 - P[x \leq 4] = 1 - .440 = .560$ from Table 2.

5.45 Let x be the number of fatalities for the current year, with $\mu = 2.5$.

    a $P[x = 2] = P[x \leq 2] - P[x \leq 1] = .544 - .287 = .257$

    b $P[x \geq 2] = 1 - P[x \leq 1] = 1 - .287 = .713$

    c $P[x \leq 1] = .287$

    d $P[x \geq 1] = 1 - P[x \leq 0] = 1 - .082 = .918$

5.47 The random variable x, number of bacteria, has a Poisson distribution with $\mu = 2$. The probability of interest is
$$P[x \text{ exceeds maximum count}] = P[x > 5]$$
Using the fact that $\mu = 2$ and $\sigma = \sqrt{2} = 1.414$, most of the observations should fall within $\mu \pm 2\sigma$ or 0 to 4. Hence, it is unlikely that x will exceed 5. In fact, the exact Poisson probability is
$P[x > 5] = 1 - P[x \leq 5] = 1 - .983 = .017$.

5.49 **a** $\dfrac{C_1^2 C_1^2}{C_2^5} = \dfrac{3(2)}{10} = .6$

    b $\dfrac{C_2^4 C_1^3}{C_3^7} = \dfrac{6(3)}{35} = .5143$

    c $\dfrac{C_4^5 C_0^3}{C_4^8} = \dfrac{5(1)}{70} = .0714$

5.51 The formula for $p(x)$ is $p(x) = \dfrac{C_x^4 C_{3-x}^{11}}{C_3^{15}}$ for $x = 0, 1, 2, 3$

    a $p(0) = \dfrac{C_0^4 C_3^{11}}{C_3^{15}} = \dfrac{165}{455} = .36$ $p(1) = \dfrac{C_1^4 C_2^{11}}{C_3^{15}} = \dfrac{220}{455} = .48$

          $p(2) = \dfrac{C_2^4 C_1^{11}}{C_3^{15}} = \dfrac{66}{455} = .15$ $p(3) = \dfrac{C_3^4 C_0^{11}}{C_3^{15}} = \dfrac{4}{455} = .01$

    b   The probability histogram is shown below.

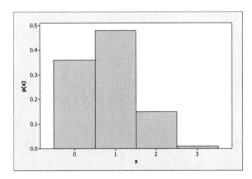

    c   Using the formulas given in Section 5.4.

$$\mu = E(x) = n\left(\frac{M}{N}\right) = 3\left(\frac{4}{15}\right) = .8$$

$$\sigma^2 = n\left(\frac{M}{N}\right)\left(\frac{N-M}{N}\right)\left(\frac{N-n}{N-1}\right) = 3\left(\frac{4}{15}\right)\left(\frac{15-4}{15}\right)\left(\frac{15-3}{15-1}\right) = .50286$$

    d   Calculate the intervals

$\mu \pm 2\sigma = .8 \pm 2\sqrt{.50286} = .8 \pm 1.418$ or $-.618$ to 2.218

$\mu \pm 3\sigma = .8 \pm 3\sqrt{.50286} = .8 \pm 1.418$ or -1.327 to 2.927

Then,

$P[-.618 \le x \le 2.218] = p(0) + p(1) + p(2) = .99$

$P[-1.327 \le x \le 2.927] = p(0) + p(1) + p(2) = .99$

These results agree with Tchebysheff's Theorem.

5.53   The formula for $p(x)$ is $p(x) = \dfrac{C_x^2 C_{3-x}^4}{C_3^6}$ for x = number of defectives = $0, 1, 2$. Then

$p(0) = \dfrac{C_0^2 C_3^4}{C_3^6} = \dfrac{4}{20} = .2 \qquad p(1) = \dfrac{C_1^2 C_2^4}{C_3^6} = \dfrac{12}{20} = .6 \qquad p(2) = \dfrac{C_2^2 C_1^4}{C_3^6} = \dfrac{4}{20} = .2$

These results agree with the probabilities calculated in Exercise 4.90.

5.55   **a**   The random variable x has a hypergeometric distribution with $N = 8, M = 5$ and $n = 3$. Then

$$p(x) = \frac{C_x^5 C_{3-x}^3}{C_3^8} \text{ for } x = 0, 1, 2, 3$$

    b $P(x = 3) = \dfrac{C_3^5 C_0^3}{C_3^8} = \dfrac{10}{56} = .1786$

    c $P(x = 0) = \dfrac{C_0^5 C_3^3}{C_3^8} = \dfrac{1}{56} = .01786$

    d $P(x \le 1) = \dfrac{C_0^5 C_3^3}{C_3^8} + \dfrac{C_1^5 C_2^3}{C_3^8} = \dfrac{1+15}{56} = .2857$

5.57   See Section 5.2 in the text.

5.59   The hypergeometric distribution is appropriate when sampling from a finite rather than an infinite population of successes and failures. In this case, the probability of p of a success is not constant from trial to trial and the binomial distribution is not appropriate.

5.61   Refer to Exercise 5.60 and assume that $p = .1$ instead of $p = .5$.

    a $P[x = 0] = p(0) = C_0^3(.1)^0(.9)^3 = .729 \qquad\qquad P[x = 1] = p(1) = C_1^3(.1)^1(.9)^2 = .243$

          $P[x = 2] = p(2) = C_2^3(.1)^2(.9)^1 = .027 \qquad\qquad P[x = 3] = p(3) = C_3^3(.1)^3(.9)^0 = .001$

b Note that the probability distribution is no longer symmetric; that is, since the probability of observing a head is so small, the probability of observing a small number of heads on three flips is increased (see the figure below).

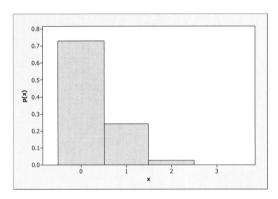

c $\mu = np = 3(.1) = .3$ and $\sigma = \sqrt{npq} = \sqrt{3(.1)(.9)} = .520$

d The desired intervals are
$$\mu \pm \sigma = .3 \pm .520 \quad \text{or} \quad -.220 \text{ to } .820$$
$$\mu \pm 2\sigma = .3 \pm 1.04 \quad \text{or} \quad -.740 \text{ to } 1.34$$

The only value of x which falls in this first interval is $x = 0$, and the fraction of measurements in this interval will be .729. The values of $x = 0$ and $x = 1$ are enclosed by the second interval, so that $.729 + .243 = .972$ of the measurements fall within two standard deviations of the mean, consistent with both Tchebysheff's Theorem and the Empirical Rule.

5.63 Define x to be the number supporting the administrator's claim, with $p = .8$ and $n = 25$.

a Using the binomial tables for $n = 25$, $P[x \geq 22] = 1 - P[x \leq 21] = 1 - .766 = .234$

b $P[x = 22] = P[x \leq 22] - P[x \leq 21] = .902 - .766 = .136$

c The probability of observing an event as extreme as $x = 22$ (or more extreme) is quite high assuming that $p = .8$. Hence, this is not an unlikely event and we would not doubt the claim.

5.65 Refer to Exercise 5.64. Redefine x to be the number of people who choose an interior number in the sample of $n = 20$. Then x has a binomial distribution with $p = .3$.

a $P[x \geq 8] = 1 - P[x \leq 7] = 1 - .772 = .228$

b Observing eight or more people choosing an interior number is not an unlikely event, assuming that the integers are all equally likely. Therefore, there is no evidence to indicate that people are more likely to choose the interior numbers than any others.

5.67 Let x be the number of Canadians who rank owning a vacation home as the number one status symbol. Then x has a binomial distribution with $n = 400$ and $p = .6$.

a $\mu = np = 400(.6) = 240$

b $\sigma = \sqrt{npq} = \sqrt{400(.6)(.4)} = 9.798$

c Since n is large and $p = .6$, this binomial distribution will be relatively mound-shaped, and approximately 95% of the values of x should lie in the interval
$$\mu \pm 2\sigma \Rightarrow 240 \pm 19.596 \Rightarrow 220.404 \text{ to } 259.596$$

d The value $x = 200$ lies $(200 - 240)/9.798 = -4.08$ standard deviations below the mean. This is an unusual result. Perhaps the sample was not randomly selected, or perhaps the 60% figure is no longer accurate, and in fact this percentage has decreased.

5.69 It is given that x = number of patients with a psychosomatic problem, $n = 25$, and $p = P[\text{patient has psychosomatic problem}]$. A psychiatrist wishes to determine whether or not $p = .8$.

a Assuming that the psychiatrist is correct (that is, $p = .8$), the expected value of x is
$E(x) = np = 25(.8) = 20$.

b $\sigma^2 = npq = 25(.8)(.2) = 4$

c Given that $p = .8$, $P[x \le 14] = .006$ from Table 1 in Appendix I.

d Assuming that the psychiatrist is correct, the probability of observing $x = 14$ or the more unlikely values, $x = 0, 1, 2, \ldots, 13$ is very unlikely. Hence, one of two conclusions can be drawn. Either we have observed a very unlikely event, or the psychiatrist is incorrect and p is actually less than .8. We would probably conclude that the psychiatrist is incorrect. The probability that we have made an incorrect decision is $P[x \le 14 \text{ given } p = .8] = .006$ which is quite small.

5.71 Define x to be the number of students 30 years or older, with $n = 200$ and $p = P[\text{student is 30+ years}] = .25$.

a Since x has a binomial distribution, $\mu = np = 200(.25) = 50$ and $\sigma = \sqrt{npq} = \sqrt{200(.25)(.75)} = 6.124$.

b The observed value, $x = 35$, lies $\dfrac{35 - 50}{6.124} = -2.45$ standard deviations below the mean. It is unlikely that $p = .25$.

5.73 **a** If there is no preference for either design, then $p = P[\text{choose the second design}] = .5$.

b Using the results of part **a** and $n = 25$, $\mu = np = 25(.5) = 12.5$ and $\sigma = \sqrt{npq} = \sqrt{6.25} = 2.5$.

c The observed value, $x = 20$, lies $z = \dfrac{20 - 12.5}{2.5} = 3$ standard deviations above the mean. This is an unlikely event, assuming that $p = .5$. We would probably conclude that there is a preference for the second design and that $p > .5$.

5.75 **a** The random variable x, the number of plants with red petals, has a binomial distribution with $n = 10$ and $p = P[\text{red petals}] = .75$.

b Since the value $p = .75$ is not given in Table 1, you must use the binomial formula to calculate
$P(x \ge 9) = C_9^{10}(.75)^9(.25)^1 + C_{10}^{10}(.75)^{10}(.25)^0 = .1877 + .0563 = .2440$

c $P(x \le 1) = C_0^{10}(.75)^0(.25)^{10} + C_1^{10}(.75)^1(.25)^9 = .0000296$.

d Refer to part **c**. The probability of observing $x = 1$ or something even more unlikely $(x = 0)$ is very small – .0000296. This is a highly unlikely event if in fact $p = .75$. Perhaps there has been a nonrandom choice of seeds, or the 75% figure is not correct for this particular genetic cross.

5.77 Let x be the number of lost calls in a series of $n = 11$ trials. If the coin is fair, then $p = \dfrac{1}{2}$ and x has a binomial distribution.

a $P[x = 11] = \left(\dfrac{1}{2}\right)^{11} = \dfrac{1}{2048}$ which is the same as odds of 1:2047.

b If $n = 13$, $P[x = 13] = \left(\dfrac{1}{2}\right)^{13} = \dfrac{1}{8192}$ which is a very unlikely event.

Introduction to Probability and Statistics, 2ce

5.79 **a** The distribution of x is actually hypergeometric, with $N = 1200$, $n = 20$ and $M =$ number of defectives in the lot. However, since N is so large in comparison to n, the distribution of x can be closely approximated by the binomial distribution with $n = 20$ and $p = P[\text{defective}]$.

    b If p is small, with $np < 7$, the Poisson approximation can be used.

    c If there are 10 defectives in the lot, then $p = 10/1200 = .008333$ and $\mu = .1667$. The probability that the lot is shipped is

$$P(x=0) \approx \frac{(.1667)^0 e^{-.1667}}{0!} = .85$$

If there are 20 defectives, $p = 20/1200$ and $\mu = .3333$. Then

$$P(x=0) \approx \frac{(.3333)^0 e^{-.3333}}{0!} = .72$$

If there are 30 defectives, $p = 30/1200$ and $\mu = .5$. Then

$$P(x=0) \approx \frac{(.5)^0 e^{-.5}}{0!} = .61$$

5.81 The random variable x, the number of offspring with Tay-Sachs disease, has a binomial distribution with $n = 3$ and $p = .25$. Use the binomial formula.

    a $P(x=3) = C_3^3 (.25)^3 (.75)^0 = (.25)^3 = .015625$

    b $P(x=1) = C_1^3 (.25)^1 (.75)^2 = 3(.25)(.75)^2 = .421875$

    c Remember that the trials are independent. Hence, the occurrence of Tay-Sachs in the first two children has no affect on the third child, and $P(\text{third child develops Tay-Sachs}) = .25$.

5.83 **a** The random variable x, the number of tasters who pick the correct sample, has a binomial distribution with $n=5$ and, if there is no difference in the taste of the three samples,

$$p = P(\text{taster picks the correct sample}) = \frac{1}{3}$$

    b The probability that exactly one of the five tasters chooses the latest batch as different from the others is $P(x=1) = C_1^5 \left(\frac{1}{3}\right)^1 \left(\frac{2}{3}\right)^4 = .3292$

    c The probability that at least one of the tasters chooses the latest batch as different from the others is

$$P(x \leq 1) = 1 - P(x=0) = 1 - C_0^5 \left(\frac{1}{3}\right)^0 \left(\frac{2}{3}\right)^5 = .8683$$

5.85 Refer to Exercise 5.84.
    a The average value of x is $\mu = np = 20(.7) = 14$.
    b The standard deviation of x is $\sigma = \sqrt{npq} = \sqrt{20(.7)(.3)} = \sqrt{4.2} = 2.049$.
    c The z-score corresponding to $x = 10$ is $z = \frac{x-\mu}{\sigma} = \frac{10-14}{2.049} = -1.95$.

Since this z-score does not exceed 3 in absolute value, we would not consider the value $x = 10$ to be an unusual observation.

5.87 The random variable x has a Poisson distribution with $\mu = 2$. Use Table 2 in Appendix I or the Poisson formula to find the following probabilities.

a $P(x = 0) = \dfrac{2^0 e^{-2}}{0!} = e^{-2} = .135335$

b $P(x \leq 2) = \dfrac{2^0 e^{-2}}{0!} + \dfrac{2^1 e^{-2}}{1!} + \dfrac{2^2 e^{-2}}{2!}$
$= .135335 + .270671 + .270671 = .676676$

5.89 The random variable x, the number of subjects who revert to their first learned method under stress, has a binomial distribution with $n = 6$ and $p = .8$. The probability that at least five of the six subjects revert to their first learned method is $P(x \geq 5) = 1 - P(x \leq 4) = 1 - .345 = .655$

5.91 The random variable x, the number of Kobo homeowners with earthquake insurance, has a binomial distribution with $n = 15$ and $p = .1$.

a $P(x \geq 1) = 1 - P(x = 0) = 1 - .206 = .794$

b $P(x \geq 4) = 1 - P(x \leq 3) = 1 - .944 = .056$

c Calculate $\mu = np = 15(.1) = 1.5$ and $\sigma = \sqrt{npq} = \sqrt{15(.1)(.9)} = 1.1619$. Then approximately 95% of the values of x should lie in the interval
$\mu \pm 2\sigma \Rightarrow 1.5 \pm 2(1.1619) \Rightarrow -.82$ to 3.82.
or between 0 and 3.

5.93 The random variable x, the number of Canadians who drive less to cope with the high gas prices, has a binomial distribution with $n = 100$ and $p = .46$.

a The average value of x is $\mu = np = 100(.46) = 46$.

b The standard deviation of x is $\sigma = \sqrt{npq} = \sqrt{100(.46)(.54)} = \sqrt{24.84} = 4.984$.

c The z-score corresponding to $x = 59$ is $z = \dfrac{x - \mu}{\sigma} = \dfrac{59 - 46}{4.984} = 2.61$

Since this value is not greater than 3 in absolute value, it is not an extremely unusual observation. It is between 2 and 3, however, so it is somewhat unusual.

5.95 Use the **Calculating Binomial Probabilities** applet. The correct answers are given below.

a $P(x < 6) = 0.00006$

b $P(x = 8) = .042$

c $P(x > 14) = .0207$

d $P(2 < x < 6) = .5948$

e $P(x \geq 6) = 1$

5.97 It is known (part **c**, Exercise 5.96) that the probability of less than two successful operations is .0067. This is a very rare event, and the fact that it has occurred leads to one of two conclusions. Either the success rate is in fact 80% and a very rare occurrence has been observed, or the success rate is less than 80% and we are observing a likely event under the latter assumption. The second conclusion is more likely. We would probably conclude that the success rate for this team is less than 80% and would put little faith in the team.

5.99 Define x to be the number of young adults who prefer McDonald's. Then x has a binomial distribution with $n = 100$ and $p = .5$. Use the **Calculating Binomial Probabilities** applet.

    a $P(61 \leq x \leq 100) = .0176$

    b $P(40 \leq x \leq 60) = .9648$

    c   If 40 prefer Burger King, then 60 prefer McDonalds, and vice versa. The probability is the same as that calculated in part **b**, since $p = .5$.

Case Study:
How Safe is Plastic Surgery? Myth versus Fact!

Do breast implants increase the suicide rate?

1. Let x be the number of suicide cases for breast-implant recipients, and let μ be the average number of suicide cases for Ontario and Quebec. Then, the reasonable estimate of μ is given by
$$58 = .75\mu + \mu \text{ or } \mu = 58/1.75 = 33.14.$$
Yes, 75% is correct percentage.

2. Since x has a Poisson distribution with $\mu = 33.14$, the standard deviation of x is $\sigma = \sqrt{\mu} = \sqrt{33.14} = 5.76$.

3. The z-score for the observed value of x is $z = \dfrac{x - \mu}{\sigma} = \dfrac{58 - 33.14}{5.76} = 4.32$
This is a very large number. It is very likely that the breast-implants do increase the suicide rate.

Do breast implants reduce cancer risk?

1. Let x be the number of cancer deaths for breast-implant recipients, and let μ be the average number of cancer deaths among the general female population in Ontario and Quebec. Note that (303 - 229)/303 = 24.42; the given percentage of 24.5 is correct.

2. The reasonable estimate of μ is 303.

3. Since x has a Poisson distribution with $\mu = 303$, the standard deviation of x is $\sigma = \sqrt{\mu} = \sqrt{303} = 17.41$.

4. The z-score for the observed value of x is $z = \dfrac{x - \mu}{\sigma} = \dfrac{58 - 303}{17.41} = -14.07$.
This is a very small number. It is very likely that the breast implants do reduce cancer risk.

Project 5: Relations Among Useful Discrete Probability Distributions

a   (i) The probability distribution can be written as: $p(x) = \dfrac{C_x^7 C_{5-x}^{13}}{C_5^{20}}$, for x = 0, 1, 2, 3, 4, 5.

(ii) The mean of x is $\mu = n\left(\dfrac{M}{N}\right) = 5\left(\dfrac{7}{20}\right) = 1.75$, and the variance of x is $\sigma^2 = n\left(\dfrac{M}{N}\right)\left(\dfrac{N-M}{N}\right)\left(\dfrac{N-n}{N-1}\right)$
$= 5\left(\dfrac{7}{20}\right)\left(\dfrac{20-7}{20}\right)\left(\dfrac{20-5}{20-1}\right) = 0.898026$.

(iii) For 2 standard deviations from the mean, the interval is $1.75 \pm 2(0.898026)^{.5} = (-0.14529, 3.64529)$. This would encompass x = 0, 1, 2, 3. Now, $p(0) + p(1) + p(2) + p(3) = 0.969298$. This result agrees with Tchebysheff's Theorem (see page 71 in the text) that at least a 3/4 proportion of the data lie within two standard deviations of the mean. For 3 standard deviations from the mean, the interval is $.75 \pm 3(0.898026)^{.5} = (-1.09293, 4.59293)$. This would encompass x = 0, 1, 2, 3, 4. Now, $p(0) + p(1) + p(2) + p(3) + p(4) = 0.998646$. This result agrees with Tchebysheff's Theorem (see page 71 in the text) that at least an 8/9 proportion of the data lie within three standard deviations of the mean.

(iv) The answer is $1 - p(0) = 0.916989$.

b   (i) Using the hypergeometric distribution, $p(2) = \dfrac{C_2^{80} C_{5-2}^{100}}{C_5^{180}} = 0.343200$.

(ii) Using the binomial distribution $p(2) = C_2^5 \left(\dfrac{80}{180}\right)^2 \left(1 - \dfrac{80}{180}\right)^{5-2} = 0.338702$; the approximation is decent, as the number selected (5) is small compared to the total sample size (180). On page 230 of the textbook, it states that if $\left(\dfrac{n}{N}\right)$ is less than or equal to 0.1, the hypergeometric distribution is well approximated by the binomial. Clearly, $\left(\dfrac{5}{180}\right)$ satisfies this criteria.

(iii) Hypergeometric distribution: The mean of x is $\mu = n\left(\dfrac{M}{N}\right) = 5\left(\dfrac{80}{180}\right) = 2.222222$, and the variance of x is $\sigma^2 = n\left(\dfrac{M}{N}\right)\left(\dfrac{N-M}{N}\right)\left(\dfrac{N-n}{N-1}\right) = 5\left(\dfrac{80}{180}\right)\left(\dfrac{180-80}{180}\right)\left(\dfrac{180-5}{180-1}\right) = 1.206980$.

Binomial distribution: The mean of x is $\mu = np = 5\left(\dfrac{80}{180}\right) = 2.222222$, and the variance of x is $\sigma^2 = npq = 5\left(\dfrac{80}{180}\right)\left(1 - \dfrac{80}{180}\right) = 1.234568$.

In summary, the mean for the two distributions are identical. The variances are only slightly different.

c   (i) In this case, $\mu = 28{,}572(0.00007) = 2.00004$ for the Poisson distribution, and thus the probability that no children will have cancer is $p(0) = \dfrac{2.00004^0 e^{-2.00004}}{0!} = 0.135330$.

(ii) The probability that at most two will have cancer is $P(x \leq 2) = p(0) + p(1) + p(2) = \dfrac{2.00004^0 e^{-2.00004}}{0!} + \dfrac{2.00004^1 e^{-2.00004}}{1!} + \dfrac{2.00004^2 e^{-2.00004}}{2!} = 0.676666$.

(iii) The probability that at least 7 will not have cancer is equal to the probability that up to 3 do develop cancer: that is, $p(0) + p(1) + p(2) + p(3)$. In a sample of 10 children, $\mu = np = (10)(0.00007) = 0.0007$. Thus, using the Poisson distribution to approximate each of the four binomial probabilities, we obtain:

$p(0) + p(1) + p(2) + p(3) = \dfrac{0.0007^0 e^{-0.0007}}{0!} + \dfrac{0.0007^1 e^{-0.0007}}{1!} + \dfrac{0.0007^2 e^{-0.0007}}{2!} + \dfrac{0.0007^3 e^{-0.0007}}{3!} =$
$0.999300245 + 0.000699510 + 0.000000245 + 0.0000000001 = 0.9999999999999$, a virtual certainty.

Chapter 6: The Normal Probability Distribution

6.1 The Exercise Reps are designed to provide practice for the student in evaluating areas under the normal curve. The following notes may be of some assistance.

    a Table 3, Appendix I tabulates the cumulative area under a standard normal curve to the left of a specified value of z.

    b Since the total area under the curve is one, the total area lying to the right of a specified value of z and the total area to its left must add to 1. Thus, in order to calculate a "tail area", such as the one shown in Figure 6.1, the value of $z = z_0$ will be indexed in Table 3, and the area that is obtained will be subtracted from 1. Denote the area obtained by indexing $z = z_0$ in Table 3 by $A(z_0)$ and the desired area by A. Then, in the above example, $A = 1 - A(z_0)$.

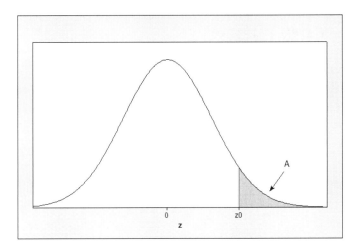

    c To find the area under the standard normal curve between two values, z_1 and z_2, calculate the difference in their cumulative areas, $A = A(z_2) - A(z_1)$.

    d Note that z, similar to x, is actually a random variable which may take on an infinite number of values, both positive and negative. Negative values of z lie to the left of the mean, $z = 0$, and positive values lie to the right.

Reread the instructions in the My Personal Trainer section if necessary. The answers are shown in the table.

The Interval	Write the probability	Rewrite the Probability (if needed)	Find the probability
Less than -2	$P(z < -2)$	not needed	.0228
Greater than 1.16	$P(z > 1.16)$	$1 - P(z \leq 1.16)$	$1 - .8770 = .1230$
Greater than 1.645	$P(z > 1.645)$	$1 - P(z \leq 1.645)$	$1 - .9500 = .0500$
Between -2.33 and 2.33	$P(-2.33 < z < 2.33)$	$P(z < 2.33) - P(z < -2.33)$	$.9901 - .0099 = .9802$
Between 1.24 and 2.58	$P(1.24 < z < 2.58)$	$P(z < 2.58) - P(z < 1.24)$	$.9951 - .8925 = .1026$
Less than or equal to 1.88	$P(z \leq 1.88)$	not needed	.9699

6.3   **a**   It is necessary to find the area to the left of $z = 1.6$. That is, $A = A(1.6) = .9452$.

    b   The area to the left of $z = 1.83$ is $A = A(1.83) = .9664$.

    c $A = A(.90) = .8159$

    d $A = A(4.58) \approx 1$. Notice that the values in Table 3 approach 1 as the value of z increases. When the value of z is larger than $z = 3.49$ (the largest value in the table), we can assume that the area to its left is approximately 1.

6.5   **a** $P(-1.43 < z < .68) = A(.68) - A(-1.43) = .7517 - .0764 = .6753$

    b $P(.58 < z < 1.74) = A(1.74) - A(.58) = .9591 - .7190 = .2401$

    c $P(-1.55 < z < -.44) = A(-.44) - A(-1.55) = .3300 - .0606 = .2694$

    d $P(z > 1.34) = 1 - A(1.34) = 1 - .9099 = .0901$

    e   Since the value of $z = -4.32$ is not recorded in Table 3, you can assume that the area to the left of $z = -4.32$ is very close to 0. Then $P(z < -4.32) \approx 0$

6.7   Now we are asked to find the z-value corresponding to a particular area.

    a   We need to find a z_0 such that $P(z > z_0) = .025$. This is equivalent to finding an indexed area of $1 - .025 = .975$. Search the interior of Table 3 until you find the four-digit number **.9750**. The corresponding z-value is **1.96**; that is, $A(1.96) = .9750$. Therefore, $z_0 = 1.96$ is the desired z-value (see the figure below).

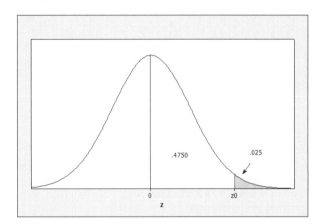

    b   We need to find a z_0 such that $P(z < z_0) = .9251$ (see below). Using Table 3, we find a value such that the indexed area is .9251. The corresponding z-value is $z_0 = 1.44$.

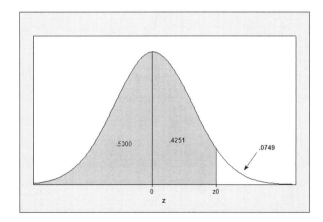

6.9 **a** Similar to Exercise 6.7b. The value of z_0 must be positive and $A(z_0) = .9505$. Hence, $z_0 = 1.65$.

b It is given that the area to the left of z_0 is .0505, shown as A_1 in the figure below. The desired value is not tabulated in Table 3 but falls between two tabulated values, .0505 and .0495. Hence, using linear interpolation (as you did in Exercise 6.6b) z_0 will lie halfway between −1.64 and −1.65, or $z_0 = -1.645$.

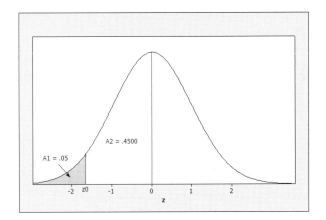

6.11 The pth percentile of the standard normal distribution is a value of z which has area $p/100$ to its left. Since all four percentiles in this exercise are greater than the 50th percentile, the value of z will all lie to the right of $z = 0$, as shown for the 90th percentile in the figure below.

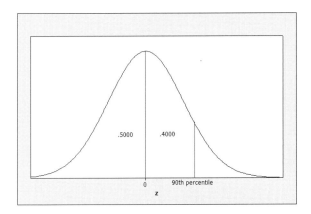

a From the figure, the area to the left of the 90th percentile is .9000. From Table 3, the appropriate value of z is closest to $z = 1.28$ with area .8997. Hence the 90th percentile is approximately $z = 1.28$.

b As in part **a**, the area to the left of the 95th percentile is .9500. From Table 3, the appropriate value of z is found using linear interpolation (see Exercise 6.9b) as $z = 1.645$. Hence the 95th percentile is $z = 1.645$.

c The area to the left of the 98th percentile is .9800. From Table 3, the appropriate value of z is closest to $z = 2.05$ with area .9798. Hence the 98th percentile is approximately $z = 2.05$.

d The area to the left of the 99th percentile is .9900. From Table 3, the appropriate value of z is closest to $z = 2.33$ with area .9901. Hence the 99th percentile is approximately $z = 2.33$.

6.13 Similar to Exercise 6.12.

a Calculate $z_1 = \dfrac{1.00 - 1.20}{.15} = -1.33$ and $z_2 = \dfrac{1.10 - 1.20}{.15} = -.67$. Then
$$P(1.00 < x < 1.10) = P(-1.33 < z < -.67) = .2514 - .0918 = .1596$$

Introduction to Probability and Statistics, 2e

b Calculate $z = \dfrac{x-\mu}{\sigma} = \dfrac{1.38-1.20}{.15} = 1.2$. Then
$$P(x > 1.38) = P(z > 1.2) = 1 - .8849 = .1151$$

c Calculate $z_1 = \dfrac{1.35-1.20}{.15} = 1$ and $z_2 = \dfrac{1.50-1.20}{.15} = 2$. Then
$$P(1.35 < x < 1.50) = P(1 < z < 2) = .9772 - .8413 = .1359$$

6.15 The 99th percentile of the standard normal distribution was found in Exercise 6.11d to be $z = 2.33$. Since the relationship between the general normal random variable x and the standard normal z is $z = \dfrac{x-\mu}{\sigma}$, the corresponding percentile for this general normal random variable is found by solving for $x = \mu + z\sigma$;

$$2.33 = \dfrac{x-35}{10}$$
$$x - 35 = 23.3 \quad \text{or} \quad x = 58.3$$

6.17 The random variable x is normal with unknown μ and σ. However, it is given that
$$P(x > 4) = P\left(z > \dfrac{4-\mu}{\sigma}\right) = .9772 \text{ and } P(x > 5) = P\left(z > \dfrac{5-\mu}{\sigma}\right) = .9332.$$ These probabilities are shown in the figure below.

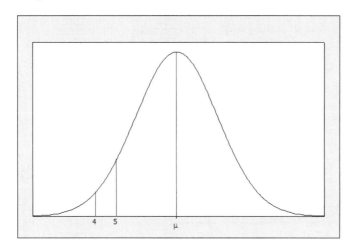

The value $\dfrac{4-\mu}{\sigma}$ is negative, with $A\left(\dfrac{4-\mu}{\sigma}\right) = 1 - .9772 = .0228$ or $\dfrac{4-\mu}{\sigma} = -2$ (i)

The value $\dfrac{5-\mu}{\sigma}$ is also negative, with $A\left(\dfrac{5-\mu}{\sigma}\right) = 1 - .9332 = .0668$ or $\dfrac{5-\mu}{\sigma} = -1.5$ (ii)

Equations (i) and (ii) provide two equations in two unknowns which can be solved simultaneously for μ and σ. From (i), $\sigma = \dfrac{\mu - 4}{2}$ which, when substituted into (ii) yields

$$5 - \mu = -1.5\left(\dfrac{\mu - 4}{2}\right)$$
$$10 - 2\mu = -1.5\mu + 6$$
$$\mu = 8$$

and from (i), $\sigma = \dfrac{8-4}{2} = 2$.

6.19 The random variable x, the height of a Canadian male, has a normal distribution with $\mu = 177$ and $\sigma = 8$.

a $P(x > 185) = P\left(z > \dfrac{185 - 177}{8}\right) = P(z > 1.00) = 1 - .8413 = .1587$

b $P(170 < x < 185) = P\left(\dfrac{170 - 177}{8} < z < \dfrac{185 - 177}{8}\right) = P(-.875 < z < 1.00) = .8413 - .1908 = .6505$

c $z = \dfrac{188 - 177}{8} = 1.375$

This would not be considered an unusually large value, since it is less than two standard deviations from the mean.

6.21 The random variable x, cerebral blood flow, has a normal distribution with $\mu = 74$ and $\sigma = 16$.

a $P(60 < x < 80) = P\left(\dfrac{60 - 74}{16} < z < \dfrac{80 - 74}{16}\right) = P(-.88 < z < .38) = .6480 - .1894 = .4586$

b $P(x > 100) = P\left(z > \dfrac{100 - 74}{16}\right) = P(z > 1.62) = 1 - .9474 = .0526$

c $P(x < 40) = P\left(z < \dfrac{40 - 74}{16}\right) = P(z < -2.12) = .0170$

6.23 The random variable x, total weight of 8 people, has a mean of $\mu = 550$ and a variance $\sigma^2 = 445$. It is necessary to find $P(x > 590)$ and $P(x > 680)$ if the distribution of x is approximately normal. The z-value corresponding to $x_1 = 590$ is $z_1 = \dfrac{x_1 - \mu}{\sigma} = \dfrac{590 - 550}{\sqrt{445}} = \dfrac{40}{21.095} = 1.90$. Hence,

$$P(x > 590) = P(z > 1.90) = 1 - A(1.90) = 1 - .9713 = .0287.$$

Similarly, the z-value corresponding to $x_2 = 680$ is $z_2 = \dfrac{x_2 - \mu}{\sigma} = \dfrac{680 - 550}{\sqrt{445}} = 6.16$.

and $P(x > 680) = P(z > 6.16) = 1 - A(6.16) \approx 1 - 1 \approx 0$.

6.25 It is given that x, the unsupported stem diameter of a sunflower plant, is normally distributed with $\mu = 35$ and $\sigma = 3$.

a $P(x > 40) = P\left(z > \dfrac{40 - 35}{3}\right) = P(z > 1.67) = 1 - .9525 = .0475$

b From part **a**, the probability that one plant has stem diameter of more than 40 mm is .0475. Since the two plants are independent, the probability that two plants both have diameters of more than 40 mm is
$$(.0475)(.0475) = .00226$$

c Since 95% of all measurements for a normal random variable lie within 1.96 standard deviations of the mean, the necessary interval is
$$\mu \pm 1.96\sigma \;\Rightarrow\; 35 \pm 1.96(3) \;\Rightarrow\; 35 \pm 5.88$$
or in the interval 29.12 to 40.88.

d The 90th percentile of the standard normal distribution was found in Exercise 6.11a to be $z = 1.28$. Since the relationship between the general normal random variable x and the standard normal z is $z = \dfrac{x - \mu}{\sigma}$, the corresponding percentile for this general normal random variable is found by solving for $x = \mu + z\sigma$.

$$x = 35 + 1.28(3) \quad \text{or} \quad x = 38.84$$

6.27 **a** It is given that the prime interest rate forecasts, x, are approximately normal with mean $\mu = 4.5$ and standard deviation $\sigma = 0.1$. It is necessary to determine the probability that x exceeds 4.75. Calculate $z = \dfrac{x-\mu}{\sigma} = \dfrac{4.75-4.5}{0.1} = 2.5$. Then $P(x > 4.75) = P(z > 2.5) = 1 - .9938 = .0062$.

b Calculate $z = \dfrac{x-\mu}{\sigma} = \dfrac{4.375-4.5}{0.1} = -1.25$. Then $P(x < 4.375) = P(z < -1.25) = .1056$.

6.29 It is given that the counts of the number of bacteria are normally distributed with $\mu = 85$ and $\sigma = 9$. The z-value corresponding to $x = 100$ is $z = \dfrac{x-\mu}{\sigma} = \dfrac{100-85}{9} = 1.67$ and
$$P(x > 100) = P(z > 1.67) = 1 - .9525 = .0475$$

6.31 Let w be the number of words specified in the contract. Then x, the number of words in the manuscript, is normally distributed with $\mu = w + 20,000$ and $\sigma = 10,000$. The publisher would like to specify w so that
$$P(x < 100,000) = .95.$$
As in Exercise 6.30, calculate
$$z = \dfrac{100,0000 - (w+20,000)}{10,000} = \dfrac{80,000 - w}{10,000}.$$
Then $P(x < 100,000) = P\left(z < \dfrac{80,000-w}{10,000}\right) = .95$. It is necessary that $z_0 = (80,000 - w)/10,000$ be such that
$$P(z < z_0) = .95 \Rightarrow A(z_0) = .9500 \quad \text{or} \quad z_0 = 1.645.$$
Hence,
$$\dfrac{80,000-w}{10,000} = 1.645 \quad \text{or} \quad w = 63,550.$$

6.33 The amount of money spent at shopping centres on Sundays is normally distributed with $\mu = 85$ and $\sigma = 10$.

a The z-value corresponding to $x = 90$ is $z = \dfrac{x-\mu}{\sigma} = \dfrac{90-85}{10} = 0.5$. Then
$$P(x > 90) = P(z > 0.5) = 1 - .6915 = .3085$$

b The z-value corresponding to $x = 100$ is $z = \dfrac{x-\mu}{\sigma} = \dfrac{100-85}{10} = 1.5$. Then
$$P(90 < x < 100) = P(0.5 < z < 1.5) = .9332 - .6915 = .2417$$

c First, find $P(x > 100) = P(z > 1.5) = 1 - .9332 = .0668$ for a single shopper. For two shoppers, use the Multiplication Rule.
P(both shoppers spend more than $100) = P(1st spends more than $100) $\times$ P(2nd spends more than $100)
$$= (.0668)(.0668) = .0045$$

6.35 Follow the instructions in the My Personal Trainer section. The blanks are filled in below.
Consider a binomial random variable with $n = 25$ and $p = .6$.
a Can we use the normal approximation? Calculate $np = $ **15** and $nq = $ **10.**
b Are np and nq both greater than 5?   Yes__**X**__   No_____
c If the answer to part b is yes, calculate $\mu = np = $ **15** and $\sigma = \sqrt{npq} = $ **2.449**
d To find the probability of more than 9 successes, what values of x should be included?
$$x = \mathbf{10, 11, \ldots 25.}$$
e To include the entire block of probability for the first value of $x = 10$, start at **9.5**.

108 *Student's Solutions Manual to accompany*

f Calculate $z = \dfrac{x \pm .5 - np}{\sqrt{npq}} = \dfrac{9.5 - 15}{2.449} = -2.25$

g Calculate $P(x > 9) \approx P(z > -2.25) = 1 - .0122 = .9878$.

6.37

a The normal approximation will be appropriate if both np and nq are greater than 5. For this binomial experiment, $np = 25(.3) = 7.5$ and $nq = 25(.7) = 17.5$ and the normal approximation is appropriate.

b For the binomial random variable, $\mu = np = 7.5$ and $\sigma = \sqrt{npq} = \sqrt{25(.3)(.7)} = 2.291$.

c The probability of interest is the area under the binomial probability histogram corresponding to the rectangles $x = 6, 7, 8$ and 9 in the figure below.

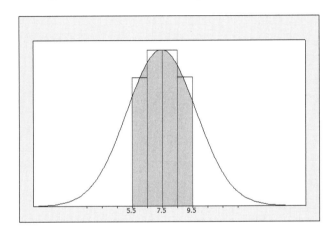

To approximate this area, use the "correction for continuity" and find the area under a normal curve with mean $\mu = 7.5$ and $\sigma = 2.291$ between $x_1 = 5.5$ and $x_2 = 9.5$. The z-values corresponding to the two values of x are

$$z_1 = \dfrac{5.5 - 7.5}{2.291} = -.87 \quad \text{and} \quad z_2 = \dfrac{9.5 - 7.5}{2.291} = .87$$

The approximating probability is $P(5.5 < x < 9.5) = P(-.87 < z < .87) = .8078 - .1922 = .6156$.

d From Table 1, Appendix I,
$$P(6 \leq x \leq 9) = P(x \leq 9) - P(x \leq 5) = .811 - .193 = .618$$
which is not too far from the approximate probability calculated in part **c**.

6.39 Similar to Exercise 6.38.

a The approximating probability will be $P(x > 22.5)$ where x has a normal distribution with $\mu = 100(.2) = 20$ and $\sigma = \sqrt{100(.2)(.8)} = 4$. Then
$$P(x > 22.5) = P\left(z > \dfrac{22.5 - 20}{4}\right) = P(z > .62) = 1 - .7324 = .2676$$

b The approximating probability is now $P(x > 21.5)$ since the entire rectangle corresponding to $x = 22$ must be included.
$$P(x > 21.5) = P\left(z > \dfrac{21.5 - 20}{4}\right) = P(z > .38) = 1 - .6480 = .3520$$

c To include the entire rectangles for $x = 21$ and $x = 24$, the approximating probability is
$$P(20.5 < x < 24.5) = P(.12 < z < 1.12) = .8686 - .5478 = .3208$$

d To include the entire rectangle for $x = 25$, the approximating probability is
$$P(x < 25.5) = P(z < 1.38) = .9162$$

Introduction to Probability and Statistics, 2e

6.41 Using the binomial tables for $n = 20$ and $p = .3$, you can verify that

a $P(x = 5) = P(x \leq 5) - P(x \leq 4) = .416 - .238 = .178$

b $P(x \geq 7) = 1 - P(x \leq 6) = 1 - .608 = .392$

6.43 Similar to previous exercises.

a With $n = 20$ and $p = .4$, $P(x \geq 10) = 1 - P(x \leq 9) = 1 - .755 = .245$.

b To use the normal approximation, find the mean and standard deviation of this binomial random variable: $\mu = np = 20(.4) = 8$ and $\sigma = \sqrt{npq} = \sqrt{20(.4)(.6)} = \sqrt{4.2} = 2.191$.

Using the continuity correction, it is necessary to find the area to the right of 9.5. The z-value corresponding to $x = 9.5$ is

$$z = \frac{9.5 - 8}{2.191} = .68 \quad \text{and} \quad P(x \geq 10) \approx P(z > .68) = 1 - .7517 = .2483.$$

Note that the normal approximation is very close to the exact binomial probability.

6.45 **a** The aproximating probability will be $P(x > 20.5)$ where x has a normal distribution with $\mu = 50(.731) = 36.55$ and $\sigma = \sqrt{50(.731)(.269)} = 3.136$. Then

$$P(x > 20.5) = P\left(z > \frac{20.5 - 36.55}{3.136}\right) = P(z > -5.12) \approx 1 - 0 = 1.00$$

b The approximating probability is

$$P(x < 14.5) = P\left(z < \frac{14.5 - 36.55}{3.136}\right) = P(z < -7.03) = 0$$

c If fewer than 28 students *do not* support same sex marriage, then $50 - 28 = 22$ or more do support same sex marriage. The approximating probability is

$$P(x > 21.5) = P\left(z > \frac{21.5 - 36.55}{3.136}\right) = P(z > -4.80) \approx 1 - 0 = 1$$

d As long as your class can be assumed to be a representative sample of all Canadians, the probabilities in parts **a-c** will be accurate.

6.47 Define x to be the number of guests claiming a reservation at the motel. Then $p = P[\text{guest claims reservation}] = 1 - .1 = .9$ and $n = 215$. The motel has only 200 rooms. Hence, if $x > 200$, a guest will not receive a room. The probability of interest is then $P(x \leq 200)$. Using the normal approximation, calculate

$$\mu = np = 215(.9) = 193.5 \quad \text{and} \quad \sigma = \sqrt{215(.9)(.1)} = \sqrt{19.35} = 4.399$$

The probability $P(x \leq 200)$ is approximated by the area under the appropriate normal curve to the left of 200.5. The z-value corresponding to $x = 200.5$ is $z = \dfrac{200.5 - 193.5}{\sqrt{19.35}} = 1.59$ and

$$P(x \leq 200) \approx P(z < 1.59) = .9441$$

6.49 Define x to be the number of Peruvian adults who support death penalty. Then the random variable x has a binomial distribution with $n = 503$ and $p = .81$. Calculate

$$\mu = np = 503(.81) = 407.43 \quad \text{and} \quad \sigma = \sqrt{503(.81)(.19)} = \sqrt{77.4117} = 8.8$$

a Using the normal approximation with correction for continuity, we find

$$P(x = 420) = P(419.5 < x < 420.5) = P\left(\frac{419.5 - 407.43}{8.8} < z < \frac{420.5 - 407.43}{8.8}\right)$$

$$= P(1.37 < z < 1.49) = .9319 - .9147 = .0172$$

b The approximating probability is
$$P(x < 419.5) = P\left(z < \frac{419.5 - 407.43}{8.8}\right) = P(z < 1.37) = .9147$$

c The approximating probability is
$$P(x > 420.5) = P\left(z > \frac{420.5 - 407.43}{8.8}\right) = P(z > 1.49) = 1 - .9319 = .0681$$

d You can only assume that Peruvian adults in Lima are representative sample of all adults in the capital city, but not of all adults in the entire nation.

6.51 Define x to be the number of consumers who preferred a *Pepsi* product. Then the random variable x has a binomial distribution with $n = 500$ and $p = .28$, if *Pepsi's* market share is indeed 28%. Calculate
$$\mu = np = 500(.28) = 140 \text{ and } \sigma = \sqrt{500(.28)(.72)} = \sqrt{100.8} = 10.0399$$

a Using the normal approximation with correction for continuity, we find the area between $x = 159.5$ and $x = 160.5$:
$$P(159.5 < x < 160.5) = P\left(\frac{159.5 - 140}{10.0399} < z < \frac{160.5 - 140}{10.0399}\right) = P(1.94 < z < 2.04) = .9793 - .9738 = .0055$$

b Find the area between $x = 119.5$ and $x = 150.5$:
$$P(119.5 < x < 150.5) = P\left(\frac{119.5 - 140}{10.0399} < z < \frac{150.5 - 140}{10.0399}\right) = P(-2.04 < z < 1.05) = .8531 - .0207 = .8324$$

c Find the area to the left of $x = 149.5$:
$$P(x < 149.5) = P\left(z < \frac{149.5 - 140}{10.0399}\right) = P(z < .95) = .8289$$

d The value $x = 232$ lies $z = \frac{232 - 140}{10.0399} = 9.16$ above the mean, if *Pepsi's* market share is indeed 28%. This is such an unusual occurrence that we would conclude that *Pepsi's* market share is higher than claimed.

6.53 Refer to Exercise 6.52, and let x be the number of Canadians who feel they will have to cut back on spending. Then x has a binomial distribution with $n = 100$ and $p = .37$.

a The average value of x is $\mu = np = 100(.37) = 37$.

b The standard deviation of x is $\sigma = \sqrt{npq} = \sqrt{100(.37)(.63)} = 4.828$.

c The z-score for $x = 50$ is $z = \frac{x - \mu}{\sigma} = \frac{50 - 37}{4.828} = 2.69$ which lies between two and three standard deviations away from the mean. This is considered a somewhat unusual occurrence.

6.55   **a**   The desired are A_1, as shown in the figure on the next page, is found by subtracting the cumulative areas corresponding to $z = 1.56$ and $z = 0.3$, respectively.

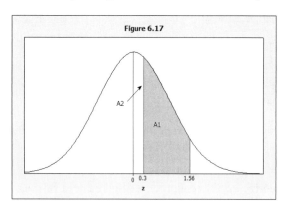

$$A_1 = A(1.56) - A(.3) = .9406 - .6179 = .3227.$$

b   The desired area is shown below:

$$A_1 + A_2 = A(.2) - A(-.2) = .5793 - .4207 = .1586$$

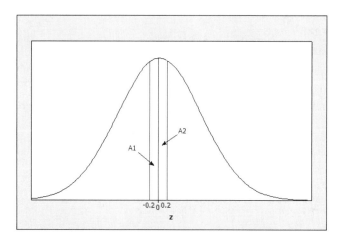

6.57   The procedure is reversed now, because the area under the curve is known. The objective is to determine the particular value, z_0, which will yield the given probability. In this exercise, it is necessary to find a z_0 such that $P(z > z_0) = .5000$. Then $P(z < z_0) = 1 - .5000 = .5000$, and from Table 3, the desired value of z_0 is 0.

6.59 $P(-z_0 < z < z_0) = 1 - 2A(-z_0) = .5000$. Hence, $A(-z_0) = \frac{1}{2}(1 - .5000) = .2500$. The desired value, z_0, will be between $z_1 = .67$ and $z_2 = .68$ with associated probabilities $P_1 = .2514$ and $P_2 = .2483$. Since the desired tail area, .2500, is closer to $P_1 = .2514$, we approximate z_0 as $z_0 = .67$. The values $z = -.67$ and $z = .67$ represent the 25th and 75th percentiles of the standard normal distribution.

6.61   The range of faculty ages should be approximately from 25 to 65 (25 for a new PhD). However, the distribution will not be normal (with average value 45 and symmetric) since there will be an overabundance of older tenured faculty. The distribution will probably be skewed to the right.

6.63 It is given that x is normally distributed with $\mu = 10$ and $\sigma = 3$. Let t be the guarantee time for the car. It is necessary that only 5% of the cars fail before time t (see below). That is,

$$P(x < t) = .05 \quad \text{or} \quad P\left(z < \frac{t-10}{3}\right) = .05$$

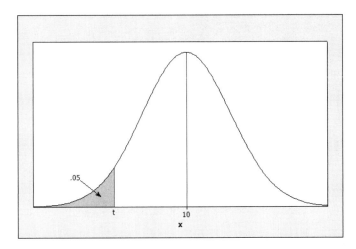

From Table 3, we know that the value of z that satisfies the above probability statement is $z = -1.645$. Hence,

$$\frac{t-10}{3} = -1.645 \quad \text{or} \quad t = 5.065 \text{ months}.$$

6.65 It is given that $\mu = 3.1$ and $\sigma = 1.2$. A washer must be replaced if its lifetime is less than one year, so that the desired fraction is $A = P(x < 1)$. The corresponding z-value is $z = \frac{x - \mu}{\sigma} = \frac{1 - 3.1}{1.2} = -1.75$ and $P(x < 1) = P(z < -1.75) = .0401$.

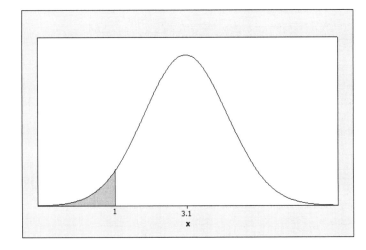

6.67 For this exercise $\mu = 70$ and $\sigma = 12$. The object is to determine a particular value, x_0, for the random variable x so that $P(x < x_0) = .90$ (that is, 90% of the students will finish the examination before the set time limit). Refer to the figure below.

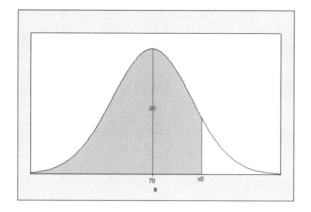

We must have

$$P(x < x_0) = P\left(z \le \frac{x_0 - 70}{12}\right) = .90$$

$$A\left(\frac{x_0 - 70}{12}\right) = .90$$

Consider $z_0 = \frac{x_0 - 70}{12}$. Without interpolating, the approximate value for z_0 is

$$z_0 = \frac{x_0 - 70}{12} = 1.28 \quad \text{or} \quad x_0 = 85.36$$

6.69 For the binomial random variable x, the mean and standard deviation are calculated under the assumption that the advertiser's claim is correct and $p = .2$. Then

$$\mu = np = 1000(.2) = 200 \text{ and } \sigma = \sqrt{npq} = \sqrt{1000(.2)(.8)} = 12.649$$

If the advertiser's claim is correct, the z-score for the observed value of x, $x = 184$, is

$$z = \frac{x - \mu}{\sigma} = \frac{184 - 200}{12.649} = -1.26$$

That is, the observed value lies 1.26 standard deviations below the mean. This is not an unlikely occurrence. Hence, we would have no reason to doubt the advertiser's claim.

6.71 It is given that the random variable x (millilitres of fill) is normally distributed with mean μ and standard deviation $\sigma = 10$. It is necessary to find a value of μ so that $P(x > 250) = .01$. That is, an 250- millilitre cup will overflow when $x > 250$, and this should happen only 1% of the time. Then

$$P(x > 250) = P\left(z > \frac{250 - \mu}{10}\right) = .01.$$

From Table 3, the value of z corresponding to an area (in the upper tail of the distribution) of .01 is $z_0 = 2.33$. Hence, the value of μ can be obtained by solving for μ in the following equation:

$$2.33 = \frac{250 - \mu}{10} \quad \text{or} \quad \mu = 226.7$$

6.73   The random variable x is the size of the first year class. That is, the admissions office will send letters of acceptance to (or accept deposits from) a certain number of qualified students. Of these students, a certain number will actually enter the first year class. Since the experiment results in one of two outcomes (enter or not enter), the random variable x, the number of students entering the first year class, has a binomial distribution with

n = number of deposits accepted and
$p = P$[student, having been accepted, enters first year class] = .8

a   It is necessary to find a value for n such $P(x \leq 120) = .95$. Note that,

$$\mu = np = .8n \text{ and } \sigma = \sqrt{npq} = \sqrt{.16n}$$

Using the normal approximation, we need to find a value of n such that $P(x \leq 120) = .95$, shown below. The z-value corresponding to $x = 120.5$ is $z = \dfrac{x - \mu}{\sigma} = \dfrac{120.5 - .8n}{\sqrt{.16n}}$

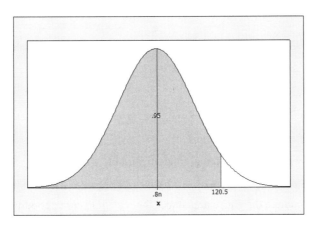

From Table 3, the z-value corresponding to an area of .05 in the right tail of the normal distribution is 1.645. Then, $\dfrac{120.5 - .8n}{\sqrt{.16n}} = 1.645$

Solving for n in the above equation, we obtain the following quadratic equation:

$$.8n + .658\sqrt{n} - 120.5 = 0$$

Let $x = \sqrt{n}$. Then the equation takes the form $ax^2 + bx + c = 0$ which can be solved using the quadratic formula, $x = \dfrac{-b \pm \sqrt{b^2 - 4ac}}{2a}$, or $x = \dfrac{-.658 \pm \sqrt{.433 + 4(96.4)}}{1.6} = \dfrac{-.658 \pm 19.648}{2}$

Since x must be positive, the desired root is $x = \sqrt{n} = \dfrac{18.990}{1.6} = 11.869$ or $n = (11.869)^2 = 140.86$.

Thus, 141 deposits should be accepted.

b   Once $n = 141$ has been determined, the mean and standard deviation of the distribution are

$$\mu = np = 141(.8) = 112.8 \text{ and } \sigma = \sqrt{npq} = \sqrt{22.56} = 4.750.$$

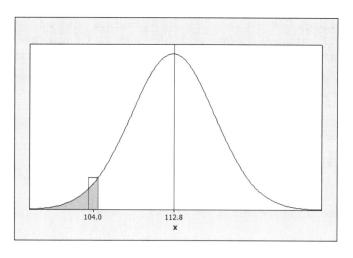

Then the approximation for $P(x<105)$, shown above, is

$$P(x \leq 104.5) = P\left(z \leq \frac{104.5-112.8}{4.750}\right) = P(z \leq -1.75) = .5 - .4599 = .0401$$

6.75 Define x = number of incoming calls that are long distance
$p = P$[incoming call is long distance] = .3
$n = 200$

The desired probability is $P(x \geq 50)$, where x is a binomial random variable with

$$\mu = np = 200(.3) = 60 \quad \text{and} \quad \sigma = \sqrt{npq} = \sqrt{200(.3)(.7)} = \sqrt{42} = 6.481$$

A correction for continuity is made to include the entire area under the rectangle corresponding to $x = 50$ and hence the approximation will be

$$P(x \geq 49.5) = P\left(z \geq \frac{49.5-60}{6.481}\right) = P(z \geq -1.62) = 1 - .0526 = .9474$$

6.77 The following information is available:
x = number of relays from supplier A
$p = P$[relay comes from supplier A] = 2/3
$n = 75$

Calculate $\mu = np = 75(2/3) = 50$ and $\sigma = \sqrt{npq} = \sqrt{16.667} = 4.082$. The probability of interest is $P(x \leq 48)$, which is approximated by $P(x \leq 48.5)$ using the normal approximation to the binomial distribution. The z-value corresponding to $x = 48.5$ is $z = \frac{x-\mu}{\sigma} = \frac{48.5-50}{\sqrt{16.667}} = -.37$ and

$$P(x \leq 48) \approx P(z \leq -.37) = .3557$$

6.79 The random variable x, the gestation time for a human baby is normally distributed with $\mu = 278$ and $\sigma = 12$.

a From Exercise 6.59, the values (rounded to two decimal places) $z = -.67$ and $z = .67$ represent the 25th and 75th percentiles of the standard normal distribution. Converting these values to their equivalents for the general random variable x using the relationship $x = \mu + z\sigma$, you have:

The lower quartile: $x = -.67(12) + 278 = 269.96$ and

The upper quartile: $x = .67(12) + 278 = 286.04$

b If you consider a month to be approximately 30 days, the value $x = 6(30) = 180$ is unusual, since it lies

$$z = \frac{x - \mu}{\sigma} = \frac{180 - 278}{12} = -8.167$$

standard deviations below the mean gestation time.

6.81 Define x to be the number of men who have fished in the last year. Then x has a binomial distribution with $n = 180$ and $p = .41$. Calculate

$$\mu = np = 180(.41) = 73.8 \quad \text{and} \quad \sigma = \sqrt{npq} = \sqrt{180(.41)(.59)} = 6.5986$$

a Using the normal approximation with correction for continuity,

$$P(x < 50) \approx P\left(z < \frac{49.5 - 73.8}{6.5986}\right) = P(z < -3.68) \approx 0$$

b $P(50 \leq x \leq 75) \approx P\left(\frac{49.5 - 73.8}{6.5986} \leq z \leq \frac{75.5 - 73.8}{6.5986}\right) = P(-3.68 \leq z \leq .26) = .6026 - .0000 = .6026$

c The sample is not random, since mailing lists for a sporting goods company will probably contain more fishermen than the population in general. Since the sampling was not random, the survey results are not reliable.

6.83 In order to implement the traditional interpretation of "curving the grades", the proportions shown in the table need to be applied to the normal curve, as shown in the figure below.

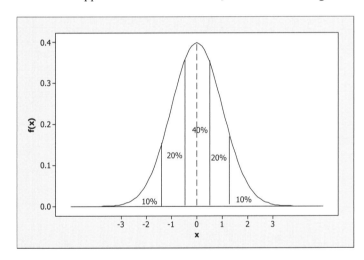

a The C grades constitute the middle 40%, that is, 20% on either side of the mean. The lower boundary has area .3000 to its left. From Table 3, we need to find a value of z such that $A(z) = .3000$. The closest value in the table is .3015 with $z = -.52$. The upper boundary is then $z = +.52$.

b The cutoff for the lowest D and highest B grades constitute the lower and upper boundaries of the middle 80%, that is, 40% on either side of the mean. The lower boundary has area .1000 to its left, so we need to find a value of z such that $A(z) = .1000$. The closest value in the table is .1003 with $z = -1.28$. The upper boundary is then $z = 1.28$.

6.85 Assume that the temperatures of healthy humans is approximately normal with $\mu = 37.1$ and $\sigma = 0.2$.

a $P(x > 37.22) = P(z > \frac{37.22 - 37.1}{0.2}) = P(z > .6) = 1 - .7257 = .2743$

b From Exercise 6.11**b**, we found that the 95th percentile of the *standard* normal (z) distribution is $z = 1.645$. Since $z = \frac{x - \mu}{\sigma} = \frac{x - 37.1}{0.2}$, solve for x to find the 95th percentile for the temperatures:

$$1.645 = \frac{x - 37.1}{0.2} \Rightarrow x = 37.1 + 1.645(0.2) = 37.429 \text{ degrees.}$$

6.87 Use either the **Normal Distribution Probabilities** or the **Normal Probabilities and z-scores** applets.
   a $P(-2.0 < z < 2.0) = .9772 - .0228 = .9544$
   b $P(-2.3 < z < -1.5) = .0668 - .0107 = .0561$

6.89 Use the **Normal Probabilities and z-scores** applet. Enter 0 as the mean and 1 as the standard deviation. Choose **One-tail** from the dropdown list and enter the desired probability in the box marked "prob". The necessary value of z_0 will appear in the box marked "z".
   a-b

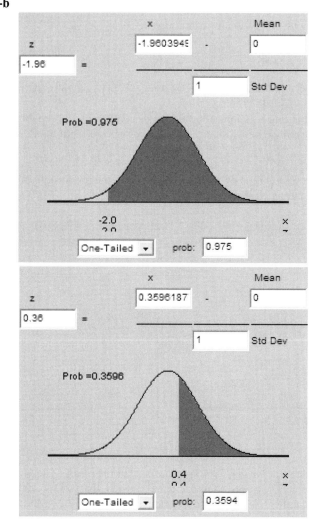

6.91 **a** Use the **Normal Distribution Probabilities** applet. Enter 5 as the mean and 2 as the standard deviation, and the appropriate lower and upper boundaries for the probabilities you need to calculate. The probability is read from the applet as **Prob = 0.9651**.
   b Use the **Normal Probabilities and z-scores** applet. Enter 5 as the mean and 2 as the standard deviation and $x = 7.5$. Choose **One-tail** from the dropdown list and read the probability as **Prob = 0.1056**.
   c Use the **Normal Probabilities and z-scores** applet. Enter 5 as the mean and 2 as the standard deviation and $x = 0$. Choose **Cumulative** from the dropdown list and read the probability as **Prob = 0.0062**.

6.93 **a-b** Using either the **Calculating Binomial Probabilities** applet or **Normal Approximation to Binomial Probabilities** applet with $n = 100$ and $p = .18$, we find $P(x \geq 60) = 1.2\,e-20 \approx 0$. The normal approximation is $P(x \geq 60) = 1 - P(x \leq 59) \approx 1 - 1 = 0$.

6.95 **a** It is given that the scores on a national achievement test were approximately normally distributed with a mean of 540 and standard deviation of 110. It is necessary to determine how far, in standard deviations, a score of 680 departs from the mean of 540. Calculate $z = \dfrac{x - \mu}{\sigma} = \dfrac{680 - 540}{110} = 1.27$.

b To find the percentage of people who scored higher than 680, we find the area under the standardized normal curve greater than 1.27. Using Table 3, this area is equal to
$$P(x > 680) = P(z > 1.27) = 1 - .8980 = .1020$$
Thus, approximately 10.2% of the people who took the test scored higher than 680. (The applet uses three decimal place accuracy and shows $z = \mathbf{1.273}$ with **Prob = 0.1016**.)

6.97 It is given that the probability of a successful single transplant from the early gastrula stage is .65. In a sample of 100 transplants, the mean and standard deviation of the binomial distribution are
$$\mu = np = 100(.65) = 65 \quad \text{and} \quad \sigma = \sqrt{npq} = \sqrt{100(.65)(.35)} = 4.770$$
It is necessary to find the probability that more than 70 transplants will be successful. This is approximated by the area under a normal curve with $\mu = 65$ and $\sigma = 4.77$ to the right of 70.5. The z-value corresponding to $x = 70.5$ is $z = \dfrac{x - \mu}{\sigma} = \dfrac{70.5 - 65}{4.77} = 1.15$ and $P(x > 70) \approx P(z > 1.15) = 1 - .8749 = .1251$

(The applet uses more decimal place accuracy and shows **Prob = 0.8756**, so that $P(x > 70) \approx .1244$.)

Case Study:
The Long and the Short of It

1 If x, the height of an adult Chinese male is normally distributed, with $\mu = 167.64$ and $\sigma = 6.86$,

$$P(x \leq 152.4) = P\left(z \leq \frac{152.4 - 167.64}{6.86}\right) = P(z \leq -2.22) = .0132$$

2 No. Seligman's odds of 40 to 1 (1 in 41 or .0244) are greater than the odds calculated in part **1**.

3 Seligman's assumption that the heights of Chinese males are approximately normal is correct, as you will learn in Chapter 7. Also, the claim that the mean height is 167.64 centimetres or taller is plausible, since the historical trend is towards an increase in height. In fact, if the mean is greater than 167.64, the probability calculated in part **1** would be even smaller. The assumption that $\sigma = 6.86$ "because it looks about right for that mean" is highly suspect, and greatly affects the calculations in part **1**.

Another flaw in the assumptions is that the distribution of the heights of all Chinese males may not be a good model for the distribution of heights of potential candidates for Deng Xiaoping's replacement. Since candidates would likely be a very select group of older Chinese males, and since human heights decrease as they age, the mean height of this group is likely less than the population of adult Chinese males in general. This would increase the probability calculated in part **1**.

4 The answer will depend on whether you are willing to accept Seligman's assumptions or not. See part **3** and draw your own conclusions.

Project 6-A: The Spectrum of Prematurity

a   (i) The probability of having a birth with mild prematurity is $P(33 \leq x < 37)$, since mild prematurity extends to 36 *completed weeks*. Now, $P(33 \leq x < 37) = P\left(\frac{33-40}{2} < z < \frac{37-40}{2}\right) = P(-3.5 < z < -1.5) = P(z < -1.5) - P(z < -3.5) = 0.0668 - 0.0002 = 0.0666$.

(ii) The probability of having a birth with extreme prematurity is $P(x < 28) = P\left(z < \frac{28-40}{2}\right) = P(z < -6) \approx 0$.

(iii) For the lower quartile, $z = -0.675$, and for the upper quartile, $z = 0.675$. Since $z = \frac{x-\mu}{\sigma}$, $x = \mu + z\sigma$. Thus, the lower quartile is $z = 40 + (-0.675)(2) = 38.65$ and the upper quartile is $z = 40 + 0.675(2) = 41.35$.

(iv) Yes, this would be very unusual. After only 24 weeks of gestation before birth, we would consider this to be *extreme prematurity*, which, as shown in part (ii) above, is extremely unlikely.

(v) For the 20th percentile, $z = -0.84$. Since $z = \frac{x-\mu}{\sigma}$, $x = \mu + z\sigma$. Thus, $x = 40 + (-0.84)(2) = 38.32$.

(vi) 83.4% corresponds to $z = 0.97$. Since $z = \frac{x-\mu}{\sigma}$, $x = \mu + z\sigma$. Thus, 83.4% of the gestational time occurs before $x = 40 + 0.97(2) = 41.94$.

b   (i) The probability of having a birth with moderate prematurity is $P(999.5 \text{ g} \leq x < 1500 \text{ g})$. Note that it is ambiguous from the question whether or not 1500 g exactly is mild or moderate prematurity. For this problem, $\mu = 3400$ g and $\sigma = 800$ g. Notice that it is necessary to use the same units (grams or kilograms) consistently. Therefore, $P(999.5 \text{ g} \leq x < 1500 \text{ g}) = P\left(\frac{999.5-3400}{800} < z < \frac{1500-3400}{800}\right) = P(-3 < z < -2.375) = P(z < -2.375) - P(z < -3) = 0.0088 - 0.0013 = 0.0075$.

(ii) The probability of having birth with extreme prematurity is $P(x < 1000) = P\left(z < \frac{999.5-3400}{800}\right) = P(z < -3) = 0.0013$.

(iii) The probability of having a baby weighing at least 6 kg (6000 g) is $P(x > 5999.5) = P\left(z > \frac{5999.5-3400}{800}\right) = P(z > 3.249) = 0.0006$. Since this probability is so low, having a baby weighing this much is highly unlikely.

(iv) To be in the top 5% corresponds to the 95th percentile, where $z = 1.645$. Since $z = \frac{x-\mu}{\sigma}$, $x = \mu + z\sigma$. Thus, the value of x we need is $x = 3.4$ kg $+ 1.645(.8$ kg$) = 4.716$ kg.

(v) 87.70% above is the same as 12.3% below, corresponding to $z = -1.16$. Since $x = \mu + z\sigma$, $x = 3.4 - 1.16(.8) = 2.472$ kg. Thus, 87.70% of the birth weights will be above 2.472 kg.

(vi) The information will result in two equations with two unknowns. The probability that x exceeds 4kg with probability 0.975 implies that $z = -1.96$ at $x = 4$. The probability that x exceeds 5 kg with probability 0.95 implies that $z = -1.645$ at $x = 5$. Since $x = \mu + z\sigma$, we get two equations: $4 = \mu - 1.96\,\sigma$, $5 = \mu - 1.645\sigma$. Subtracting the first equation from the second yields: $1 = 0.315\sigma$, or $\sigma = 3.1746$ kg. Then, $\mu = 10.2222$ kg. Such a high average weight on this planet is not comparable to the average weight of humans. Their babies are almost three times as heavy, on average.

Project 6-B: Premature Babies in Canada

c The normal approximation to the Binomial would not be appropriate in this instance. The rule of thumb illustrated on page 255 of the text states that the normal approximation will be adequate if both np and nq are greater than 5. In our case, $np = 25(0.086) = 2.15$, which is clearly less than 5.

d Since Alberta has an SGA rate of 8.7%, we would expect $.087(200) = 17.4$ SGA births in a random sample of 200.

e The SGA rate for Ontario is 8.9%. Since $np = 200(.089) = 17.8 > 5$, and $nq = 200(.911) > 5$, we may use the normal approximation to the Binomial, with $\mu = np = 17.8$ and $\sigma = \sqrt{npq} = \sqrt{200 \cdot (0.089) \cdot (0.911)}$ $= 4.026885$. And so, $P(x \geq 60) \approx P(x > 59.5) = P\left(z > \frac{x-\mu}{\sigma}\right) = P\left(z > \frac{59.5-17.8}{4.026885}\right) = P(z > 10.355) \approx 0$.

f The SGA rate for Canada was 8.3%. Since $np = 500(.083) = 41.5 > 5$, and $nq = 500(.917) > 5$, we may use the normal approximation to the Binomial, with $\mu = np = 41.5$ and $\sigma = \sqrt{npq} = \sqrt{500 \cdot (0.083) \cdot (0.917)}$ $= 6.168914$. Thus, the probability that at most 50 of these 500 will be declared SGA is $P(x \leq 50) \approx P(x < 50.5) = \left(z < \frac{50.5-41.5}{6.168914}\right) = P(z < 1.459) \approx 0.9279$. Furthermore, the probability that more than 10% of the births will be SGA is equivalent to more than $500(10\%) = 50$ births being SGA. But this is just the compliment of the probability of having at most 50 SGA births. Thus, the probability that more than 10% of the births will be SGA is simply $1 - 0.9279 = 0.0721$.

g The probability that at least 1 birth out of 25 is SGA is $1 - P$(none are SGA). The exact binomial probability that none are SGA is $p(0) = C_0^{25}(0.083)^0(0.917)^{25} = 0.114613$. Thus, the exact probability that at least 1 birth is SGA is $1 - 0.114613 = 0.885387$. For the approximate probability, we note that $\mu = np = 25(0.083)$ $= 2.075$. Thus, we don't expect our approximation to be that good. Proceeding, we also note that $\sigma = \sqrt{npq} = \sqrt{25 \cdot (0.083) \cdot (0.917)} = 1.379411$. Thus, the approximate probability that at least 1 of these 25 will be SGA is $P(x > 0.5) = \left(z > \frac{0.5-2.075}{1.379411}\right) = P(z > -1.142) \approx 0.8729$. The approximate probability is not as bad as anticipated, with a discrepancy of only $0.8854 - 0.8729 = 0.0125$ or 1.25%.

Chapter 7: Sampling Distributions

7.1 You can select a simple random sample of size $n = 20$ using Table 10 in Appendix I. First choose a starting point and consider the first three digits in each number. Since the experimental units have already been numbered from 000 to 999, the first 20 can be used. The three digits OR the (three digits – 500) will identify the proper experimental unit. For example, if the three digits are 742, you should select the experimental unit numbered $742 - 500 = 242$. The probability that any three digit number is selected is $2/1000 = 1/500$. One possible selection for the sample size $n = 20$ is

242	134	173	128	399
056	412	188	255	388
469	244	332	439	101
399	156	028	238	231

7.3 Each student will obtain a different sample, using Table 10 in Appendix I.

7.5 If all of the town citizenry is likely to pass this corner, a sample obtained by selecting every tenth person is probably a fairly random sample.

7.7 Since the enumeration list is updated only once every three years and will not reflect changes in personal circumstances within that period, it may not cover the Canadian citizens over 18 years of age who are residing in Ontario for less than three years. The list also excludes the youth group who receives elector eligibility within the last three year period.

7.9 Use a randomization scheme similar to that used in Exercise 7.1. Number each of the 50 rats from 01 to 50. To choose the 25 rats who will receive the dose of MX, select 25 two-digit random numbers from Table 10. Each two-digit number OR the (two digits – 50) will identify the proper experimental unit.

7.11   **a** The sample was chosen from patients in the clinical practice of one of the authors who were willing to participate. This sample is not randomly selected; it is a convenience sample.
     b Valid inferences can be made from this study *only if* the convenience sample chosen by the researcher *behaves like a random sample*. That is, the patients in this particular clinical practice must be representative of the population of patients as a whole.
     c In order to increase the chances of obtaining a sample that is representative of the population of patients as a whole, the researcher might try to obtain a larger base of patients to choose from. Perhaps there is a computerized database from which he or she might select a random sample.

7.13   **a** The first question is more unbiased.
     b Notice that the percentage for the Health Care Category drops dramatically when the phrase "that is one about which you are most concerned" is added to the question.

7.15 Follow the instructions in the My Personal Trainer section. The blanks have been filled in below.

The sampling distribution of $\bar{x}$ will be approximately **normal** with mean **53** and standard deviation (or standard error) **3**.

7.17 Follow the instructions in the My Personal Trainer section. The blanks have been filled in below.
The sampling distribution of $\bar{x}$ will be approximately **normal** with mean **100** and standard deviation (or standard error) **3.16**.

7.19 Regardless of the shape of the population from which we are sampling, the sampling distribution of the sample mean will have a mean μ equal to the mean of the population from which we are sampling, and a standard deviation equal to $\sigma/\sqrt{n}$.

    a $\mu = 10$; $\sigma/\sqrt{n} = 3/\sqrt{36} = .5$

    b $\mu = 5$; $\sigma/\sqrt{n} = 2/\sqrt{100} = .2$

    c $\mu = 120$; $\sigma/\sqrt{n} = 1/\sqrt{8} = .3536$

7.21   **a**   The sketch of the normal distribution with mean $\mu = 5$ and standard deviation $\sigma/\sqrt{n} = .2$ is left to the student. The interval $5 \pm .4$ or 4.6 to 5.4 should be located on the $\bar{x}$ axis.

    b   The probability of interest $P(-.15 < (\bar{x} - \mu) < .15)$ and is shown below.

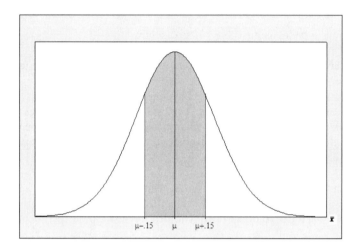

    c $P(-.15 < (\bar{x} - \mu) < .15) = P\left(\dfrac{-.15}{.2} < \dfrac{(\bar{x} - \mu)}{\sigma/\sqrt{n}} < \dfrac{.15}{.2}\right)$

        $= P(-.75 < z < .75) = .7734 - .2266 = .5468$

7.22-23 For a population with $\sigma = 1$, the standard error of the mean is $\sigma/\sqrt{n} = 1/\sqrt{n}$

The values of $\sigma/\sqrt{n}$ for various values of n are tabulated and plotted below and on the next page. Notice that the standard error *decreases* as the sample size *increases*.

n	1	2	4	9	16	25	100
$SE(\bar{x}) = \sigma/\sqrt{n}$	1.00	.707	.500	.333	.250	.200	.100

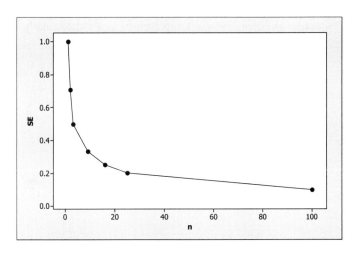

7.25 **a** Age of equipment, technician error, technician fatigue, equipment failure, difference in chemical purity, contamination from outside sources, and so on.

b The variability in the average measurement is measured by the standard error, $\sigma/\sqrt{n}$. In order to decrease this variability you should increase the sample size n.

7.27 The number of bacteria in one cubic metre of water can be thought of as the sum of 1,000,000 random variables, each of which is the number of bacteria in a particular cubic centimetre of water. Hence, the Central Limit Theorem insures the approximate normality of the sum.

7.29 **a** The population from which we are randomly sampling $n = 35$ measurements is not necessarily normally distributed. However, the sampling distribution of $\bar{x}$ does have an approximate normal distribution, with mean μ and standard deviation $\sigma/\sqrt{n}$. The probability of interest is

$$P(|\bar{x}-\mu|<1) = P(-1<(\bar{x}-\mu)<1).$$

Since $z = \dfrac{\bar{x}-\mu}{\sigma/\sqrt{n}}$ has a standard normal distribution, we need only find $\sigma/\sqrt{n}$ to approximate the above probability. Though σ is unknown, it can be approximated by $s = 12$ and $\sigma/\sqrt{n} \approx 12/\sqrt{35} = 2.028$. Then $P(|\bar{x}-\mu|<1) = P(-1/2.028 < z < 1/2.028)$

$$= P(-.49 < z < .49) = .6879 - .3121 = .3758$$

b No. There are many possible values for x, the actual percent tax savings, as given by the probability distribution for x.

7.31 **a** The random variable $T = \sum x_i$, were x_i is normally distributed with mean $\mu = 630$ and standard deviation $\sigma = 40$ for $i = 1, 2, 3$. The Central Limit Theorem states that T is normally distributed with mean $n\mu = 3(630) = 1890$ and standard deviation $\sigma\sqrt{n} = 40\sqrt{3} = 69.282$.

b Calculate $z = \dfrac{T - 1890}{69.282} = \dfrac{2000 - 1890}{69.282} = 1.59$.

Then $P(T > 2000) = P(z > 1.59) = 1 - .9441 = .0559$.

Introduction to Probability and Statistics, 2ce

7.33 **a** Since the original population is normally distributed, the sample mean $\bar{x}$ is also normally distributed (for any sample size) with mean $\mu = 37$ and standard deviation

$$\sigma/\sqrt{n} = 0.2/\sqrt{130} = .0175$$

The z-value corresponding to $\bar{x} = 36.81$ is $z = \dfrac{\bar{x} - \mu}{\sigma/\sqrt{n}} = \dfrac{36.81 - 37}{.0175} = -10.86$ and

$$P(\bar{x} < 36.81) = P(z < -10.86) \approx 0$$

b Since the probability is extremely small, the average temperature of 36.81 degrees is very unlikely.

7.35 Follow the instructions in the My Personal Trainer section. The blanks have been filled in below.
The sampling distribution of $\hat{p}$ will be approximately **normal** with mean $p = .7$ and standard deviation (or standard error) $\sqrt{\dfrac{pq}{n}} = \sqrt{\dfrac{.7(.3)}{50}} = \textbf{.0648}$.

7.36 Follow the instructions in the My Personal Trainer section. The blanks have been filled in below.
To find the probability that the sample proportion is less than .8, write down the event of interest __$P(\hat{p} < .8)$__. When $\hat{p} = .8$,

$$z = \dfrac{\hat{p} - p}{\sqrt{\dfrac{pq}{n}}} = \dfrac{.8 - .7}{.0648} = 1.54$$

Find the probability: $P(\hat{p} < .8) = P(z < 1.54) = .9382$

7.38 Since the sample sizes are very large, the sampling distributions in Exercise 7.37 will each be approximately normal, with appropriate means and variances. The interval $p \pm 2 SE(\hat{p})$ should cover 95% of the measurements.

7.39 **a** The sampling distribution will be approximately normal with mean p=.3 and SE=.0458. The interval that lies within .08 of the population proportion is (.3-.08, .3+.08) or (.22, .38).

b The probability of interest is $P(|\hat{p} - p| \leq .08) = P(-.08 \leq (\hat{p} - p) \leq .08)$

Since $\hat{p}$ is approximately normal, with standard deviation $SE(\hat{p}) = .0458$ from Exercise 7.37,

$$P(-.08 \leq (\hat{p} - p) \leq .08) = P\left[\dfrac{-.08}{.0458} \leq z \leq \dfrac{.08}{.0458}\right]$$
$$= P(-1.75 \leq z \leq 1.75) = .9599 - .0401 = .9198$$

7.41 The values $SE = \sqrt{pq/n}$ for $n = 100$ and various values of p are tabulated and graphed below and on the next page. Notice that SE is maximum for $p = .5$ and becomes very small for p near zero and one.

p	.01	.10	.30	.50	.70	.90	.99
$SE(\hat{p})$	.0099	.03	.0458	.05	.0458	.03	.0099

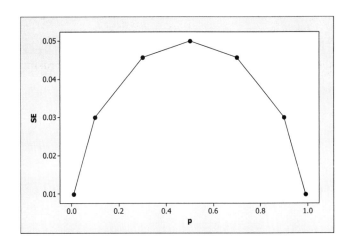

7.43 a For $n = 100$ and $p = .46$, $np = 46$ and $nq = 54$ are both greater than 5. Therefore, the normal approximation will be appropriate, with mean $p = .46$ and $SE = \sqrt{\frac{pq}{n}} = \sqrt{\frac{.46(.54)}{100}} = .0498$.

b $P(\hat{p} > .5) = P\left(z > \frac{.5 - .46}{.0498}\right) = P(z > .80) = 1 - .7881 = .2119$

c $P(.35 < \hat{p} < .55) = P\left(\frac{.35 - .46}{.0498} < z < \frac{.55 - .46}{.0498}\right) = P(-2.21 < z < 1.81) = .9649 - .0136 = .9513$

d The value $\hat{p} = .3$ lies $z = \frac{\hat{p} - p}{\sqrt{\frac{pq}{n}}} = \frac{.3 - .46}{.0498} = -3.21$ standard deviations from the mean. This is an unlikely occurrence, assuming that $p = .46$. Perhaps the sampling was not random, or the 46% figure is not correct.

7.45 a The random variable $\hat{p}$, the sample proportion of brown M&Ms in a package of $n = 55$, has a binomial distribution with $n = 55$ and $p = .13$. Since $np = 7.15$ and $nq = 47.85$ are both greater than 5, this binomial distribution can be approximated by a normal distribution with mean $p = .13$ and $SE = \sqrt{\frac{.13(.87)}{55}} = .04535$.

b $P(\hat{p} < .2) = P\left(z < \frac{.2 - .13}{.04535}\right) = P(z < 1.54) = .9382$

c $P(\hat{p} > .35) = P\left(z > \frac{.35 - .13}{.04535}\right) = P(z > 4.85) \approx 1 - 1 = 0$

d From the Empirical Rule (and the general properties of the normal distribution), approximately 95% of the measurements will lie within 2 (or 1.96) standard deviations of the mean:
$p \pm 2SE \implies .13 \pm 2(.04535)$
$.13 \pm .09$ or $.04$ to $.22$

7.47 a The random variable $\hat{p}$, the sample proportion of consumers who like nuts or caramel in their chocolate, has a binomial distribution with $n = 200$ and $p = .75$. Since $np = 150$ and $nq = 50$ are both greater than 5, this binomial distribution can be approximated by a normal distribution with mean $p = .75$ and $SE = \sqrt{\frac{.75(.25)}{200}} = .03062$

b $P(\hat{p} > .80) = P\left(z > \dfrac{.80 - .75}{.03062}\right) = P(z > 1.63) = 1 - .9484 = .0516$

c From the Empirical Rule (and the general properties of the normal distribution), approximately 95% of the measurements will lie within 2 (or 1.96) standard deviations of the mean:
$$p \pm 2SE \quad \Rightarrow \quad .75 \pm 2(.03062)$$
$$.75 \pm .06 \quad \text{or} \quad .69 \text{ to } .81$$

7.49 Similar to Exercise 7.48.

a The upper and lower control limits are
$$UCL = \bar{\bar{x}} + 3\dfrac{s}{\sqrt{n}} = 155.9 + 3\dfrac{4.3}{\sqrt{5}} = 155.9 + 5.77 = 161.67$$
$$LCL = \bar{\bar{x}} - 3\dfrac{s}{\sqrt{n}} = 155.9 - 3\dfrac{4.3}{\sqrt{5}} = 155.9 - 5.77 = 150.13$$

b The control chart is constructed by plotting two horizontal lines, one the upper control limit and one the lower control limit (see Figure 7.15 in the text). Values of $\bar{x}$ are plotted, and should remain within the control limits. If not, the process should be checked.

7.51 **a** The upper and lower control limits for a p chart are
$$UCL = \bar{p} + 3\sqrt{\dfrac{\bar{p}(1-\bar{p})}{n}} = .035 + 3\sqrt{\dfrac{.035(.965)}{100}} = .035 + .055 = .090$$
$$LCL = \bar{p} - 3\sqrt{\dfrac{\bar{p}(1-\bar{p})}{n}} = .035 - 3\sqrt{\dfrac{.035(.965)}{100}} = .035 - .055 = -.020$$
or LCL = 0 (since p cannot be negative).

b The control chart is constructed by plotting two horizontal lines, one the upper control limit and one the lower control limit (see Figure 7.16 in the text). Values of $\hat{p}$ are plotted, and should remain within the control limits. If not, the process should be checked.

7.53 **a** The upper and lower control limits are
$$UCL = \bar{\bar{x}} + 3\dfrac{s}{\sqrt{n}} = 10{,}752 + 3\dfrac{1605}{\sqrt{5}} = 10{,}752 + 2153.3 = 12{,}905.3$$
$$LCL = \bar{\bar{x}} - 3\dfrac{s}{\sqrt{n}} = 10{,}752 - 3\dfrac{1605}{\sqrt{5}} = 10{,}752 - 2153.3 = 8598.7$$

b The $\bar{x}$ chart will allow the manager to monitor daily gains or losses to see whether there is a problem with any particular table.

7.55 Calculate $\bar{p} = \dfrac{\sum \hat{p}_i}{k} = \dfrac{.14 + .21 + \cdots + .26}{30} = .197$. The upper and lower control limits for the p chart are then
$$UCL = \bar{p} + 3\sqrt{\dfrac{\bar{p}(1-\bar{p})}{n}} = .197 + 3\sqrt{\dfrac{.197(.803)}{100}} = .197 + .119 = .316$$
$$LCL = \bar{p} - 3\sqrt{\dfrac{\bar{p}(1-\bar{p})}{n}} = .197 - 3\sqrt{\dfrac{.197(.803)}{100}} = .197 - .119 = .078$$

7.57 Using all 104 measurements, the value of s is calculated to be $s = .006717688$ and $\bar{\bar{x}} = .0256$. Then the upper and lower control limits are
$$UCL = \bar{\bar{x}} + 3\dfrac{s}{\sqrt{n}} = .0256 + 3\dfrac{.06717688}{\sqrt{4}} = .0357$$
$$LCL = \bar{\bar{x}} - 3\dfrac{s}{\sqrt{n}} = .0256 - 3\dfrac{.06717688}{\sqrt{4}} = .0155$$

7.59 Refer to Exercise 7.58. The sample mean is outside the control limits in hours 1, 2, 3, 4 and 5. The process should be checked.

7.61 Refer to Exercise 7.60. If samples of size $n = 3$ are drawn without replacement, there are 4 possible samples with sample means shown below.

Sample	Observations	$\bar{x}$
1	6, 1, 3	3.333
2	6, 1, 2	3
3	6, 3, 2	3.667
4	1, 3, 2	2

The sampling distribution of $\bar{x}$ is then $p(\bar{x}) = \dfrac{1}{4}$ for $\bar{x} = 2, 3, 3.333, 3.667$

The graph of the sampling distribution is shown below.

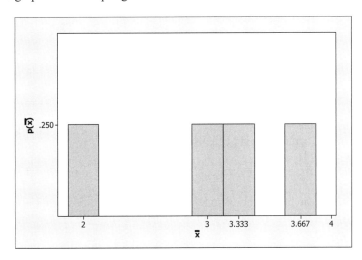

7.63 **a** Using the range approximation, the standard deviation σ can be approximated as

$$\sigma \approx \frac{R}{4} = \frac{55-5}{4} = 12.5$$

b The sampling distribution of $\bar{x}$ is approximately normal with mean μ and standard error

$$\sigma/\sqrt{n} \approx 12.5/\sqrt{400} = .625$$

Then $P(|\bar{x} - \mu| \le 2) = P\left(\dfrac{-2}{.625} \le z \le \dfrac{2}{.625}\right) = P(-3.2 \le z \le 3.2) = .9993 - .0007 = .9986$.

c If the scientists are worried that the estimate of $\mu = 35$ is too high, and if your estimate is $\bar{x} = 31.75$, then your estimate lies $z = \dfrac{\bar{x} - \mu}{\sigma/\sqrt{n}} = \dfrac{31.75 - 35}{.625} = -5.2$ standard deviations below the mean. This is a very unlikely event if in fact $\mu = 35$. It is more likely that the scientists are correct in assuming that the mean is an overestimate of the mean biomass for tropical woodlands.

7.65 **a** To divide a group of 20 people into two groups of 10, use Table 10 in Appendix I. Assign an identification number from 01 to 20 to each person. Then select ten two digit numbers from the random number table to identify the ten people in the first group. (If the number is greater than 20, subtract multiples of 20 from the random number until you obtain a number between 01 and 20.)

b Although it is not possible to select an actual random sample from this hypothetical population, the researcher must obtain a sample that *behaves like* a random sample. A large database of some sort should be used to ensure a fairly representative sample.

	c	The researcher has actually selected a *convenience sample*; however, it will probably behave like a simple random sample, since a person's enthusiasm for a paid job should not affect his response to this psychological experiment.
7.67	a	From the 50 lettuce seeds, the researcher must choose a group of 26 and a group of 13 for the experiment. Identify each seed with a number from 01 to 50 and then select random numbers from Table 10. The first 26 numbers chosen will identify the seeds in the first Petri dish, and the next 13 will identify the seeds in the third Petri dish. If the same number is picked twice, simply ignore it and go on to the next number. Use a similar procedure to choose the groups of 26 and 13 radish seeds.
	b	The seeds in these two packages must be representative of all seeds in the general population of lettuce and radish seeds.

7.69 Referring to Table 10 in Appendix I, we will select 20 numbers. First choose a starting point and consider the first four digits in each number. If the four digits are a number greater than 7000, discard it. Continue until 20 numbers have been chosen. The customers have already been numbered from 0001 to 7000. One possible selection for the sample size $n = 20$ is

1048	2891	5108	4866
2236	6355	0236	5416
2413	0942	0101	3263
4216	1036	5216	2933
3757	0711	0705	0248

7.71	a	Since a census is done on each cluster (a city block), this is an example of cluster sampling.
	b	This is a 1-in-10 systematic sample.
	c	The wards are the strata, and the sample is a stratified sample.
	d	This is a 1-in-10 systematic sample.
	e	This is a simple random sample from the population of all tax returns filed in the city of Halifax.

7.73	a	The number of packages which can be assembled in 8 hours is the sum of 8 observations on the random variable described here. Hence, its mean is $n\mu = 8(16.4) = 131.2$ and its standard deviation is $\sigma\sqrt{n} = 1.3\sqrt{8} = 3.677$.
	b	If the original population is approximately normal, the sampling distribution of a sum of 8 normal random variables will also be approximately normal. Since the original population is exactly normal, so will be the sampling distribution of the sum.
	c	$P(x > 135) = P\left(z > \dfrac{135 - 131.2}{3.677}\right) = P(z > 1.03) = 1 - .8485 = .1515$

7.75	a	The average proportion of defectives is $\overline{p} = \dfrac{.04 + .02 + \cdots + .03}{25} = .032$

and the control limits are

$$UCL = \overline{p} + 3\sqrt{\dfrac{\overline{p}(1-\overline{p})}{n}} = .032 + 3\sqrt{\dfrac{.032(.968)}{100}} = .0848$$

and $\quad LCL = \overline{p} - 3\sqrt{\dfrac{\overline{p}(1-\overline{p})}{n}} = .032 - 3\sqrt{\dfrac{.032(.968)}{100}} = -.0208$

If subsequent samples do not stay within the limits, $UCL = .0848$ and $LCL = 0$, the process should be checked.

	b	From part **a**, we must have $\hat{p} > .0848$.
	c	An erroneous conclusion will have occurred if in fact $p < .0848$ and the sample has produced $\hat{p} = .15$ by chance. One can obtain an upper bound on the probability of this particular type of error by calculating $P(\hat{p} \geq .15$ when $p = .0848)$.

7.77 Refer to Exercise 7.75, in which $UCL = .0848$ and $LCL = 0$. For the next 5 samples, the values of $\hat{p}$ are .02, .04, .09, .07, .11. Hence, samples 3 and 5 are producing excess defectives. The process should be checked.

7.79 Answers will vary from student to student. Paying cash for opinions will not necessarily produce a random sample of opinions of all Pepsi and Coke drinkers.

7.81 a Since the fill per can is normal with mean 355 and standard deviation 5.91, the total fill for a case of 24 cans will also have a normal distribution with mean $n\mu = 24(355) = 8520$ and standard deviation $\sigma\sqrt{n} = 5.91\sqrt{24} = 28.953$

b Let T be the total fill for the case of pop. Then
$$P(T < 8457) = P(z < \frac{8457 - 8520}{28.953}) = P(z < -2.18) = .0146$$

c $P(\bar{x} < 348.93) = P(z < \frac{348.93 - 355}{5.91/\sqrt{6}}) = P(z < -2.52) = .0059$

7.83 a The average proportion of inoperable components is $\bar{p} = \frac{6+7+\cdots+5}{50(15)} = \frac{75}{750} = .10$

and the control limits are

$$UCL = \bar{p} + 3\sqrt{\frac{\bar{p}(1-\bar{p})}{n}} = .10 + 3\sqrt{\frac{.10(.90)}{50}} = .2273$$

and $$LCL = \bar{p} - 3\sqrt{\frac{\bar{p}(1-\bar{p})}{n}} = .10 - 3\sqrt{\frac{.1(.9)}{50}} = -.0272$$

If subsequent samples do not stay within the limits, $UCL = .2273$ and $LCL = 0$, the process should be checked.

7.85 a The theoretical mean and standard deviation of the sampling distribution of $\bar{x}$ when $n = 2$ are
$$\mu = 3.5 \quad \text{and} \quad SE = \sigma/\sqrt{n} = 1.708/\sqrt{2} = 1.208$$

b-c Answers will vary from student to student. The distribution should be relatively uniform with mean and standard deviation close to those given in part **a**.

7.87 a The theoretical mean and standard deviation of the sampling distribution of $\bar{x}$ when $n = 4$ are
$$\mu = 3.5 \quad \text{and} \quad \sigma/\sqrt{n} = 1.708/\sqrt{4} = .854$$

b-c Answers will vary from student to student. The distribution should be relatively uniform with mean and standard deviation close to those given in part **a**.

7.89 Use the **Normal Probabilities for Means** applet. Enter the values of μ, σ, n.

a Enter $\bar{x} = 1100$, choosing **Cumulative** from the dropdown list. The applet shows $P(\bar{x} < 1100) = .0062$. Since the entire area to the left of 1110 is 0.5, the area between 1100 and 1110 is $.5 - .0062 = .4938$.

b Enter $\bar{x} = 1120$, choosing **One-Tailed** from the dropdown list. The applet shows $P(\bar{x} > 1120) = .0062$.

c Enter $\bar{x} = 900$, choosing **Cumulative** from the dropdown list. The applet shows $P(\bar{x} < 900) = .0000$.

Case Study:
Sampling the Roulette at Monte Carlo

1   Each bet results in a gain of (– \$5) if he loses and \$175 if he wins. Thus, the probability distribution of the gain x on a single \$5 bet is

X	p(x)
–5	37/38
175	1/38

2   Then
$$E(x) = \sum xp(x) = (-5)(37/38) + 175(1/38) = -.2632$$
$$\sigma_x^2 = \sum x^2 p(x) - \mu^2 = (-5)^2(37/38) + (175)^2(1/38) - (-.2632)^2 = 830.1939$$

3   The gain for the evening is the sum $S = \sum x_i$ of the gains or losses for the 200 bets of \$5 each. When the sample size is large, the Central Limit Theorem assures that this sum will be approximately normal with mean
$$\mu_S = n\mu = 200(-.2632) = -\$52.64$$
and variance
$$\sigma_S^2 = n\sigma_x^2 = 200(830.1939) = 166,038.78$$
$$\sigma_S = \sqrt{166,038.78} = 407.48$$

The total winnings will vary from –\$1000 (if the gambler loses all 200 bets) to \$35,000 (if the gambler wins all 200 bets), a range of \$36,000. However, most of the winnings (95%) will fall in the interval
$$\mu_S \pm 2\sigma_S = -52.64 \pm 2(407.43)$$

or –\$867.50 to \$762.22. The large gains are highly improbable.

4   The loss of \$1000 on any one night will occur only if there are no wins in 200 bets of \$5. The probability of this event is $\left(\frac{37}{38}\right)^{200} = .005$. Define y to be the number of evenings on which a loss of \$1000 occurs. Then y has a binomial distribution with $p = .005$ and $n = 365$. Using the Poisson approximation to the binomial with $\mu = np = 1.825$, the probability of interest is approximately
$$p(7) \approx \frac{(1.825)^7 e^{-1.825}}{7!} = .002$$
which is highly improbable.

5   The largest evening's winnings, \$1160, is not surprising. It lies $z = \frac{1160 - (-52.64)}{407.43} = 2.98$ standard deviations above the mean, so that $P(\text{winnings} \geq 1160) = P(z \geq 2.98) = 1 - .9986 = .0014$ for any one evening. The probability of observing winnings of \$1160 or greater on one evening out of 365 is then approximated using the Poisson approximation with $\mu = 365(.0014) = .511$ or $p(1) \approx \frac{(.511)^1 e^{-.511}}{1!} = .3065$

Project 7-A: Canada's Average IQ Just Jumped a Bunch- Stephen Hawking's Coming to Canada!

a We first need to assume that the population is normal, with a mean of 140 and standard deviation of 5. Then, the probability a random professor has an IQ below 130 can be found as follows:
Calculate $z = \frac{x-\mu}{\sigma} = \frac{130-140}{5} = -2$, so that $P(z < 130) = 0.0228$.

b (i) Since we cannot assume that the population is normal, we must invoke the Central Limit Theorem. The CLT tells us, when n is large ($n > 30$), that sampling distribution of $\overline{x}$ is approximately normal.

(ii) The sampling distribution is normal with a mean of 140 and a standard deviation of $\frac{\sigma}{\sqrt{n}} = \frac{5}{\sqrt{81}} = 0.\overline{5}$.

(iii) We calculate $z = \frac{\bar{x}-\mu}{\sigma/\sqrt{n}} = \frac{130-140}{0.\overline{5}} = -18$, so that $P(\bar{x} < 130) = P(z < -18) \approx 0$. The sampling distribution of $\bar{x}$ is much less variable than the single observation in part (a) above, which accounts for the fact that it is much less likely to get an average IQ less than 130 with a large sample than with a single observation.

(iv) No. Since our sample size is already relatively large, Stephen Hawking's score of 187 will not have much of an effect on the average IQ score.

(v) No. Even if we consider the probability of a professors' IQ score being below 130.5, we still obtain a z-score well under -3.5. This means that the probability that the average IQ score of selected Canadian professors is less than or equal to 130 would still remain essentially zero.

(vi) No, there is no way of calculating the exact probability that the sample average IQ is 135, due to the fact that the normal distribution is a continuous distribution.

(vii) Given that we have a relatively large sample (81), a sample mean of 130 would make the claim that μ = 140 suspect. A formal hypothesis test should be conducted to test this. However, it is very possible that μ is actually less than 140.

(viii) If we want to calculate the probability that the sample mean differs by more than 2 from the population mean, we must calculate two things; $P(\bar{x} > 142)$ and $P(\bar{x} < 138)$. First, calculate $z = \frac{\bar{x}-\mu}{\sigma/\sqrt{n}} = \frac{142-140}{0.\overline{5}} = 3.6$, so that $P(z > 142) \approx 0$, and $z = \frac{\bar{x}-\mu}{\sigma/\sqrt{n}} = \frac{138-140}{0.\overline{5}} = -3.6$, so that $P(z < 138) \approx 0$. Hence, the probability of the sample mean differing from the population mean by more than 2 is $P(138 > \bar{x} > 142) = P(\bar{x} < 138) + P(\bar{x} > 142) \approx 0$.

(ix) Since the normal distribution is a symmetric distribution, we can consider only finding an x such that $P(x < \bar{x}) = P(z < \bar{z}) = .025$. The other half must be symmetric. Looking at Table 3, the value of z corresponding the probability $P(\bar{z} < z) = .025$ is $z = -1.96$. Thus, for the lower bound, we calculate $\bar{x} = z\frac{\sigma}{\sqrt{n}} + \mu = (-1.96)(0.\overline{5}) + 140 = 138.9\overline{1}$; for the upper bound $(1.96)(0.\overline{5}) + 140 = 138.9\overline{1}$. Hence, we would expect the sample average to be within $(138.9\overline{1} < \bar{x} < 141.0\overline{8})$ with probability 0.95.

c **(i)** If for all i, x_i is normally distributed with a mean of μ and a standard deviation of σ, the total score is normally distributed.

(ii) By the Central Limit Theorem, we know that the total score $\sum x_i$ is normally distributed with mean $n\mu = 81(140) = 11{,}340$ and standard deviation $\sigma\sqrt{n} = 5\sqrt{81} = 45$.

(iii) We can calculate the probability of finding the total score between 11,000 and 11,400 (inclusive) as follows: $P(11{,}000 \leq \sum x_i \leq 11{,}400) = P(\sum x_i < 11{,}400.5) - P(\sum x_i < 10{,}999.5)$. Then,

$$z = \frac{\sum x_i - n\mu}{\sigma\sqrt{n}} = \frac{11{,}400.5 - 11{,}340}{45} = 1.34\text{, so that } P(\sum x_i < 11{,}400.5) = 0.9099\text{, and}$$

$$z = \frac{\sum x_i - n\mu}{\sigma\sqrt{n}} = \frac{10{,}999.5 - 11{,}340}{45} = -7.57\text{, so that } P(\sum x_i < 10{,}999.5) \approx 0.$$

Hence, the probability that the total score will be between 11,000 and 11,400 inclusive is 0.9099.

Project 7-B: Test the Nation on CBC

a   The mean of the sample proportion of individuals who scored above average is $p = 0.2$ and the standard deviation is $SE(\hat{p}) = \sqrt{\frac{pq}{n}} = \sqrt{\frac{(0.2)(0.8)}{64}} = 0.05$.

b   From page 293 of the text, we know that the normal distribution will be appropriate if $np > 5$ and $nq > 5$. First, $np = 64(0.2) = 18.2 > 5$, and $nq = 64(0.8) = 51.2 > 5$. This confirms that the distribution of the sample proportion is approximately normal.

c   First, calculate $z = \frac{\hat{p}-p}{\sqrt{\frac{pq}{n}}} = \frac{0.18-0.20}{0.05} = -0.4$, so that $P(\hat{p} > 18\%) = P(z > -0.4) = 0.6554$.

d   To calculate the probability that the sample proportion $\hat{p}$ lies between 19% and 23%, we must first find $P(\hat{p} < 23\%)$ and $P(\hat{p} < 19\%)$. First, calculate $z = \frac{\hat{p}-p}{\sqrt{\frac{pq}{n}}} = \frac{0.23-0.20}{0.05} = 0.6$, so that $P(\hat{p} < 23\%) = P(z < 0.6) = 0.7257$. Also, $z = \frac{\hat{p}-p}{\sqrt{\frac{pq}{n}}} = \frac{0.19-0.20}{0.05} = -0.2$, so that $P(\hat{p} < 19\%) = P(z < -0.2) = 0.4207$. Thus, $P(19\% < \hat{p} < 23\%) = 0.7257 - 0.4207 = 0.305$.

e   Since the normal distribution is a symmetric distribution, we only need to find a $\hat{p}$ such that $P(p < \hat{p}) = 0.005$. The other half must be symmetric. Looking at Table 3 in the text, the value of z corresponding to this probability is approximately $z = -2.575$. By symmetry, the other z we need is 2.575. Thus, for the lower bound, we calculate $\hat{p} = z\sqrt{\frac{pq}{n}} + p = (-2.575)(0.05) + 0.20 = 0.07125$, and for the upper bound, we calculate $\hat{p} = z\sqrt{\frac{pq}{n}} + p = (2.575)(0.05) + 0.20 = 0.32875$. Hence, 99% of the time, the sample proportion $\hat{p}$ will lie between $(0.07125 < \hat{p} < 0.32875)$.

f   There are multiple factors that can contribute to a sample proportion as small as 10%. However, the most probable reason for this would be that sampling was not done randomly, leading to a biased $\hat{p}$.

g   With a mean of $p = 0.2$ and a standard deviation of $SE(\hat{p}) = 0.05$, we notice that a value of $\hat{p} = 0.35$ is three times the standard deviation away from the mean. When this occurs, we would consider the value of $\hat{p}$ to be an outlier. Therefore, the value of $\hat{p} = 0.35$ would certainly be considered unusual.

Chapter 8: Large-Sample Estimation

8.1 The margin of error in estimation provides a practical upper bound to the difference between a particular estimate and the parameter which it estimates. In this chapter, the margin of error is $1.96 \times$ (standard error of the estimator).

8.3 For the estimate of μ given as $\bar{x}$, the margin of error is $1.96 \, SE = 1.96 \dfrac{\sigma}{\sqrt{n}}$.

  a $\quad 1.96\sqrt{\dfrac{0.2}{30}} = .160$

  b $\quad 1.96\sqrt{\dfrac{0.9}{30}} = .339$

  c $\quad 1.96\sqrt{\dfrac{1.5}{30}} = .438$

8.5 The margin of error is $1.96 \, SE = 1.96 \dfrac{\sigma}{\sqrt{n}}$, where σ can be estimated by the sample standard deviation s for large values of n.

  a $\quad 1.96\sqrt{\dfrac{4}{50}} = .554$

  b $\quad 1.96\sqrt{\dfrac{4}{500}} = .175$

  c $\quad 1.96\sqrt{\dfrac{4}{5000}} = .055$

8.7 For the estimate of p given as $\hat{p} = x/n$, the margin of error is $1.96 \, SE = 1.96\sqrt{\dfrac{pq}{n}}$.

  a $\quad 1.96\sqrt{\dfrac{(.5)(.5)}{30}} = .179$

  b $\quad 1.96\sqrt{\dfrac{(.5)(.5)}{100}} = .098$

  c $\quad 1.96\sqrt{\dfrac{(.5)(.5)}{400}} = .049$

  d $\quad 1.96\sqrt{\dfrac{(.5)(.5)}{1000}} = .031$

8.9 For the estimate of p given as $\hat{p} = x/n$, the margin of error is $1.96 \, SE = 1.96\sqrt{\dfrac{pq}{n}}$. Use the estimated value given in the exercise for p.

  a $\quad 1.96\sqrt{\dfrac{(.1)(.9)}{100}} = .0588$

  b $\quad 1.96\sqrt{\dfrac{(.3)(.7)}{100}} = .0898$

c $1.96\sqrt{\dfrac{(.5)(.5)}{100}} = .098$

d $1.96\sqrt{\dfrac{(.7)(.3)}{100}} = .0898$

e $1.96\sqrt{\dfrac{(.9)(.1)}{100}} = .0588$

f The largest margin of error occurs when $p = .5$.

8.11 The point estimate for p is given as $\hat{p} = \dfrac{x}{n} = \dfrac{655}{900} = .728$ and the margin of error is approximately

$$1.96\sqrt{\dfrac{\hat{p}\hat{q}}{n}} = 1.96\sqrt{\dfrac{.728(.272)}{900}} = .029$$

8.13 The point estimate of μ is $\bar{x} = 39.8°$ and the margin of error with $s = 17.2$ and $n = 50$ is

$$1.96\, SE = 1.96\dfrac{\sigma}{\sqrt{n}} \approx 1.96\dfrac{s}{\sqrt{n}} = 1.96\dfrac{17.2}{\sqrt{50}} = 4.768$$

8.15 The point estimate of μ is $\bar{x} = 7.2\%$ and the margin of error with $s = 5.6\%$ and $n = 200$ is

$$1.96\, SE = 1.96\dfrac{\sigma}{\sqrt{n}} \approx 1.96\dfrac{s}{\sqrt{n}} = 1.96\dfrac{5.6}{\sqrt{200}} = .776$$

8.17 **a** The point estimate for p is given as $\hat{p} = \dfrac{x}{n} = .69$ and the margin of error is approximately

$$1.96\sqrt{\dfrac{\hat{p}\hat{q}}{n}} = 1.96\sqrt{\dfrac{.69(.31)}{1275}} = 1.96\sqrt{.0002} \approx .027$$

b The poll's margin of error does agree with the results of part **a.**

8.19 **a** This method of sampling would not be random, since only interested viewers (those who were adamant in their approval or disapproval) would reply.

b The results of such a survey will not be valid, and a margin or error would be useless, since its accuracy is based on the assumption that the sample was random.

8.21 A 95% confidence interval for the population mean μ is given by

$$\bar{x} \pm 1.96\dfrac{\sigma}{\sqrt{n}}$$

where σ can be estimated by the sample standard deviation s for large values of n.

a $13.1 \pm 1.96\sqrt{\dfrac{3.42}{36}} = 13.1 \pm .604$ or $12.496 < \mu < 13.704$

b $2.73 \pm 1.96\sqrt{\dfrac{.1047}{64}} = 2.73 \pm .079$ or $2.651 < \mu < 2.809$

c Intervals constructed in this manner will enclose the true value of μ 95% of the time in repeated sampling. Hence, we are fairly confident that these particular intervals will enclose μ.

8.23 **a** $\bar{x} \pm z_{.005} \dfrac{\sigma}{\sqrt{n}} = \bar{x} \pm 2.58 \dfrac{\sigma}{\sqrt{n}} \approx 34 \pm 2.58 \sqrt{\dfrac{12}{38}} = 34 \pm 1.450$ or $32.550 < \mu < 35.450$.

b $\bar{x} \pm z_{.05} \dfrac{\sigma}{\sqrt{n}} = \bar{x} \pm 1.645 \dfrac{\sigma}{\sqrt{n}} \approx 1049 \pm 1.645 \sqrt{\dfrac{51}{65}} = 1049 \pm 1.457$ or $1047.543 < \mu < 1050.457$.

c $\bar{x} \pm z_{.025} \dfrac{\sigma}{\sqrt{n}} = \bar{x} \pm 1.96 \dfrac{\sigma}{\sqrt{n}} \approx 66.3 \pm 1.96 \sqrt{\dfrac{2.48}{89}} = 66.3 \pm .327$ or $65.973 < \mu < 66.627$.

8.25 Calculate $\hat{p} = \dfrac{x}{n} = \dfrac{27}{500} = .054$. Then an approximate 95% confidence interval for p is

$$\hat{p} \pm 1.96 \sqrt{\dfrac{\hat{p}\hat{q}}{n}} = .054 \pm 1.96 \sqrt{\dfrac{.054(.946)}{500}} = .054 \pm .020 \qquad \text{or} \qquad .034 < p < .074.$$

Notice that the interval is narrower than the one calculated in Exercise 8.25, even though the confidence coefficient is larger and n is larger. This is because the value of p (estimated by $\hat{p}$) is quite close to zero, causing $\sigma_{\hat{p}}$ to be small.

8.27 Refer to Exercise 8.26.
   a When the sample size is doubled, the width is decreased by $1/\sqrt{2}$.
   b When the sample size is quadrupled, the width is decreased by $1/\sqrt{4} = 1/2$.

8.29 With $n = 30$, $\bar{x} = .145$ and $s = .0051$, a 90% confidence interval for μ is approximated by

$$\bar{x} \pm 1.645 \dfrac{s}{\sqrt{n}} = .145 \pm 1.645 \dfrac{.0051}{\sqrt{30}} = .145 \pm .0015 \text{ or } .1435 < \mu < .1465$$

8.31 **a** An approximate 95% confidence interval for p is

$$\hat{p} \pm 1.96 \sqrt{\dfrac{\hat{p}\hat{q}}{n}} = .78 \pm 1.96 \sqrt{\dfrac{.78(.22)}{1030}} = .78 \pm .025 \text{ or } .755 < p < .805.$$

b An approximate 95% confidence interval for p is

$$\hat{p} \pm 1.96 \sqrt{\dfrac{\hat{p}\hat{q}}{n}} = .69 \pm 1.96 \sqrt{\dfrac{.69(.31)}{1030}} = .69 \pm .028 \text{ or } .662 < p < .718.$$

8.33 The 90% confidence interval for p is

$$\hat{p} \pm 1.645 \sqrt{\dfrac{\hat{p}\hat{q}}{n}} = .90 \pm 1.645 \sqrt{\dfrac{.90(.10)}{5756}} = .90 \pm .0065 \text{ or } .8935 < p < .9065.$$

8.35 **a** The time to complete an online order is probably not mound-shaped. The minimum value of x is zero, and there is an average time of $\mu = 4.5$, with a standard deviation of $\sigma = 2.7$. If we calculate $\mu - 2\sigma = -.9$, leaving no possibility for a measurement to fall more than two standard deviations below the mean. For a mound-shaped distribution, approximately 2.5% should fall in that range. The distribution is probably skewed to the right.

b Since n is large, the Central Limit Theorem ensures that the sample mean $\bar{x}$ is approximately normal, and the standard normal distribution can be used to construct a confidence interval for μ.

c The 95% confidence interval for μ is

$$\bar{x} \pm 1.96 \dfrac{s}{\sqrt{n}} = 4.5 \pm 1.96 \dfrac{2.7}{\sqrt{50}} = 4.5 \pm .478 \text{ or } 4.022 < \mu < 4.978$$

8.37 **a** The point estimate of p is $\hat{p} = \dfrac{x}{n} = \dfrac{192}{300} = .64$, and the approximate 95% confidence interval for p is

$$\hat{p} \pm 1.96\sqrt{\dfrac{\hat{p}\hat{q}}{n}} = .64 \pm 1.96\sqrt{\dfrac{.64(.36)}{300}} = .64 \pm .054 \text{ or } .586 < p < .694.$$

b Since the possible values for p given in the confidence interval does not includes the value $p = .71$, we would disagree with the reported percentage.

c The confidence interval in part **a** only estimates the proportion who say they always vote in federal general elections. This is different from the proportion who actually *do* vote in the next federal general election.

8.39 **a** 90% confidence interval for $\mu_1 - \mu_2$ is

$$(\bar{x}_1 - \bar{x}_2) \pm 1.645\sqrt{\dfrac{\sigma_1^2}{n_1} + \dfrac{\sigma_2^2}{n_2}}$$

Estimating σ_1^2 and σ_2^2 with s_1^2 and s_2^2, the approximate interval is

$$(2.9 - 5.1) \pm 1.645\sqrt{\dfrac{.83}{64} + \dfrac{1.67}{64}} = -2.2 \pm .325$$

or $-2.525 < (\mu_1 - \mu_2) < -1.875$. In repeated sampling, 90% of all intervals constructed in this manner will enclose $\mu_1 - \mu_2$. Hence, we are fairly certain that this particular interval contains $\mu_1 - \mu_2$.

b The 99% confidence interval for $\mu_1 - \mu_2$ is approximately

$$(\bar{x}_1 - \bar{x}_2) \pm 2.58\sqrt{\dfrac{s_1^2}{n_1} + \dfrac{s_2^2}{n_2}}$$

$$(2.9 - 5.1) \pm 2.58\sqrt{\dfrac{.83}{64} + \dfrac{1.67}{64}}$$

$$-2.2 \pm .510 \quad \text{or} \quad -2.710 < (\mu_1 - \mu_2) < -1.690$$

Since the value $\mu_1 - \mu_2 = 0$ is not in the confidence interval, it is not likely that $\mu_1 = \mu_2$. You should conclude that there is a difference in the two population means.

8.41 Similar to previous exercises. The 90% confidence interval for $\mu_1 - \mu_2$ is approximately

$$(\bar{x}_1 - \bar{x}_2) \pm 1.645\sqrt{\dfrac{s_1^2}{n_1} + \dfrac{s_2^2}{n_2}}$$

$$(2.4 - 3.1) \pm 1.645\sqrt{\dfrac{1.44}{100} + \dfrac{2.64}{100}}$$

$$-0.7 \pm .332 \quad \text{or} \quad -1.032 < (\mu_1 - \mu_2) < -0.368$$

Intervals constructed in this manner will enclose $\mu_1 - \mu_2$ 90% of the time. Hence, we are fairly certain that this particular interval encloses $(\mu_1 - \mu_2)$.

8.43 Similar to previous exercises. The 95% confidence interval for $\mu_1 - \mu_2$ is approximately

$$(\bar{x}_1 - \bar{x}_2) \pm 1.96 \sqrt{\frac{s_1^2}{n_1} + \frac{s_2^2}{n_2}}$$

$$(21.3 - 13.4) \pm 1.96 \sqrt{\frac{(2.6)^2}{30} + \frac{(1.9)^2}{30}}$$

$$7.9 \pm 1.152 \quad \text{or} \quad 6.748 < (\mu_1 - \mu_2) < 9.052$$

Intervals constructed in this manner will enclose $(\mu_1 - \mu_2)$ 95% of the time in repeated sampling. Hence, we are fairly certain that this particular interval encloses $(\mu_1 - \mu_2)$.

8.45 **a** The 95% confidence interval is approximately

$$\bar{x} \pm 1.96 \frac{s}{\sqrt{n}} = 14.06 \pm 1.96 \frac{5.65}{\sqrt{376}} = 14.06 \pm .571 \quad \text{or} \quad 13.489 < \mu < 14.631$$

b The 95% confidence interval is approximately

$$\bar{x} \pm 1.96 \frac{s}{\sqrt{n}} = 12.96 \pm 1.96 \frac{5.93}{\sqrt{308}} = 12.96 \pm .662 \quad \text{or} \quad 12.298 < \mu < 13.622$$

c The 95% confidence interval for $\mu_1 - \mu_2$ is approximately

$$(\bar{x}_1 - \bar{x}_2) \pm 1.96 \sqrt{\frac{s_1^2}{n_1} + \frac{s_2^2}{n_2}}$$

$$(14.06 - 12.96) \pm 1.96 \sqrt{\frac{(5.65)^2}{376} + \frac{(5.93)^2}{308}}$$

$$1.10 \pm .875 \quad \text{or} \quad .225 < (\mu_1 - \mu_2) < 1.975$$

d Since the confidence interval in part **c** has two positive endpoints, it does not contain the value $\mu_1 - \mu_2 = 0$. Hence, it is not likely that the means are equal. It appears that there is a real difference in the mean scores.

8.47 **a** The 99% confidence interval for $\mu_1 - \mu_2$ is approximately

$$(\bar{x}_1 - \bar{x}_2) \pm 2.58 \sqrt{\frac{s_1^2}{n_1} + \frac{s_2^2}{n_2}}$$

$$(15 - 23) \pm 2.58 \sqrt{\frac{4^2}{30} + \frac{10^2}{40}}$$

$$-8 \pm 4.49 \quad \text{or} \quad -12.49 < (\mu_1 - \mu_2) < -3.51$$

b Since the confidence interval in part **a** has two negative endpoints, it does not contain the value $\mu_1 - \mu_2 = 0$. Hence, it is not likely that the means are equal. It appears that there is a real difference in the mean times to completion for the two groups.

8.49   **a**   The best estimate of $p_1 - p_2$ is $\hat{p}_1 - \hat{p}_2$ where $\hat{p}_1 = \dfrac{x_1}{n_1} = \dfrac{120}{500} = .24$ and $\hat{p}_2 = \dfrac{x_2}{n_2} = \dfrac{147}{500} = .294$.

       b   The standard error is calculated by estimating p_1 and p_2 with $\hat{p}_1$ and $\hat{p}_2$ in the formula:

$$SE = \sqrt{\dfrac{p_1 q_1}{n_1} + \dfrac{p_2 q_2}{n_2}} \approx \sqrt{\dfrac{\hat{p}_1 \hat{q}_1}{n_1} + \dfrac{\hat{p}_2 \hat{q}_2}{n_2}} = \sqrt{\dfrac{.24(.76)}{500} + \dfrac{.294(.706)}{500}} = .0279$$

       c   From part **b**, the approximate margin of error is

$$1.96\sqrt{\dfrac{.24(.76)}{500} + \dfrac{.294(.706)}{500}} = 1.96(.0279) = .055$$

8.51   **a**   Calculate $\hat{p}_1 = \dfrac{x_1}{n_1} = \dfrac{849}{1265} = .671$ and $\hat{p}_2 = \dfrac{x_2}{n_2} = \dfrac{910}{1688} = .539$. The approximate 99% confidence interval

           is

$$(\hat{p}_1 - \hat{p}_2) \pm 2.58 \sqrt{\dfrac{\hat{p}_1 \hat{q}_1}{n_1} + \dfrac{\hat{p}_2 \hat{q}_2}{n_2}}$$

$$(.671 - .539) \pm 2.58 \sqrt{\dfrac{.671(.329)}{1265} + \dfrac{.539(.461)}{1688}}$$

$$.132 \pm .046 \quad \text{or} \quad .086 < (p_1 - p_2) < .178$$

           In repeated sampling, 99% of all intervals constructed in this manner will enclose $p_1 - p_2$. Hence, we are fairly certain that this particular interval contains $p_1 - p_2$.

       b   Since the value $p_1 - p_2 = 0$ is not in the confidence interval, it is not likely that $p_1 = p_2$. You should conclude that there is a difference in the two population proportions.

8.53   Calculate $\hat{p}_1 = \dfrac{x_1}{250} = .70$ and $\hat{p}_2 = \dfrac{x_2}{250} = .86$. The approximate 95% confidence interval is

$$(\hat{p}_1 - \hat{p}_2) \pm 1.96 \sqrt{\dfrac{\hat{p}_1 \hat{q}_1}{n_1} + \dfrac{\hat{p}_2 \hat{q}_2}{n_2}}$$

$$(.70 - .86) \pm 1.96 \sqrt{\dfrac{.70(.30)}{250} + \dfrac{.86(.14)}{250}}$$

$$-.16 \pm .071 \quad \text{or} \quad -.231 < (p_1 - p_2) < -.089$$

Since the value $p_1 - p_2 = 0$ is not in the confidence interval, it is not likely that $p_1 = p_2$. You should conclude that there is a difference in the proportion of Conservatives and Liberals who favor the new initiative. It appears that the percentage of Liberal voters is higher than the Conservative percentage.

8.55   **a**   Calculate $\hat{p}_1 = \dfrac{390}{430} = .907$ and $\hat{p}_2 = \dfrac{100}{570} = .175$. The approximate 80% confidence interval is

$$(\hat{p}_1 - \hat{p}_2) \pm 1.28 \sqrt{\dfrac{\hat{p}_1 \hat{q}_1}{n_1} + \dfrac{\hat{p}_2 \hat{q}_2}{n_2}}$$

$$(.907 - .175) \pm 1.28 \sqrt{\dfrac{.907(.093)}{430} + \dfrac{.175(.825)}{570}}$$

$$.732 \pm .027 \quad \text{or} \quad .705 < (p_1 - p_2) < .759$$

       b   Since the value $p_1 - p_2 = 0$ is not in the confidence interval in part **a**, it is not likely that $p_1 = p_2$.

8.57 The following sample information is available:

$$n_1 = n_2 = 200 \quad \hat{p}_1 = \frac{142}{200} = .71 \quad \hat{p}_2 = \frac{120}{200} = .60.$$

The approximate 95% confidence interval is

$$(\hat{p}_1 - \hat{p}_2) \pm 1.96\sqrt{\frac{\hat{p}_1\hat{q}_1}{n_1} + \frac{\hat{p}_2\hat{q}_2}{n_2}}$$

$$(.71 - .60) \pm 1.96\sqrt{\frac{.71(.29)}{200} + \frac{.60(.40)}{200}}$$

$$.11 \pm .093 \quad \text{or} \quad .017 < (p_1 - p_2) < .203$$

Intervals constructed in this manner will enclose the true value of $p_1 - p_2$ 95% of the time in repeated sampling. Hence, we are fairly certain that this particular interval encloses $p_1 - p_2$.

8.59 **a** The approximate 98% confidence interval is

$$(\hat{p}_1 - \hat{p}_2) \pm 2.33\sqrt{\frac{\hat{p}_1\hat{q}_1}{n_1} + \frac{\hat{p}_2\hat{q}_2}{n_2}}$$

$$(.20 - .26) \pm 2.33\sqrt{\frac{.20(.80)}{500} + \frac{.26(.74)}{500}}$$

$$-.06 \pm .062 \quad \text{or} \quad -.122 < (p_1 - p_2) < .002$$

b Intervals constructed in this manner enclose the true value of $p_1 - p_2$ 98% of the time in repeated sampling. Hence, we are fairly certain that this particular interval encloses $p_1 - p_2$.

c Since the value $p_1 - p_2 = 0$ is in the confidence interval, it is possible that $p_1 = p_2$. You should not conclude that there is a difference in the proportion of men and women who think that space should remain commercial free.

8.61 **a** The point estimate for p is given as $\hat{p} = \frac{x}{n} = \frac{23}{41} = .561$ and the margin of error is approximately

$$1.96\sqrt{\frac{\hat{p}\hat{q}}{n}} = 1.96\sqrt{\frac{.56(.44)}{41}} = .152$$

b Calculate $\hat{p}_1 = \frac{10}{32} = .3125$ and $\hat{p}_2 = \frac{23}{41} = .561$. The approximate 95% confidence interval is

$$(\hat{p}_1 - \hat{p}_2) \pm 1.96\sqrt{\frac{\hat{p}_1\hat{q}_1}{n_1} + \frac{\hat{p}_2\hat{q}_2}{n_2}}$$

$$(.3125 - .561) \pm 1.96\sqrt{\frac{.3125(.6875)}{32} + \frac{.561(.439)}{41}}$$

$$-.2485 \pm .2211 \quad \text{or} \quad -.4696 < (p_1 - p_2) < -.0274$$

8.63 Follow the instructions in the My Personal Trainer section. The answers are shown in the table below.

Type of Data	One or Two Samples	Margin of error	p or σ	Bound, B	Solve this inequality	Sample size
Quantitative	Two	$1.96\sqrt{\dfrac{\sigma_1^2}{n_1}+\dfrac{\sigma_2^2}{n_2}}$	$\sigma_1 \approx \sigma_2 \approx 10$	4	$1.96\sqrt{\dfrac{10^2}{n}+\dfrac{10^2}{n}} \le 4$	$n_1 = n_2 \ge 49$
Binomial	Two	$1.96\sqrt{\dfrac{p_1 q_1}{n_1}+\dfrac{p_2 q_2}{n_2}}$	$p_1 \approx p_2 \approx .5$	.10	$1.96\sqrt{\dfrac{.5(.5)}{n}+\dfrac{.5(.5)}{n}}$	$n_1 = n_2 \ge 193$

8.65 The 99% upper confidence bound is calculated using $\hat{p} = x/n = 196/400 = .49$ and a value $z_\alpha = z_{.01} = 2.33$. The lower confidence bound for the binomial parameter p is approximately

$$\hat{p} - 2.33\sqrt{\dfrac{\hat{p}\hat{q}}{n}} = .49 - 2.33\sqrt{\dfrac{.49(.51)}{400}} = .49 - .058 \quad \text{or} \quad p > .432$$

8.67 It is necessary to find the sample size required to estimate a certain parameter to within a given bound with confidence $(1-\alpha)$. Recall from Section 8.5 that we may estimate a parameter with $(1-\alpha)$ confidence within the interval (estimator) $\pm z_{\alpha/2} \times$ (std error of estimator). Thus, $z_{\alpha/2} \times$ (std error of estimator) provides the margin of error with $(1-\alpha)$ confidence. The experimenter will specify a given bound B. If we let $z_{\alpha/2} \times$ (std error of estimator) $\le B$, we will be $(1-\alpha)$ confident that the estimator will lie within B units of the parameter of interest.

For this exercise, the parameter of interest is μ, B = 1.6 and $1-\alpha = .95$. Hence, we must have

$$1.96\dfrac{\sigma}{\sqrt{n}} \le 1.6 \Rightarrow 1.96\dfrac{12.7}{\sqrt{n}} \le 1.6$$

$$\sqrt{n} \ge \dfrac{1.96(12.7)}{1.6} = 15.5575$$

$$n \ge 242.04 \quad \text{or} \quad n \ge 243$$

8.69 In this exercise, the parameter of interest is $\mu_1 - \mu_2$, $n_1 = n_2 = n$, and $\sigma_1^2 \approx \sigma_2^2 \approx 27.8$. Then we must have

$$z_{\alpha/2} \times (\text{std error of } \bar{x}_1 - \bar{x}_2) \le B$$

$$1.645\sqrt{\dfrac{\sigma_1^2}{n_1}+\dfrac{\sigma_2^2}{n_2}} \le .17 \Rightarrow 1.645\sqrt{\dfrac{27.8}{n}+\dfrac{27.8}{n}} \le .17$$

$$\sqrt{n} \ge \dfrac{1.645\sqrt{55.6}}{.17} \Rightarrow n \ge 5206.06 \quad \text{or} \quad n_1 = n_2 = 5207$$

8.71 The parameter to be estimated is the population mean μ and the 90% upper confidence bound is calculated using a value $z_\alpha = z_{.10} = 1.28$. The upper bound is approximately

$$\bar{x} + 1.28\dfrac{s}{\sqrt{n}} = 5474 + 1.28\dfrac{764}{\sqrt{36}} = 5474 + 162.99 \quad \text{or} \quad \mu < 5636.99$$

8.73 Similar to Exercise 8.71.

$$z_{.025}\sqrt{\frac{p_1q_1}{n_1}+\frac{p_2q_2}{n_2}} \le .03 \Rightarrow 1.96\sqrt{\frac{(.5)(.5)}{n}+\frac{(.5)(.5)}{n}} \le .03$$

$$\sqrt{n} \ge \frac{1.96\sqrt{.5}}{.03} \Rightarrow n \ge 2134.2 \text{ or } n_1 = n_2 = 2135$$

8.75 a For the difference $\mu_1 - \mu_2$ in the population means this year and ten years ago, the 99% lower confidence bound uses $z_{.01} = 2.33$ and is calculated as

$$(\bar{x}_1 - \bar{x}_2) - 2.33\sqrt{\frac{s_1^2}{n_1}+\frac{s_2^2}{n_2}} = (33.1 - 28.6) - 2.33\sqrt{\frac{11.3^2}{400}+\frac{12.7^2}{400}}$$

$$4.5 - 1.980 \text{ or } (\mu_1 - \mu_2) > 2.52$$

b Since the difference in the means is positive, you can conclude that there has been a decrease in the average per-capita beef consumption over the last ten years.

8.77 Similar to previous exercises with B = .1 and $\sigma \approx .5$. The required sample size is obtained by solving

$$1.96\frac{\sigma}{\sqrt{n}} \le B \Rightarrow 1.96\frac{.5}{\sqrt{n}} \le .1$$

$$\sqrt{n} \ge \frac{1.96(.5)}{.1} = 9.8 \Rightarrow n \ge 96.04 \text{ or } n \ge 97$$

Notice that water specimens should be selected randomly and not necessarily from the same rainfall, in order that all observations are independent.

8.79 The parameter of interest is $\mu_1 - \mu_2$, the difference in grade-point averages for the two populations of students. Assume that $n_1 = n_2 = n$, and $\sigma_1^2 \approx \sigma_2^2 \approx (.6)^2 = .36$ and that the desired bound is .2. Then

$$1.96\sqrt{\frac{\sigma_1^2}{n_1}+\frac{\sigma_2^2}{n_2}} \le .2 \Rightarrow 1.96\sqrt{\frac{.36}{n}+\frac{.36}{n}} \le .2$$

$$\sqrt{n} \ge \frac{1.96\sqrt{.72}}{.2} \Rightarrow n \ge 69.149$$

or $n_1 = n_2 = 70$ students should be included in each group.

8.81 See Section 7.4 of the text.

8.83 The 90% confidence interval for $\mu_1 - \mu_2$ is approximately

$$(\bar{x}_1 - \bar{x}_2) \pm 1.645\sqrt{\frac{s_1^2}{n_1}+\frac{s_2^2}{n_2}}$$

$$(100.4 - 96.2) \pm 1.645\sqrt{\frac{.8^2}{50}+\frac{1.3^2}{60}}$$

$$4.2 \pm .333 \text{ or } 3.867 < (\mu_1 - \mu_2) < 4.533$$

8.85 a The point estimate of p is $\hat{p} = \frac{x}{n} = \frac{240}{500} = .48$ with approximate margin of error

$$1.96\sqrt{\frac{\hat{p}\hat{q}}{n}} = 1.96\sqrt{\frac{.48(.52)}{500}} = .044$$

b An approximate 90% confidence interval for p is

$$\hat{p} \pm 1.645\sqrt{\frac{\hat{p}\hat{q}}{n}} = .48 \pm 1.645\sqrt{\frac{.48(.52)}{500}} = .48 \pm .037$$

or $.433 < p < .517$. Intervals constructed in this manner enclose the true value of p 90% of the time. Hence, we are fairly certain that this particular interval encloses p.

8.87 Calculate $\hat{p}_1 = \frac{17}{40} = .425$, $\hat{p}_2 = \frac{23}{80} = .2875$. The approximate 99% confidence interval for $p_1 - p_2$ is

$$(\hat{p}_1 - \hat{p}_2) \pm 2.58\sqrt{\frac{\hat{p}_1\hat{q}_1}{n_1} + \frac{\hat{p}_2\hat{q}_2}{n_2}}$$

$$(.425 - .2875) \pm 2.58\sqrt{\frac{.425(.575)}{40} + \frac{.2875(.7125)}{80}}$$

$$.1375 \pm .240 \quad \text{or} \quad -.1025 < (p_1 - p_2) < .3775$$

Intervals constructed in this manner will enclose the true value of $p_1 - p_2$ 99% of the time in repeated sampling. Hence, we are fairly certain that this particular interval encloses $p_1 - p_2$.

8.89 The parameter of interest is $p_1 - p_2$, $n_1 = n_2 = n$, and $B = .03$. Since no prior knowledge is available about p_1 and p_2, we assume the largest possible variation, which occurs if $p_1 = p_2 = .5$. Then

$$z_{.025}\sqrt{\frac{p_1q_1}{n_1} + \frac{p_2q_2}{n_2}} \leq .03 \Rightarrow 1.96\sqrt{\frac{(.5)(.5)}{n} + \frac{(.5)(.5)}{n}} \leq .03$$

$$\sqrt{n} \geq \frac{1.96\sqrt{.5}}{.03} \Rightarrow n \geq 2134.2 \quad \text{or} \quad n_1 = n_2 = 2135$$

8.91 The best sample estimator for the mean μ is $\bar{x} = \sum x_i/n$. Therefore, the estimated value of μ is $\bar{x} = 9.7$. The margin of error is

$$1.96\sigma_{\bar{x}} = 1.96\frac{\sigma}{\sqrt{n}} \approx 1.96\frac{s}{\sqrt{n}} = 1.96\left(\frac{5.8}{\sqrt{35}}\right) = 1.92$$

The population in this exercise is the difference in blood pressure for all college-aged smokers between the beginning of the experiment and five years later.

8.93 **a** The "error margin" is equivalent to the margin of error, the upper limit to the difference between the estimate and the parameter to be estimated.

b If $n = 30$ and $s = .017$, the margin of error is

$$1.96\frac{s}{\sqrt{n}} = 1.96\frac{.017}{\sqrt{30}} = .00608$$

and the chemist is correct.

8.95

The following sample information is available:

$$n_1 = n_2 = 100 \quad \hat{p}_1 = \frac{13}{100} = .13 \quad \hat{p}_2 = \frac{6}{100} = .06.$$

a The approximate 98% confidence interval for $p_1 - p_2$ is

$$(\hat{p}_1 - \hat{p}_2) \pm 2.33 \sqrt{\frac{\hat{p}_1 \hat{q}_1}{n_1} + \frac{\hat{p}_2 \hat{q}_2}{n_2}}$$

$$(.13 - .06) \pm 2.33 \sqrt{\frac{.13(.87)}{100} + \frac{.06(.94)}{100}}$$

$$.07 \pm .096 \quad \text{or} \quad -.026 < (p_1 - p_2) < .166$$

b Since the value $p_1 - p_2 = 0$ does lie in the interval in part **a**, it is possible that $p_1 = p_2$. You should not conclude that there is a difference in mortality rates for the two rations.

8.97 The 95% confidence intervals for the average cheese and beer consumptions are:

Cheese: $\bar{x} \pm 1.96 \frac{s}{\sqrt{n}} = 16.1 \pm 1.96 \frac{2.1}{\sqrt{40}} = 16.1 \pm .65$ or $15.45 < \mu < 16.75$

Beer: $\bar{x} \pm 1.96 \frac{s}{\sqrt{n}} = 81 \pm 1.96 \frac{10.2}{\sqrt{40}} = 81 \pm 3.16$ or $77.84 < \mu < 84.16$

The 83.9 litre figure appears to be a possible value for the average beer consumption, but the 14.2 kilograms of cheese does not fall among the possible values for the average cheese consumption.

8.99 Answers will vary from student to student. Comparisons should be made within the two types of plants for the two treatments (free-standing versus supported). For example, 95% confidence intervals for the two comparisons are calculated as:

Sunflower: $(\bar{x}_1 - \bar{x}_2) \pm 1.96 \sqrt{\frac{s_1^2}{n_1} + \frac{s_2^2}{n_2}} \Rightarrow (\bar{x}_1 - \bar{x}_2) \pm 1.96 \sqrt{SE_1^2 + SE_2^2}$

$(35.3 - 32.1) \pm 1.96 \sqrt{.72^2 + .72^2} \Rightarrow 3.2 \pm 2.00$ or $1.2 < (\mu_1 - \mu_2) > 5.2$

Maize: $(\bar{x}_1 - \bar{x}_2) \pm 1.96 \sqrt{\frac{s_1^2}{n_1} + \frac{s_2^2}{n_2}} \Rightarrow (\bar{x}_1 - \bar{x}_2) \pm 1.96 \sqrt{SE_1^2 + SE_2^2}$

$(16.2 - 14.6) \pm 1.96 \sqrt{.41^2 + .40^2} \Rightarrow 1.6 \pm 1.12$ or $.48 < (\mu_1 - \mu_2) > 2.72$

There is a difference in the mean basal diameters for both types of plant.

8.101 Using the range approximation to obtain an estimate of σ, we have $\sigma \approx \frac{R}{4} = \frac{13,000 - 4800}{4} = 2050$

and the desired value of *n* is obtained:

$$1.96 \frac{\sigma}{\sqrt{n}} \leq B \Rightarrow 1.96 \frac{2050}{\sqrt{n}} \leq 500 \Rightarrow n \geq 64.58 \text{ or } n \geq 65$$

8.103 The approximate 90% confidence interval for $p_1 - p_2$ is

$$(\hat{p}_1 - \hat{p}_2) \pm 1.645 \sqrt{\frac{\hat{p}_1 \hat{q}_1}{n_1} + \frac{\hat{p}_2 \hat{q}_2}{n_2}}$$

$$(.2 - .1) \pm 1.645 \sqrt{\frac{.2(.8)}{400} + \frac{.1(.9)}{400}}$$

$$.1 \pm .041 \quad \text{or} \quad .059 < (p_1 - p_2) < .141$$

8.105 **a** Since $n = 69$ for each of the variables recorded, the 95% confidence interval is approximated by

$$\bar{x} \pm 1.96 \frac{s}{\sqrt{69}}$$

b A 95% confidence interval for μ, the mean time to skate, is approximated as

$$1.953 \pm 1.96 \frac{.131}{\sqrt{69}} = 1.953 \pm .031 \text{ or } 1.922 < \mu < 1.984$$

8.107 **a** The approximate 95% confidence interval for μ is

$$\bar{x} \pm 1.96 \frac{s}{\sqrt{n}} = 5.753 \pm 1.96 \frac{.892}{\sqrt{69}} = 5.753 \pm .210 \text{ or } 5.543 < \mu < 5.963.$$

b Using $s = .892$ as our estimate of σ, we must have

$$2.58 \frac{.892}{\sqrt{n}} \leq .1 \implies \sqrt{n} \geq \frac{2.58(.892)}{.1}$$

$$n \geq 529.63 \text{ or } n \geq 530$$

8.109 Calculate $\hat{p} = \frac{79}{121} = .653$. Then the approximate 95% confidence interval for p is

$$\hat{p} \pm 1.96 \sqrt{\frac{\hat{p}\hat{q}}{n}} = .653 \pm 1.96 \sqrt{\frac{.653(.347)}{121}} = .653 \pm .085 \text{ or } .568 < p < .738.$$

8.111 The point estimate of μ is $\bar{x} = 21.6$ and the margin of error in estimation is

$$1.96 \, SE = 1.96 \frac{\sigma}{\sqrt{n}} \approx 1.96 \frac{s}{\sqrt{n}} = 1.96 \frac{2.1}{\sqrt{70}} = .49$$

8.113 For this exercise, the parameter of interest is μ, $B = 0.5$ and $1 - \alpha = .95$. Hence, we must have

$$1.96 \frac{\sigma}{\sqrt{n}} \leq 0.5 \implies 1.96 \frac{4}{\sqrt{n}} \leq 0.5$$

$$\sqrt{n} \geq \frac{1.96(4)}{0.5} = 15.68$$

$$n \geq 245.86 \text{ or } n \geq 246$$

8.115 **a** The half-width of a 95% confidence interval is

$$1.96 \frac{\sigma}{\sqrt{n}} \implies 1.96 \frac{35}{\sqrt{50}} \implies 9.702$$

b Use the **Interpreting Confidence Intervals** applet. The intervals will vary from student to student, but the half-width of all intervals is shown at the bottom of the applet to be $1.96(4.95) = 9.702$.

8.117 Use the **Interpreting Confidence Intervals** applet. Answers will vary, but the widths of all the intervals should be the same. Most of the simulations will show between 90 and 99 intervals that work correctly.

8.119 Use the **Exploring Confidence Intervals** applet.
a Move the slider at the top of the applet to change the confidence level. For 99% confidence, the applet shows **z = 2.58**; for 95% confidence, the applet shows **z = 1.96**; for 90% confidence, the applet shows **z = 1.65** (we use $z = 1.645$).
b-c Reducing the confidence level results in a narrower interval. To obtain this more precise estimate of μ, you have sacrificed the confidence that your interval works properly and actually *does* enclose the true value of μ.

8.121 Use the **Exploring Confidence Intervals** applet.
   a-b Move the slider on the left side of the applet to change the standard deviation. Increasing the variability results in a larger standard error and in a wider interval.
   c When the variability increases, you will need a wider "loop" to make sure that the interval works properly and actually *does* enclose the true value of μ. The confidence interval needs to be wider to achieve the same confidence level without increasing the sample size.

Case Study:
How Reliable Is That Poll?

1 For the total sample of size $n = 2000$ adults, the margin of error for estimating any sample proportion p is approximately

$$1.96\sqrt{\frac{pq}{n}} \approx 1.96\sqrt{\frac{.5(.5)}{2000}} = .022$$

Hence, the margin of error is the same as the margin (± 2.2) given by the survey designers. Since the split samples are smaller, the sampling error for these smaller groups will be *larger* than ± 2.2 percentage points. For each of the split samples of $n = 1000$ adults, the margin of error is

$$1.96\sqrt{\frac{pq}{n}} \approx 1.96\sqrt{\frac{.5(.5)}{1000}} = .031.$$

2 The numbers in the table are the percentages falling in a particular opinion category.

3 Rotating the order of options and the order of questions is done to avoid biases that might be caused by order of presentation.

4 **a** The approximate 95% confidence interval is

$$\hat{p} \pm 1.96\sqrt{\frac{\hat{p}\hat{q}}{n}} = .59 \pm 1.96\sqrt{\frac{.59(.41)}{1000}} = .59 \pm .030 \text{ or } .56 < p < .62.$$

b The approximate 95% confidence interval is $\hat{p} \pm 1.96\sqrt{\frac{\hat{p}\hat{q}}{n}} = .30 \pm 1.96\sqrt{\frac{.30(.70)}{1000}} = .30 \pm .028$ or $.272 < p < .328$.

5 The approximate 95% confidence interval is

$$(\hat{p}_1 - \hat{p}_2) \pm 1.96\sqrt{\frac{\hat{p}_1\hat{q}_1}{n_1} + \frac{\hat{p}_2\hat{q}_2}{n_2}}$$

$$(.79 - .87) \pm 1.96\sqrt{\frac{.79(.21)}{2000} + \frac{.87(.13)}{1000}}$$

$$-.08 \pm .027 \text{ or } -.107 < (p_1 - p_2) < -.053$$

Since zero is not in the interval, it is not possible that $p_1 = p_2$. Yes, the proportion of people who say "multiculturalism makes them proud to be Canadian," and those who say that "the fact that people from different cultural groups in Canada get along and live in peace" are different.

6 Answers will vary from student to student. Responses today will not necessarily be similar to those reported here.

Project 8-A: Saving Time and Making Patients Safer

a The point estimate for μ_2 is just the sample mean $\bar{x}_2 = 3.1$. The 95% margin of error when $n \geq 30$ is estimated as $1.96 \left(\frac{s}{\sqrt{n}}\right) = 1.96 \left(\frac{\sqrt{1.68}}{\sqrt{100}}\right) = 0.254045$.

b The 90% confidence interval is $\bar{x}_2 \pm 1.645 \left(\frac{s}{\sqrt{n}}\right) = 3.1 \pm 1.645 \left(\frac{\sqrt{1.68}}{\sqrt{100}}\right) = 3.1 \pm 0.213216$. Under repeated sampling then, the true parameter μ_2 will lie in the interval (2.886784, 3.31322) 90% of the time.

c No. Since the sample mean before the Quality Improvement is still higher than the upper boundary of our confidence interval for μ_2, we can conclude that the Quality Improvement has in fact worked.

d Yes, the quality control department should be concerned because the waiting time of 2.7 hours falls under the lower boundary of our confidence interval for μ_2.

e Yes, we can still use the standard normal distribution to construct a confidence internal for μ_2 because the Central Limit Theorem applies in this case.

f Let us find the sample size n such that the margin of error is equal to 0.5 hours:
$0.5 = 1.96 \left(\frac{1.7}{\sqrt{n}}\right) \Rightarrow n = \left[\frac{1.96(1.7)}{0.5}\right]^2 \approx 44.41$. Thus, a sample of size 45 would be required.

g The best point estimator for $(\mu_1 - \mu_2)$ is simply $(\bar{x}_1 - \bar{x}_2) = (3.5 - 3.1) = 0.4$.

h By the Central Limit Theorem, we can invoke approximate normality here. The mean of this normal distribution will be $(\bar{x}_1 - \bar{x}_2)$, and the standard deviation of this normal distribution will be $\sqrt{\frac{s_1^2}{n_1} + \frac{s_2^2}{n_2}}$. The margin of error can be found using, $1.96 \sqrt{\frac{s_1^2}{n_1} + \frac{s_2^2}{n_2}} = 1.96 \sqrt{\frac{2.82}{100} + \frac{1.68}{100}} = 0.415779$.

i A 98% confidence interval for $(\mu_1 - \mu_2)$ can be expressed as $(\bar{x}_1 - \bar{x}_2) \pm 2.33 \sqrt{\frac{s_1^2}{n_1} + \frac{s_2^2}{n_2}}$
$= (3.5 - 3.1) \pm 2.33 \sqrt{\frac{2.82}{100} + \frac{1.68}{100}} = 0.4 \pm 0.4943 = (-0.0943, 0.8943)$.

j No, since zero can be found in the interval, we cannot infer a difference in the true average waiting times.

k The phrase "98% confident" mean that under repeated sampling, 98% of constructed intervals would contain the true value of the parameter.

Project 8-B: Attitudes of Canadian Women Towards Birthing Centres and Midwife Care for Childbirth

a   The point estimate of p_2 is simply $\hat{p}_2 = 0.31$. The 95% margin of error is given by $1.96\sqrt{\frac{\hat{p}_2 \hat{q}_2}{n_2}}$
$= 1.96\sqrt{\frac{(0.31)(0.69)}{360}} = 0.048$, where $\hat{q}_2 = 1 - \hat{p}_2$.

b   The point estimate of p_3 is simply $\hat{p}_3 = 0.35$. The 95% margin of error is given by $1.96\sqrt{\frac{\hat{p}_3 \hat{q}_3}{n_3}}$
$= 1.96\sqrt{\frac{(0.35)(0.65)}{169}} = 0.072$, where $\hat{q}_3 = 1 - \hat{p}_3$.

c   The 99% confidence interval for p_3 is given by $\hat{p}_3 \pm z_{0.005}\sqrt{\frac{\hat{p}_3 \hat{q}_3}{n_3}} = 0.35 \pm 2.58\sqrt{\frac{(0.35)(0.65)}{169}}$
$= 0.35 \pm 0.09466 = (0.25534, 0.44466)$. Under repeated sampling, 99% of the time the constructed intervals for p_3 will contain the true value of p_3.

d   We need to find a sample of size n_1 such that the 90% margin of error is equal to 0.01. That is, we need, $z_{0.05}\sqrt{\frac{\hat{p}_1 \hat{q}_1}{n_1}} = 0.01$, where $z_{0.05} = 1.645$ (for a 90% confidence interval). Solving for n_1 we obtain, $n_1 = \frac{z_{0.05}^2 (\hat{p}_1 \hat{q}_1)}{(0.01)^2} = \frac{(1.645)^2 (0.28)(0.72)}{(0.01)^2} = 5455.35$. Thus, we would need to sample a minimum of 5456 women if we want the true population proportion to lie within 0.01 of our sample proportion.

e   We need to find a sample of size n_4 such that the 99% margin of error is equal to 0.1. That is, we need, $z_{0.005}\sqrt{\frac{\hat{p}_4 \hat{q}_4}{n_4}} = 0.1$, where $z_{0.005} = 2.58$ (for a 99% confidence interval). Solving for n_4 we obtain, $n_4 = \frac{z_{0.005}^2 (\hat{p}_4 \hat{q}_4)}{(0.1)^2} = \frac{(2.58)^2 (0.30)(0.70)}{(0.1)^2} = 139.78$. Thus, the researcher would need to sample a minimum of 140 women if he wants the true population proportion to lie within 0.1 of the sample proportion.

f   A 98% confidence interval is $(\hat{p}_2 - \hat{p}_3) \pm z_{0.01}\sqrt{\frac{\hat{p}_2 \hat{q}_2}{n_2} + \frac{\hat{p}_3 \hat{q}_3}{n_3}}$, where $z_{0.01} = 2.33$. Inputting the required values yields $(0.31 - 0.35) \pm 2.33\sqrt{\frac{(0.31)(0.69)}{360} + \frac{(0.35)(0.65)}{169}} = -0.4 \pm 0.1026 = (-0.5026, -0.2974)$.
Thus, we are 98% confident that the true difference in proportions is between -0.5026 and -0.2974.

g   Since zero is not in the interval, we can safely conclude that the two proportions are unequal. That is, there is a real difference in the proportion of women who would like to use a birthing centre between Ontario women and Quebec women.

h   The point estimate of the difference is simply $(\hat{p}_1 - \hat{p}_4) = 0.28 - 0.30 = -0.02$. A 95% margin of error is $1.96\sqrt{\frac{\hat{p}_1 \hat{q}_1}{n_1} + \frac{\hat{p}_4 \hat{q}_4}{n_4}} = 1.96\sqrt{\frac{(0.28)(0.72)}{49} + \frac{(0.30)(0.70)}{225}} = 0.13925$.

i   It depends on what is meant by "compare". Parameters can be compared to others in different ways. Given the methodology covered in Chapter 8, the answer is likely "no". However, there are more advanced methods, beyond the scope of this text, that would suggest the answer is "yes": it is possible to compare many proportions simultaneously.

Chapter 9: Large-Sample Tests of Hypotheses

9.1 Follow the instructions in the My Personal Trainer section. The answers are given in the table below.

Test statistic	Significance level	One or two-tailed test?	Critical value	Rejection region	Conclusion
z = 0.88	$\alpha = .05$	Two-tailed	1.96	$|z| > 1.96$	Do not reject H_0
z = −2.67	$\alpha = .05$	One-tailed (lower)	1.645	$z < -1.645$	Reject H_0
z = 5.05	$\alpha = .01$	Two-tailed	2.58	$|z| > 2.58$	Reject H_0
z = −1.22	$\alpha = .01$	One-tailed (lower)	2.33	$z < -2.33$	Do not reject H_0

9.3 a The critical value that separates the rejection and nonrejection regions for a right-tailed test based on a z-statistic will be a value of z (called z_α) such that $P(z > z_\alpha) = \alpha = .01$. That is, $z_{.01} = 2.33$ (see the figure below). The null hypothesis H_0 will be rejected if $z > 2.33$.

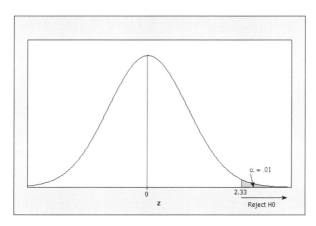

b For a two-tailed test with $\alpha = .05$, the critical value for the rejection region cuts off $\alpha/2 = .025$ in the two tails of the z distribution in Figure 9.2, so that $z_{.025} = 1.96$. The null hypothesis H_0 will be rejected if $z > 1.96$ or $z < -1.96$ (which you can also write as $|z| > 1.96$).

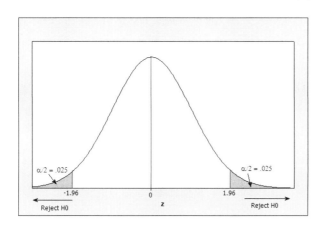

c Similar to part **a**, with the rejection region in the lower tail of the z distribution. The null hypothesis H_0 will be rejected if $z < -2.33$.

d Similar to part **b**, with $\alpha/2 = .005$. The null hypothesis H_0 will be rejected if $z > 2.58$ or $z < -2.58$ (which you can also write as $|z| > 2.58$).

9.5 Use the guidelines for statistical significance in Section 9.3. The smaller the *p*-value, the more evidence there is in favor of rejecting H_0. For part **a**, *p*-value = .1251 is not statistically significant; H_0 is not rejected. For part **b**, *p*-value = .0054 is less than .01 and the results are highly significant; H_0 should be rejected. For part **c**, *p*-value = .0351 is between .01 and .05. The results are significant at the 5% level, but not at the 1% level (P < .05).

9.7 **a** Since this is a right-tailed test, the *p*-value is the area under the standard normal distribution to the right of $z = 2.04$:
$$p\text{-value} = P(z > 2.04) = 1 - .9793 = .0207$$

b The *p*-value, .0207, is less than $\alpha = .05$, and the null hypothesis is rejected at the 5% level of significance. There is sufficient evidence to indicate that $\mu > 2.3$.

c The conclusions reached using the **critical value approach** and the ***p*-value approach** are identical.

9.9 The hypotheses to be tested are
$$H_0: \mu = 28 \quad \text{versus} \quad H_a: \mu \neq 28$$
and the test statistic is
$$z = \frac{\bar{x} - \mu_0}{\sigma/\sqrt{n}} \approx \frac{\bar{x} - \mu_0}{s/\sqrt{n}} = \frac{26.8 - 28}{6.5/\sqrt{100}} = -1.85$$
with *p*-value = $P(|z| > 1.85) = 2(.0322) = .0644$. To draw a conclusion from the *p*-value, use the guidelines for statistical significance in Section 9.3. Since the *p*-value is greater than .05, the null hypothesis should not be rejected. There is insufficient evidence to indicate that the mean is different from 28. (Some researchers might report these results as *tending towards significance*.)

9.11 **a** In order to make sure that the average weight was one kilogram, you would test
$$H_0: \mu = 1 \quad \text{versus} \quad H_a: \mu \neq 1$$
b-c The test statistic is
$$z = \frac{\bar{x} - \mu_0}{\sigma/\sqrt{n}} \approx \frac{\bar{x} - \mu_0}{s/\sqrt{n}} = \frac{1.01 - 1}{.02/\sqrt{35}} = 2.96$$
with *p*-value = $P(|z| > 2.96) = 2(.0015) = .003$. Since the *p*-value is less than .05, the null hypothesis should be rejected. The manager should report that there is sufficient evidence to indicate that the mean is different from 1.

9.13 **a-b** We want to test the null hypothesis that μ is, in fact, 80% against the alternative that it is not:
$$H_0: \mu = 80 \quad \text{versus} \quad H_a: \mu \neq 80$$
Since the exercise does not specify $\mu < 80$ or $\mu > 80$, we are interested in a two directional alternative, $\mu \neq 80$.

c The test statistic is
$$z = \frac{\bar{x} - \mu_0}{\sigma/\sqrt{n}} \approx \frac{\bar{x} - \mu_0}{s/\sqrt{n}} = \frac{79.7 - 80}{.8/\sqrt{100}} = -3.75$$

The rejection region with $\alpha = .05$ is determined by a critical value of z such that
$$P(z < -z_0) + P(z > z_0) = \frac{\alpha}{2} + \frac{\alpha}{2} = .05$$

This value is $z_0 = 1.96$ (see the figure in Exercise 9.3**b**). Hence, H_0 will be rejected if $z > 1.96$ or $z < -1.96$. The observed value, $z = -3.75$, falls in the rejection region and H_0 is rejected. There is sufficient evidence to refute the manufacturer's claim. The probability that we have made an incorrect decision is $\alpha = .05$.

9.15 a The hypothesis to be tested is
$$H_0 : \mu = 29 \quad \text{versus} \quad H_a : \mu < 29$$
and the test statistic is
$$z = \frac{\bar{x} - \mu_0}{\sigma/\sqrt{n}} \approx \frac{\bar{x} - \mu_0}{s/\sqrt{n}} = \frac{27 - 29}{5.2/\sqrt{36}} = -2.31$$

with p-value $= P(z < -2.31) = .0104$. To draw a conclusion from the p-value, use the guidelines for statistical significance in Section 9.3. Since the p-value is between .01 and .05, the test results are significant at the 5% level, but not at the 1% level.

b If $\alpha = .05$, H_0 can be rejected and you can conclude that the average score improvement is less than claimed. This would be the most beneficial way for the competitor to state these conclusions.

c If you worked for this organization, it would be more beneficial to conclude that there was *insufficient evidence at the 1% level* to conclude that the average score improvement is less than claimed.

9.17 The hypothesis to be tested is
$$H_0 : \mu = 5.97 \quad \text{versus} \quad H_a : \mu > 5.97$$
and the test statistic is
$$z = \frac{\bar{x} - \mu_0}{\sigma/\sqrt{n}} \approx \frac{\bar{x} - \mu_0}{s/\sqrt{n}} = \frac{9.8 - 5.97}{1.95/\sqrt{31}} = 10.94$$

with p-value $= P(z > 10.94) < 1 - .9998 = .0002$ (or p-value ≈ 0). Since the p-value is less than .05, the null hypothesis is rejected. There is sufficient evidence to indicate that the average diameter of the tendon for patients with AT is greater than 5.97 mm.

9.19 The hypothesis of interest is one-tailed:
$$H_0 : \mu_1 - \mu_2 = 0 \quad \text{versus} \quad H_a : \mu_1 - \mu_2 < 0$$
The test statistic, calculated under the assumption that $\mu_1 - \mu_2 = 0$, is
$$z \approx \frac{(\bar{x}_1 - \bar{x}_2) - 0}{\sqrt{\frac{s_1^2}{n_1} + \frac{s_2^2}{n_2}}} = \frac{1.24 - 1.31}{\sqrt{\frac{.056}{36} + \frac{.054}{45}}} = -1.33$$

with the unknown σ_1^2 and σ_2^2 estimated by s_1^2 and s_2^2, respectively. The student can use one of two methods for decision making:

p-value approach: Calculate p-value $= P(z < -1.33) = .0918$. Since this p-value is greater than .05, the null hypothesis is not rejected. There is insufficient evidence to indicate that the mean for population 1 is smaller than the mean for population 2.

Critical value approach: The rejection region with $\alpha = .05$, is $z < -1.645$. Since the observed value of z does not fall in the rejection region, H_0 is not rejected. There is insufficient evidence to indicate that the mean for population 1 is smaller than the mean for population 2.

9.21 **a** The hypothesis of interest is one-tailed:
$$H_0: \mu_1 - \mu_2 = 0 \quad \text{versus} \quad H_a: \mu_1 - \mu_2 > 0$$

b The test statistic, calculated under the assumption that $\mu_1 - \mu_2 = 0$, is
$$z \approx \frac{(\bar{x}_1 - \bar{x}_2) - 0}{\sqrt{\frac{s_1^2}{n_1} + \frac{s_2^2}{n_2}}} = \frac{6.9 - 5.8}{\sqrt{\frac{(2.9)^2}{35} + \frac{(1.2)^2}{35}}} = 2.074$$

The rejection region with $\alpha = .05$, is $z > 1.645$ and H_0 is rejected. There is evidence to indicate that $\mu_1 - \mu_2 > 0$, or $\mu_1 > \mu_2$. That is, there is reason to believe that Vitamin C reduces the mean time to recover.

9.23 **a** The hypothesis of interest is two-tailed:
$$H_0: \mu_1 - \mu_2 = 0 \quad \text{versus} \quad H_a: \mu_1 - \mu_2 \neq 0$$

The test statistic, calculated under the assumption that $\mu_1 - \mu_2 = 0$, is
$$z \approx \frac{(\bar{x}_1 - \bar{x}_2) - 0}{\sqrt{\frac{s_1^2}{n_1} + \frac{s_2^2}{n_2}}} = \frac{34.1 - 36}{\sqrt{\frac{(5.9)^2}{100} + \frac{(6.0)^2}{100}}} = -2.26$$

with p-value $= P(|z| > 2.26) = 2(.0119) = .0238$. Since the p-value is less than .05, the null hypothesis is rejected. There is evidence to indicate a difference in the mean lead levels for the two sections of the city.

b From Section 8.6, the 95% confidence interval for $\mu_1 - \mu_2$ is approximately
$$(\bar{x}_1 - \bar{x}_2) \pm 1.96 \sqrt{\frac{s_1^2}{n_1} + \frac{s_2^2}{n_2}}$$
$$(34.1 - 36) \pm 1.96 \sqrt{\frac{5.9^2}{100} + \frac{6.0^2}{100}}$$
$$-1.9 \pm 1.65 \quad \text{or} \quad -3.55 < (\mu_1 - \mu_2) < -.25$$

c Since the value $\mu_1 - \mu_2 = 5$ or $\mu_1 - \mu_2 = -5$ is not in the confidence interval in part **b**, it is not likely that the difference will be more than 5 ppm, and hence the statistical significance of the difference is not of practical importance to the engineers.

9.25 **a** Most people would have no preconceived idea about which of the two hotels would have higher average room rates, and a two-tailed hypothesis would be appropriate:
$$H_0: \mu_1 - \mu_2 = 0 \quad \text{versus} \quad H_a: \mu_1 - \mu_2 \neq 0$$

b The test statistic is
$$z \approx \frac{(\bar{x}_1 - \bar{x}_2) - 0}{\sqrt{\frac{s_1^2}{n_1} + \frac{s_2^2}{n_2}}} = \frac{170 - 145}{\sqrt{\frac{17.5^2}{50} + \frac{10^2}{50}}} = 8.77$$

The rejection region, with $\alpha = .01$, is $|z| > 2.58$ and H_0 is rejected. There is evidence to indicate that there is a difference in the average room rates for the Marriott and the Radisson hotels.

c The p-value for this two-tailed test is
$$p\text{-value} = P(z > 8.77) + P(z < -8.77) \approx .0000$$

Since the p-value is less than $\alpha = .01$, the null hypothesis can be rejected at the 1% level. There is sufficient evidence to conclude that $\mu_1 - \mu_2 \neq 0$.

9.27 **a**   The hypothesis of interest is two-tailed:
$$H_0: \mu_1 - \mu_2 = 0 \quad \text{versus} \quad H_a: \mu_1 - \mu_2 \neq 0$$
and the test statistic is
$$z \approx \frac{(\bar{x}_1 - \bar{x}_2) - 0}{\sqrt{\frac{s_1^2}{n_1} + \frac{s_2^2}{n_2}}} = \frac{.94 - 2.8}{\sqrt{\frac{1.2^2}{36} + \frac{2.8^2}{26}}} = -3.18$$

with p-value $= P(|z| > 3.18) = 2(.0007) = .0014$. Since the p-value is less than .05, the null hypothesis is rejected. There is evidence to indicate a difference in the mean concentrations for these two types of sites.

b   The 95% confidence interval for $\mu_1 - \mu_2$ is approximately
$$(\bar{x}_1 - \bar{x}_2) \pm 1.96\sqrt{\frac{s_1^2}{n_1} + \frac{s_2^2}{n_2}}$$
$$(.94 - 2.8) \pm 1.96\sqrt{\frac{1.2^2}{36} + \frac{2.8^2}{26}}$$
$$-1.86 \pm 1.15 \quad \text{or} \quad -3.01 < (\mu_1 - \mu_2) < -.71$$

Since the value $\mu_1 - \mu_2 = 0$ does not fall in the interval in part **b**, it is not likely that $\mu_1 = \mu_2$. There is evidence to indicate that the means are different, confirming the conclusion in part **a**.

9.29 **a**   The hypothesis of interest is two-tailed:
$$H_0: \mu_1 - \mu_2 = 0 \quad \text{versus} \quad H_a: \mu_1 - \mu_2 \neq 0$$
and the test statistic is
$$z \approx \frac{(\bar{x}_1 - \bar{x}_2) - 0}{\sqrt{\frac{s_1^2}{n_1} + \frac{s_2^2}{n_2}}} = \frac{36.72 - 36.88}{\sqrt{\frac{.7^2}{65} + \frac{.74^2}{65}}} = -1.27$$

with p-value $= P(|z| > 1.27) = 2(1 - .8980) = .2040$. Since the p-value is greater than .10, the null hypothesis is not rejected, and the results are not significant. There is insufficient evidence to indicate a difference in the mean temperatures for men versus women.

b   Since the p-value = .2040, we reject H_0 neither at the 5% level (p-value > .05) not at 1% level (p-value > .01). Using the guidelines for significance given in Section 9.3 of the text, we declare the results not statistically *significant*.

9.31 **a**   The hypothesis of interest is two-tailed:
$$H_0: p = .4 \quad \text{versus} \quad H_a: p \neq .4$$

b-c   It is given that $x = 529$ and $n = 1400$, so that $\hat{p} = \frac{x}{n} = \frac{529}{1400} = .378$. The test statistic is
$$z = \frac{\hat{p} - p_0}{\sqrt{\frac{p_0 q_0}{n}}} = \frac{.378 - .4}{\sqrt{\frac{.4(.6)}{1400}}} = -1.68$$

with p-value $= P(|z| > 1.68) = 2(.0465) = .093$. Since the p-value is not less than $\alpha = .01$, the null hypothesis is not rejected. There is insufficient evidence to indicate that p differs from .4.

9.33 **a** The two sets of hypothesis both involve a different binomial parameter p:

$$H_0: p = .75 \quad \text{versus} \quad H_a: p \neq .75 \text{ (part c)}$$
$$H_0: p = .5 \quad \text{versus} \quad H_a: p > .5 \text{ (part b)}$$

b For the second test in part **a**, $x = 47$ and $n = 100$, so that $\hat{p} = \dfrac{x}{n} = \dfrac{47}{100} = .47$, the test statistic is

$$z = \dfrac{\hat{p} - p_0}{\sqrt{\dfrac{p_0 q_0}{n}}} = \dfrac{.47 - .5}{\sqrt{\dfrac{.5(.5)}{100}}} = -.60$$

Since no value of α is specified in advance, we calculate p-value $= P(z > .60) = .2743 = .2743$. Since this p-value is greater than .10, the null hypothesis is not rejected. There is insufficient evidence to contradict the claim.

c For the first test in part **a**, $x = 68$ and $n = 100$, so that $\hat{p} = \dfrac{x}{n} = \dfrac{68}{100} = .68$, the test statistic is

$$z = \dfrac{\hat{p} - p_0}{\sqrt{\dfrac{p_0 q_0}{n}}} = \dfrac{.68 - .75}{\sqrt{\dfrac{.75(.25)}{100}}} = -1.62$$

with p-value $= P(|z| > 1.62) = 2(.0526) = .1052$. Since the p-value is greater than .10, the null hypothesis is not rejected. There is insufficient evidence to contradict the claim.

9.35 **a-b** Since the survival rate without screening is $p = 2/3$, the survival rate with an effective program may be greater than 2/3. Hence, the hypothesis to be tested is

$$H_0: p = 2/3 \quad \text{versus} \quad H_a: p > 2/3$$

c With $\hat{p} = \dfrac{x}{n} = \dfrac{164}{200} = .82$, the test statistic is

$$z = \dfrac{\hat{p} - p_0}{\sqrt{\dfrac{p_0 q_0}{n}}} = \dfrac{.82 - 2/3}{\sqrt{\dfrac{(2/3)(1/3)}{200}}} = 4.6$$

The rejection region is one-tailed, with $\alpha = .05$ or $z > 1.645$ and H_0 is rejected. The screening program seems to increase the survival rate.

d For the one-tailed test,

$$p\text{-value} = P(z > 4.6) < 1 - .9998 = .0002$$

That is, H_0 can be rejected for any value of $\alpha \geq .0002$. The results are *highly significant*.

9.37 The hypothesis of interest is

$$H_0: p = .20 \quad \text{versus} \quad H_a: p > .20$$

With $\hat{p} = \dfrac{x}{n} = \dfrac{15}{60} = .25$, the test statistic is

$$z = \dfrac{\hat{p} - p_0}{\sqrt{\dfrac{p_0 q_0}{n}}} = \dfrac{.25 - .20}{\sqrt{\dfrac{.20(.80)}{60}}} = .97$$

with p-value $= P(z > .97) < 1 - .8340 = .1660$. Since the p-value is greater than .10, H_0 is not rejected. There is insufficient evidence to dispute the claim regarding the proportion of smokers.

9.39 The hypothesis of interest is
$$H_0: p = .80 \quad \text{versus} \quad H_a: p < .80$$
with $\hat{p} = \dfrac{x}{n} = \dfrac{37}{50} = .74$, the test statistic is
$$z = \dfrac{\hat{p} - p_0}{\sqrt{\dfrac{p_0 q_0}{n}}} = \dfrac{.74 - .80}{\sqrt{\dfrac{.80(.20)}{50}}} = -1.06$$

The rejection region with $\alpha = .05$ is $z < -1.645$ and the null hypothesis is not rejected. (Alternatively, we could calculate p-value $= P(z < -1.06) = .1446$. Since this p-value is greater than .05, the null hypothesis is not rejected.) There is insufficient evidence to refute the experimenter's claim.

9.41 The hypothesis of interest is
$$H_0: p = .85 \quad \text{versus} \quad H_a: p \neq .85$$
with $\hat{p} = \dfrac{x}{n} = \dfrac{261}{300} = .87$, the test statistic is
$$z = \dfrac{\hat{p} - p_0}{\sqrt{\dfrac{p_0 q_0}{n}}} = \dfrac{.87 - .85}{\sqrt{\dfrac{.85(.15)}{300}}} = .97$$

with p-value $= 2P(|z| > .97) = 2(.1660) = .3320$. Since this p-value is greater than .10, the null hypothesis is not rejected. There is insufficient evidence to indicate that the proportion of aboriginal in Nunavut is different from that reported in census data.

9.43 a-b If p_1 cannot be larger than p_2, the only alternative to $H_0: p_1 - p_2 = 0$ is that $p_1 < p_2$, and the one-tailed alternative is $H_a: p_1 - p_2 < 0$.

 c The rejection region, with $\alpha = .05$, is $z < -1.645$ and the observed value of the test statistic is $z = -.84$. The null hypothesis is not rejected. There is no evidence to indicate that p_1 is smaller than p_2.

9.45 a The hypothesis of interest is:
$$H_0: p_1 - p_2 = 0 \quad \text{versus} \quad H_a: p_1 - p_2 < 0$$
Calculate $\hat{p}_1 = .36$, $\hat{p}_2 = .60$ and $\hat{p} = \dfrac{n_1 \hat{p}_1 + n_2 \hat{p}_2}{n_1 + n_2} = \dfrac{18 + 30}{50 + 50} = .48$. The test statistic is then
$$z = \dfrac{\hat{p}_1 - \hat{p}_2}{\sqrt{\hat{p}\hat{q}\left(\dfrac{1}{n_1} + \dfrac{1}{n_2}\right)}} = \dfrac{.36 - .60}{\sqrt{.48(.52)(1/50 + 1/50)}} = -2.40$$

The rejection region, with $\alpha = .05$, is $z < -1.645$ and H_0 is rejected. There is evidence of a difference in the proportion of survivors for the two groups.

 b From Section 8.7, the approximate 95% confidence interval is
$$(\hat{p}_1 - \hat{p}_2) \pm 1.96 \sqrt{\dfrac{\hat{p}_1 \hat{q}_1}{n_1} + \dfrac{\hat{p}_2 \hat{q}_2}{n_2}}$$
$$(.36 - .60) \pm 1.96 \sqrt{\dfrac{.36(.64)}{50} + \dfrac{.60(.40)}{50}}$$
$$-.24 \pm .19 \quad \text{or} \quad -.43 < (p_1 - p_2) < -.05$$

9.47 The hypothesis of interest is
$$H_0 : p_1 - p_2 = 0 \quad \text{versus} \quad H_a : p_1 - p_2 \neq 0$$
Calculate $\hat{p}_1 = \dfrac{12}{56} = .214$, $\hat{p}_2 = \dfrac{8}{32} = .25$, and $\hat{p} = \dfrac{x_1 + x_2}{n_1 + n_2} = \dfrac{12 + 8}{56 + 32} = .227$.

The test statistic is then
$$z = \dfrac{\hat{p}_1 - \hat{p}_2}{\sqrt{\hat{p}\hat{q}\left(\dfrac{1}{n_1} + \dfrac{1}{n_2}\right)}} = \dfrac{.214 - .25}{\sqrt{.227(.773)(1/56 + 1/32)}} = -.39$$

The rejection region, with $\alpha = .05$, is $|z| > 1.96$ and H_0 is not rejected. There is insufficient evidence to indicate a difference in the proportion of red M&Ms for the plain and peanut varieties. These results match the conclusions of Exercise 8.52.

9.49 Refer to Exercise 9.48. The 99% lower one-sided confidence bound for $p_1 - p_2$ is

$$(\hat{p}_1 - \hat{p}_2) - 2.33\sqrt{\dfrac{\hat{p}_1\hat{q}_1}{n_1} + \dfrac{\hat{p}_2\hat{q}_2}{n_2}}$$

$$(.018 - .009) - 2.33\sqrt{\dfrac{.018(.982)}{2266} + \dfrac{.009(.991)}{2266}}$$

$$.009 - .008 = .001 \quad \text{or} \quad (p_1 - p_2) > .001$$

The difference in risk between the two groups is at least 1 in 1000. This difference may not be of *practical significance* if the benefits of *Prempro* to the patient outweighs the risk.

9.51 The hypothesis of interest is
$$H_0 : p_1 - p_2 = 0 \quad \text{versus} \quad H_a : p_1 - p_2 > 0$$
Calculate $\hat{p}_1 = \dfrac{93}{121} = .769$, $\hat{p}_2 = \dfrac{119}{199} = .598$, and $\hat{p} = \dfrac{x_1 + x_2}{n_1 + n_2} = \dfrac{93 + 119}{121 + 199} = .6625$. The test statistic is then

$$z = \dfrac{\hat{p}_1 - \hat{p}_2}{\sqrt{\hat{p}\hat{q}\left(\dfrac{1}{n_1} + \dfrac{1}{n_2}\right)}} = \dfrac{.769 - .598}{\sqrt{.6625(.3375)(1/121 + 1/199)}} = 3.14$$

with p-value $= P(z > 3.14) = 1 - .9992 = .0008$. Since the p-value is less than .01, the results are reported as highly significant at the 1% level of significance. There is evidence to confirm the researcher's conclusion.

9.53 See Section 9.3 of the text.

9.55 The power of the test is $1 - \beta = P(\text{reject } H_0 \text{ when } H_0 \text{ is false})$. As μ gets farther from μ_0, the power of the test increases.

9.57 The objective of this experiment is to make a decision about the binomial parameter p, which is the probability that a customer prefers the first color. Hence, the null hypothesis will be that a customer has no preference for the first color, and the alternative will be that he does have a preference. If the null hypothesis is true, then
$$H_0 : p = P[\text{customer prefers the first color}] = 1/3$$
If the customer actually has a preference for the first color, then
$$H_a : p > 1/3$$

a   The test statistic is calculated with $\hat{p} = \dfrac{400}{1000} = .4$ as

$$z = \dfrac{\hat{p} - p_0}{\sqrt{\dfrac{p_0 q_0}{n}}} = \dfrac{.4 - 1/3}{\sqrt{\dfrac{(1/3)(2/3)}{1000}}} = 4.47$$

and the p-value is

$$p\text{-value} = P(z > 4.47) < 1 - .9998 = .0002$$

since $P(z > 4.47)$ is surely less than $P(z > 3.49)$, the largest value in Table 3.

b   Since $\alpha = .05$ is larger than the p-value, which is less than .0002, H_0 can be rejected. We conclude that customers have a preference for the first color.

9.59   **a-b**   Since it is necessary to prove that the average pH level is less than 7.5, the hypothesis to be tested is one-tailed:

$$H_0 : \mu = 7.5 \quad \text{versus} \quad H_a : \mu < 7.5$$

c   Answers will vary.

d   The test statistic is $z = \dfrac{\overline{x} - \mu}{\sigma/\sqrt{n}} \approx \dfrac{\overline{x} - \mu}{s/\sqrt{n}} = \dfrac{-.2}{.2/\sqrt{30}} = -5.477$

and the rejection region with $\alpha = .05$ is $z < -1.645$. The observed value, $z = -5.477$, falls in the rejection region and H_0 is rejected. We conclude that the average pH level is less than 7.5.

9.61   **a-b**   Since there is no prior knowledge as to which mean should be larger, the hypothesis of interest is two-tailed

$$H_0 : \mu_1 - \mu_2 = 0 \quad \text{versus} \quad H_a : \mu_1 - \mu_2 \neq 0$$

c   The test statistic is

$$z \approx \dfrac{(\overline{x}_1 - \overline{x}_2) - 0}{\sqrt{\dfrac{s_1^2}{n_1} + \dfrac{s_2^2}{n_2}}} = \dfrac{908 - 976}{\sqrt{\dfrac{347^2}{40} + \dfrac{293^2}{40}}} = -.947$$

The rejection region, with $\alpha = .05$, is two-tailed or $|z| > 1.96$. The null hypothesis is not rejected. There is insufficient evidence to indicate a difference in the two means.

9.63   Let p_1 be the proportion of defectives produced by machine A and p_2 be the proportion of defectives produced by machine B. The hypothesis to be tested is

$$H_0 : p_1 - p_2 = 0 \quad \text{versus} \quad H_a : p_1 - p_2 \neq 0$$

Calculate $\hat{p}_1 = \dfrac{16}{200} = .08$, $\hat{p}_2 = \dfrac{8}{200} = .04$, and $\hat{p} = \dfrac{x_1 + x_2}{n_1 + n_2} = \dfrac{16 + 8}{200 + 200} = .06$. The test statistic is

Then $z = \dfrac{\hat{p}_1 - \hat{p}_2}{\sqrt{\hat{p}\hat{q}\left(\dfrac{1}{n_1} + \dfrac{1}{n_2}\right)}} = \dfrac{.08 - .04}{\sqrt{.06(.94)(1/200 + 1/200)}} = 1.684$

The rejection region, with $\alpha = .05$, is $|z| > 1.96$ and H_0 is not rejected. There is insufficient evidence to indicate that the machines are performing differently in terms of the percentage of defectives being produced.

9.65 **a** The hypothesis to be tested is
$$H_0: p_1 - p_2 = 0 \quad \text{versus} \quad H_a: p_1 - p_2 > 0$$

Calculate $\hat{p}_1 = \frac{136}{200} = .68$, $\hat{p}_2 = \frac{124}{200} = .62$, and $\hat{p} = \frac{x_1 + x_2}{n_1 + n_2} = \frac{136 + 124}{200 + 200} = .65$. The test statistic is

then $z = \dfrac{\hat{p}_1 - \hat{p}_2}{\sqrt{\hat{p}\hat{q}\left(\dfrac{1}{n_1} + \dfrac{1}{n_2}\right)}} = \dfrac{.68 - .62}{\sqrt{.65(.35)(1/200 + 1/200)}} = 1.26$

and the p-value is $P(z \geq 1.26) = 1 - .8962 = .1038$.

b Since the observed p-value, .1038, is greater than $\alpha = .05$, H_0 cannot be rejected. There is insufficient evidence to support the researcher's belief.

9.67 No. The agronomist would have to show experimentally that the increase was 26453 or more cubic centimetre per quadrant in order to achieve practical importance.

9.69 The hypothesis to be tested is
$$H_0: \mu_1 - \mu_2 = 0 \quad \text{versus} \quad H_a: \mu_1 - \mu_2 \neq 0$$
and the test statistic is

$$z \approx \frac{(\bar{x}_1 - \bar{x}_2) - 0}{\sqrt{\dfrac{s_1^2}{n_1} + \dfrac{s_2^2}{n_2}}} = \frac{36 - 33}{\sqrt{\dfrac{31}{64} + \dfrac{27}{64}}} = 3.15$$

Since no value of α is specified in advance, we calculate $p\text{-value} = P(|z| > 3.15) < 2(.0008) = .0016$. Since this p-value is less than .10, you can reject H_0 at the 1% level (highly significant). There is a difference in mean stopping distances for the two models.

9.71 **a** Let p_1 be the proportion of cells in which RNA developed normally when treated with a .6 micrograms per millimeter concentration of Actinomysin-D, and p_2 be the proportion of normal cells treated with the higher concentration of Actinomysin-D. The hypothesis to be tested is
$$H_0: p_1 - p_2 = 0 \quad \text{versus} \quad H_a: p_1 - p_2 \neq 0$$

Calculate $\hat{p}_1 = \frac{55}{70} = .786$, $\hat{p}_2 = \frac{23}{70} = .329$, and $\hat{p} = \frac{x_1 + x_2}{n_1 + n_2} = \frac{55 + 23}{70 + 70} = .557$. The test statistic is

then $z = \dfrac{\hat{p}_1 - \hat{p}_2}{\sqrt{\hat{p}\hat{q}\left(\dfrac{1}{n_1} + \dfrac{1}{n_2}\right)}} = \dfrac{.786 - .329}{\sqrt{.557(.443)(1/70 + 1/70)}} = 5.44$

and the p-value is $P(|z| \geq 5.44) < 2(.0002) = .0004$.

b Since the observed p-value < .0004, it must be less than $\alpha = .05$, and H_0 is rejected. We can conclude that there is a difference in the rate of normal RNA synthesis for cells exposed to the two different concentrations of Actinomysin-D.

9.73 The hypothesis of interest is
$$H_0: p = .5 \quad \text{versus} \quad H_a: p > .5$$
with $\hat{p} = \dfrac{x}{n} = \dfrac{64}{100} = .64$, the test statistic is
$$z = \dfrac{\hat{p} - p_0}{\sqrt{\dfrac{p_0 q_0}{n}}} = \dfrac{.64 - .5}{\sqrt{\dfrac{.5(.5)}{100}}} = 2.8$$

Calculate the *p*-value $= P(z > 2.80) = 1 - .9974 = .0026$. Since the *p*-value is less than $\alpha = .01$, H_0 is rejected. The results are highly significant; there is sufficient evidence to indicate that learning is taking place.

9.75 Refer to Exercise 9.75, in which the rejection region was given as $z > 2.33$ where
$$z = \dfrac{\bar{x} - \mu_0}{s/\sqrt{n}} = \dfrac{\bar{x} - 2.3}{.29/\sqrt{35}}$$
Solving for $\bar{x}$ we obtain the critical value of $\bar{x}$ necessary for rejection of H_0.
$$\dfrac{\bar{x} - 5}{6.2/\sqrt{38}} > 2.33 \quad \Rightarrow \quad \bar{x} > 2.33 \dfrac{6.2}{\sqrt{38}} + 5 = 7.34$$
The probability of a Type II error is defined as
$$\beta = P(\text{accept } H_0 \text{ when } H_0 \text{ is false})$$
Since the acceptance region is $\bar{x} \leq 7.34$ from part **a**, β can be rewritten as
$$\beta = P(\bar{x} \leq 7.34 \text{ when } H_0 \text{ is false}) = P(\bar{x} \leq 7.34 \text{ when } \mu > 5)$$
Several alternative values of μ are given in this exercise.

a For $\mu = 6$,
$$\beta = P(\bar{x} \leq 7.34 \text{ when } \mu = 6) = P\left(z \leq \dfrac{7.34 - 6}{6.2/\sqrt{38}}\right)$$
$$= P(z \leq 1.33) = .9082$$
and $1 - \beta = 1 - .9082 = .0918$.

b For $\mu = 7$,
$$\beta = P(\bar{x} \leq 7.34 \text{ when } \mu = 7) = P\left(z \leq \dfrac{7.34 - 7}{6.2/\sqrt{38}}\right)$$
$$= P(z \leq .34) = .6331$$
and $1 - \beta = 1 - .6331 = .3669$.

c For $\mu = 8$,
$$1 - \beta = 1 - P(\bar{x} \leq 7.34 \text{ when } \mu = 8)$$
$$= 1 - P\left(z \leq \dfrac{7.34 - 8}{6.2/\sqrt{38}}\right)$$
$$= 1 - P(z \leq -.66) = .7454$$
For $\mu = 9$,
$$1 - \beta = 1 - P(\bar{x} \leq 7.34 \text{ when } \mu = 9)$$
$$= 1 - P\left(z \leq \dfrac{7.34 - 9}{6.2/\sqrt{38}}\right)$$
$$= 1 - P(z \leq -1.65) = .9505$$

For $\mu = 10$,
$$1 - \beta = 1 - P(\bar{x} \leq 7.34 \text{ when } \mu = 10)$$
$$= 1 - P\left(z \leq \frac{7.34 - 10}{6.2/\sqrt{38}}\right)$$
$$= 1 - P(z \leq -2.64) = .9959$$

For $\mu = 12$,
$$1 - \beta = 1 - P(\bar{x} \leq 7.34 \text{ when } \mu = 12)$$
$$= 1 - P\left(z \leq \frac{7.34 - 12}{6.2/\sqrt{38}}\right)$$
$$= 1 - P(z \leq -4.63) \approx 1$$

d The power curve is shown below.

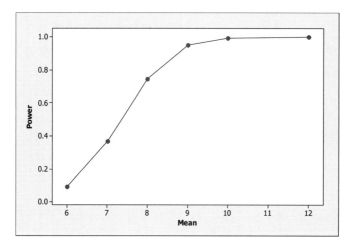

You can see that the power becomes greater than or equal to .90 for a value of μ a little smaller than $\mu = 9$. To find the exact value, we need to solve for μ in the equation:

$$1 - \beta = 1 - P(\bar{x} \leq 7.34) = 1 - P\left(z \leq \frac{7.34 - \mu}{6.2/\sqrt{38}}\right) = .90$$

or $P\left(z \leq \frac{7.34 - \mu}{6.2/\sqrt{38}}\right) = .10$

From Table 3, the value of z that cuts off .10 in the lower tail of the z-distribution is $z = -1.28$, so that

$$\frac{7.34 - \mu}{6.2/\sqrt{38}} = -1.28$$

$$\mu = 7.34 + 1.28 \frac{6.2}{\sqrt{38}} = 8.63.$$

9.77   a   The hypothesis to be tested is
$$H_0 : \mu_1 - \mu_2 = 0 \quad \text{versus} \quad H_a : \mu_1 - \mu_2 > 0$$
and the test statistic is
$$z \approx \frac{(\bar{x}_1 - \bar{x}_2) - 0}{\sqrt{\frac{s_1^2}{n_1} + \frac{s_2^2}{n_2}}} = \frac{176.7 - 163.7}{\sqrt{\frac{6.7^2}{48} + \frac{6.6^2}{77}}} = 10.61$$

The rejection region, with $\alpha = .01$, is one-tailed or $z > 2.33$ and the null hypothesis is rejected. There is sufficient evidence to indicate that the average height for males is greater than females.

   b   An approximate 99% one-sided confidence bound for $\mu_1 - \mu_2$ is approximately
$$(\bar{x}_1 - \bar{x}_2) - 2.33\sqrt{\frac{s_1^2}{n_1} + \frac{s_2^2}{n_2}}$$
$$(176.7 - 163.7) - 2.33\sqrt{\frac{6.7^2}{48} + \frac{6.6^2}{77}}$$
$$13.0 - 2.8545 = 10.15 \quad \text{or} \quad (\mu_1 - \mu_2) > 10.15$$

Males are at least 10.15 centimetres taller than females on average.

9.79   The hypothesis to be tested is
$$H_0 : \mu_1 - \mu_2 = 0 \quad \text{versus} \quad H_a : \mu_1 - \mu_2 \neq 0$$
and the test statistic is
$$z \approx \frac{(\bar{x}_1 - \bar{x}_2) - 0}{\sqrt{\frac{s_1^2}{n_1} + \frac{s_2^2}{n_2}}} = \frac{9017 - 5853}{\sqrt{\frac{7162^2}{130} + \frac{1961^2}{80}}} = 4.76$$

The rejection region, with $\alpha = .01$, is two-tailed or $|z| > 2.58$ and the null hypothesis is rejected. There is sufficient evidence to indicate that the average number of metres swum by these two groups is different.

9.81   a   The hypothesis to be tested is $H_0 : \mu = 665$ versus $H_a : \mu \neq 665$.

   b-c   The test statistic is
$$z \approx \frac{\bar{x} - \mu}{s/\sqrt{n}} = \frac{652 - 665}{32/\sqrt{40}} = -2.569$$

and the **Large-Sample Test of a Population Mean** applet gives p-value $= .0102$.

   d   The null hypothesis can be rejected at the 5% level but not at the 1% level of significance.

9.83   The **Power of a z-Test** applet displays the power correct to only two-decimal places, while the true mean μ is only displayed to the nearest integer. Therefore, as you move the slider at the bottom of the applet, several different power probabilities will appear while μ remains the same. The power values shown in the applet are all consistent with the actual values given in Table 9.2.

μ	870	875	880	885	890
Table 9.2	.9207	.3897	.0500	.3897	.9207
Applet	.90 to .94	.33 to .45	.05	.33 to .45	.90 to .94

9.85   Move the appropriate sliders on the **Power of a z-Test** applet. As the Sample size and the distance between the null and alternative values of μ increase, so does the power. If you decrease α, you will increase β, and hence the power will decrease.

Introduction to Probability and Statistics, 2ce

Case Study:
Cure for the Cold – Pooling Data: Making Sense or Folly?

1. Let p_1 be the proportion of patients who have laboratory confirmed influenza among the hypothetical population of all people who could be treated with 200 mg of ginseng extract, and let p_2 be the proportion of patients who have laboratory confirmed influenza among the hypothetical population of all people who could be treated with a placebo. The hypothesis to be tested is
$$H_0: p_1 - p_2 = 0 \quad \text{versus} \quad H_a: p_1 - p_2 < 0$$

Trial 1:

Calculate $\hat{p}_1 = \dfrac{0}{40} = 0$, $\hat{p}_2 = \dfrac{3}{49} = .0612244898$, and $\hat{p} = \dfrac{x_1 + x_2}{n_1 + n_2} = \dfrac{0+3}{40+49} = .0337078652$. The test statistic is then $z = \dfrac{\hat{p}_1 - \hat{p}_2}{\sqrt{\hat{p}\hat{q}\left(\dfrac{1}{n_1} + \dfrac{1}{n_2}\right)}} = \dfrac{0 - .0612244898}{\sqrt{.033707865(.966292134)(1/40 + 1/49)}} = -1.59$ which is not significant at $\alpha \le 0.05$, with p-value .0559.

Trial 2:

Calculate $\hat{p}_1 = \dfrac{1}{57} = .0175438596$, $\hat{p}_2 = \dfrac{6}{52} = .1153846154$, and $\hat{p} = \dfrac{x_1 + x_2}{n_1 + n_2} = \dfrac{1+6}{57+52} = .0642201835$.

The test statistic is then $z = \dfrac{\hat{p}_1 - \hat{p}_2}{\sqrt{\hat{p}\hat{q}\left(\dfrac{1}{n_1} + \dfrac{1}{n_2}\right)}} = \dfrac{.0175438596 - .1153846154}{\sqrt{.0642201835(.9357798165)(1/57 + 1/52)}} = -2.08$ which is significant at $\alpha = 0.05$, bus not significant at $\alpha = 0.01$, with p-value .0188.

Note that both studies are not significant at $\alpha = 0.01$.

2. The hypothesis to be tested is
$$H_0: p_1 - p_2 = 0 \quad \text{versus} \quad H_a: p_1 - p_2 < 0$$

Calculate $\hat{p}_1 = \dfrac{1}{97} = .0103092784$, $\hat{p}_2 = \dfrac{9}{101} = .0891089109$, and $\hat{p} = \dfrac{x_1 + x_2}{n_1 + n_2} = \dfrac{1+9}{97+101} = .0505050505$.

The test statistic is then $z = \dfrac{\hat{p}_1 - \hat{p}_2}{\sqrt{\hat{p}\hat{q}\left(\dfrac{1}{n_1} + \dfrac{1}{n_2}\right)}} = \dfrac{0.0103092784 - .0891089109}{\sqrt{.0505050505(.9494949495)(1/97 + 1/101)}} = -2.53$ which is significant, with p-value .0057.

3. The answers will vary.

Project 9-A: Proportion of "Cured" Cancer Patients: How Does Canada Compare With Europe?

a (i) The hypothesis test is $H_o: p = 0.32$ versus $H_a: p \neq 0.32$.

The test statistic is $z = \dfrac{\hat{p}-p_o}{\sqrt{\dfrac{p_o q_o}{n}}} = \dfrac{0.36-0.32}{\sqrt{\dfrac{(0.32)(0.68)}{75}}} \cong 0.7426$.

Since $z = 0.7426$ is in the acceptance region (i.e. is between -1.96 and 1.96 for a 2-sided α=5% test), we fail to reject the null hypothesis $H_o: p = 0.32$.

(ii) The p-value is equal to $P(z > 0.7426) + P(z < -0.7426) = 0.2296 + 0.2296 = 0.4592$.

The p-value is greater than the specified 5% significance level, and therefore the null hypothesis $H_o: p = 0.32$ is not rejected. Assuming H_0 were true, there would be a 0.4592 probability of observing sample proportions (of size 75) at least as large as 0.04 away from $p = 0.32$.

(iii) A 95% confidence interval is $\hat{p} \pm z_{0.025}\sqrt{\dfrac{\hat{p}\hat{q}}{n}} = 0.36 \pm 1.96\sqrt{\dfrac{(0.36)(0.64)}{75}} = (0.2514, 0.4686)$.

(iv) If we wish to test if a specific proportion, p_A say, is equal to p_0 or not, we can simply check if p_A is in this confidence interval. If it is, then we cannot reject the null hypothesis that they are equal. Note that this will only work for two-sided tests, at 5% significance.

b (i) The null hypothesis should be $H_o: p_1 = p_2$ or $H_o: (p_1 - p_2) = 0$, versus the alternative $H_a: p_1 \neq p_2$ or $H_a: (p_1 - p_2) \neq 0$.

(ii) A Type II error: the probability of accepting the null hypothesis when it is in reality false.

(iii) The standard error, using the pooled estimate, is $\sqrt{\hat{p}\hat{q}\left(\dfrac{1}{n_1}+\dfrac{1}{n_2}\right)} = \sqrt{(0.294)(0.706)\left(\dfrac{1}{150}+\dfrac{1}{1000}\right)}$
$= 0.0399$, since $\hat{p} = \dfrac{x_1+x_2}{n_1+n_2} = \dfrac{338}{1150} = 0.294$ (using the notation from the text).

(iv) The hypothesis test is $H_o: p_1 = p_2$ versus $H_a: p_1 \neq p_2$. Next, find

$z = \dfrac{\hat{p}_1-\hat{p}_2}{\sqrt{\hat{p}\hat{q}\left(\dfrac{1}{n_1}+\dfrac{1}{n_2}\right)}} = \dfrac{0.313-0.291}{\sqrt{(0.294)(0.706)\left(\dfrac{1}{150}+\dfrac{1}{1000}\right)}} = 0.5515$.

Since the z-score is low, it is relatively safe to conclude that we have a likely observation. That is, $H_o: p_1 = p_2$ may very well be true.

(v) Thus, p-value $= P(z > 0.5515) + P(z < -0.5515) = (1 - 0.7088) + (0.2912) = 0.5824$. Since the p-value is substantially larger than 1% (the significance level), $H_o: p_1 = p_2$ is not rejected.

(vi) For $\alpha = 0.01$, the rejection region is where $z > z_{0.005} = 2.58$ or where $z < -z_{0.005} = -2.58$. Since our test statistic (from part (iv) above) is $z = 0.5515$ and is not in the rejection region, our test shows no difference between the two proportions.

(vii) A 95% confidence interval for the difference $(p_1 - p_2)$ is given by,

$$(\hat{p}_1 - \hat{p}_2) \pm z_{0.025}\sqrt{\frac{\hat{p}_1 \hat{q}_1}{n_1} + \frac{\hat{p}_2 \hat{q}_2}{n_2}} = (0.313 - 0.291) \pm 1.96\sqrt{\frac{(0.313)(0.687)}{150} + \frac{(0.291)(0.709)}{1000}}$$

$= 0.022 \pm 0.079$. This implies that $-0.06 < (p_1 - p_2) < 0.10$, with probability 0.95. Since zero is contained in the interval, there is no statistical evidence that the two proportions are unequal at the 5% significance level.

c (i) The hypotheses are $H_o: p = 0.10$ versus $H_a: p > 0.10$.

(ii) The test statistic is $z = \frac{\hat{p} - p_o}{\sqrt{\frac{p_o q_o}{n}}} = \frac{0.078 - 0.10}{\sqrt{\frac{(0.10)(0.90)}{500}}} = -1.64$.

(iii) The p-value $= P(z > -1.64) = 0.9495$. Thus, based on the very large p-value, there is strong evidence that we should not reject H_0 in favor of the alternative. That is, it is very unlikely that the cure rate in Spain is actually greater than 10%.

Project 9-B: Walking and Talking: My Favorite Sport

a (i) The hypothesis test is $H_o: \mu = 4.8$ vs $H_a: \mu < 4.8$, and the test statistic is,

$z = \frac{\bar{x} - \mu_o}{s/\sqrt{n}} = \frac{4.5 - 4.8}{0.4/\sqrt{40}} = -4.74$. Assuming $\alpha = 0.01$ for this one-sided test, the critical value is $z_{0.01} = -2.33$. Thus, the null hypothesis can be rejected since $z < -2.33$.

(ii) By changing $\alpha = 0.01$ to $\alpha = 0.20$, the critical value would then be $z_{0.20} = -0.84$. Hence, we would have the same conclusion as in the first question.

(ii)* The p-value $= P(z < -4.74) \simeq 0$. Therefore, we reject H_0 as before.

(iii) For this problem, a Type I error will occur if we reject $H_o: \mu = 4.8$ in favor of $H_a: \mu < 4.8$ when in fact $H_o: \mu = 4.8$ is true. A Type II error will occur if we fail to reject $H_o: \mu = 4.8$ when it is actually false.

(iv) For this solution, we will mimic the wording, sequence, and logic of Example 9.8 from the text. The acceptance region is the area to the right of $\mu_o + 2.33\left(\frac{s}{\sqrt{n}}\right) = 4.8 + 2.33\left(\frac{0.4}{\sqrt{40}}\right) = 4.95$.

The probability of accepting H_o, given that $\mu = 4.7$, is equal to the area under the sampling distribution when the values are greater than 4.95.

The z-value that corresponds to 4.95 is, $z = \frac{\bar{x} - \mu}{s/\sqrt{n}} = \frac{4.95 - 4.7}{0.4/\sqrt{40}} = 3.95$.

Then, $\beta = P(\text{accept } H_o \text{ when } \mu = 4.7) = P(z > 3.95) = 0.00004$. Hence, the power of the test is $(1 - \beta) = 1 - 0.00004 = 0.99996$. The probability of correctly rejecting the null hypothesis, given that $\mu = 4.7$, is virtually certain.

b (i) The test is $H_o: (\mu_1 - \mu_2) = 0$ versus $H_a: (\mu_1 - \mu_2) \neq 0$, and the test statistic is

$z = \frac{(\bar{x}_1 - \bar{x}_2) - D_0}{\sqrt{\frac{s_1^2}{n_1} + \frac{s_2^2}{n_2}}} = \frac{(4.9 - 5.4) - 0}{\sqrt{\frac{0.5^2}{81} + \frac{0.2^2}{64}}} = -8.21$.

The p-value corresponding to this test statistic is basically zero. In any case, using any reasonable α, we would reject the null hypothesis and conclude that there is a difference between the average walking speed of men and women.

(ii) The 99% lower confidence bound is giving by,

$(\bar{x}_1 - \bar{x}_2) - 2.33\left(\sqrt{\frac{s_1^2}{n_1} + \frac{s_2^2}{n_2}}\right) = (4.9 - 5.4) - 2.33\left(\sqrt{\frac{0.5^2}{81} + \frac{0.2^2}{64}}\right) = (-0.5) - 0.142 = -0.642$.

The lower confidence bound does not really confirm our conclusion from the previous question, since it is only on one side, but it is consistent with our conclusions from part (i). As it states on page 397 of the text, "a 95% lower one-sided confidence bound will help you find the lowest likely value for the difference." This is therefore the additional information that is provided.

Chapter 10: Inference from Small Samples

10.1 Refer to Table 4, Appendix I, indexing *df* along the left or right margin and t_α across the top.

    a $t_{.05} = 2.015$ with 5 *df*

    b $t_{.025} = 2.306$ with 8 *df*

    c $t_{.10} = 1.330$ with 18 *df*

    d $t_{.025} \approx 1.96$ with 30 *df*

10.3   **a**   The *p*-value for a two-tailed test is defined as
$$p\text{-value} = P(|t| > 2.43) = 2P(t > 2.43)$$
so that
$$P(t > 2.43) = \frac{1}{2} p\text{-value}$$
Refer to Table 4, Appendix I, with $df = 12$. The exact probability, $P(t > 2.43)$ is unavailable; however, it is evident that $t = 2.43$ falls between $t_{.025} = 2.179$ and $t_{.01} = 2.681$. Therefore, the area to the right of $t = 2.43$ must be between .01 and .025. Since
$$.01 < \frac{1}{2} p\text{-value} < .025$$
the *p*-value can be approximated as
$$.02 < p\text{-value} < .05$$

    b   For a right-tailed test, $p\text{-value} = P(t > 3.21)$ with $df = 16$. Since the value $t = 3.21$ is larger than $t_{.005} = 2.921$, the area to its right must be less than .005 and you can bound the *p*-value as
$$p\text{-value} < .005$$

    c   For a two-tailed test, $p\text{-value} = P(|t| > 1.19) = 2P(t > 1.19)$, so that $P(t > 1.19) = \frac{1}{2} p\text{-value}$. From Table 4 with $df = 25$, $t = 1.19$ is smaller than $t_{.10} = 1.316$ so that
$$\frac{1}{2} p\text{-value} > .10 \quad \text{and} \quad p\text{-value} > .20$$

    d   For a left-tailed test, $p\text{-value} = P(t < -8.77) = P(t > 8.77)$ with $df = 7$. Since the value $t = 8.77$ is larger than $t_{.005} = 3.499$, the area to its right must be less than .005 and you can bound the *p*-value as
$$p\text{-value} < .005$$

10.5   **a**   Using the formulas given in Chapter 2, calculate $\sum x_i = 70.5$ and $\sum x_i^2 = 499.27$. Then
$$\bar{x} = \frac{\sum x_i}{n} = \frac{70.5}{10} = 7.05$$
$$s^2 = \frac{\sum x_i^2 - \frac{(\sum x_i)^2}{n}}{n-1} = \frac{499.27 - \frac{(70.5)^2}{10}}{9} = .249444 \quad \text{and} \quad s = .4994$$

b With $df = n-1 = 9$, the appropriate value of t is $t_{.01} = 2.821$ (from Table 4) and the 99% upper one-sided confidence bound is

$$\bar{x} + t_{.01}\frac{s}{\sqrt{n}} \Rightarrow 7.05 + 2.821\sqrt{\frac{.249444}{10}} \Rightarrow 7.05 + .446$$

or $\mu < 7.496$. Intervals constructed using this procedure will enclose μ 99% of the time in repeated sampling. Hence, we are fairly certain that this particular interval encloses μ.

c The hypothesis to be tested is

$$H_0 : \mu = 7.5 \quad \text{versus} \quad H_a : \mu < 7.5$$

and the test statistic is

$$t = \frac{\bar{x} - \mu}{s/\sqrt{n}} = \frac{7.05 - 7.5}{\sqrt{\frac{.249444}{10}}} = -2.849$$

The rejection region with $\alpha = .01$ and $n - 1 = 9$ degrees of freedom is located in the lower tail of the t-distribution and is found from Table 4 as $t < -t_{.01} = -2.821$. Since the observed value of the test statistic falls in the rejection region, H_0 is rejected and we conclude that μ is less than 7.5.

d Notice that the 99% upper one-sided confidence bound for μ does not include the value $\mu = 7.5$. This would confirm the results of the hypothesis test in part **c**, in which we concluded that μ is less than 7.5.

10.7 Similar to previous exercises. The hypothesis to be tested is

$$H_0 : \mu = 5 \quad \text{versus} \quad H_a : \mu < 5$$

Calculate $\bar{x} = \frac{\sum x_i}{n} = \frac{29.6}{6} = 4.933$

$$s^2 = \frac{\sum x_i^2 - \frac{(\sum x_i)^2}{n}}{n-1} = \frac{146.12 - \frac{(29.6)^2}{6}}{5} = .01867 \quad \text{and} \quad s = .1366$$

The test statistic is

$$t = \frac{\bar{x} - \mu}{s/\sqrt{n}} = \frac{4.933 - 5}{\frac{.1366}{\sqrt{6}}} = -1.195$$

The critical value of t with $\alpha = .05$ and $n - 1 = 5$ degrees of freedom is $t_{.05} = 2.015$ and the rejection region is $t < -2.015$. Since the observed value does not fall in the rejection region, H_0 is not rejected. There is no evidence to indicate that the dissolved oxygen content is less than 5 parts per million.

10.9 **a** Similar to previous exercises. The hypothesis to be tested is

$$H_0 : \mu = 100 \quad \text{versus} \quad H_a : \mu < 100$$

Calculate $\bar{x} = \frac{\sum x_i}{n} = \frac{1797.095}{20} = 89.8547$

$$s^2 = \frac{\sum x_i^2 - \frac{(\sum x_i)^2}{n}}{n-1} = \frac{165,697.7081 - \frac{(1797.095)^2}{20}}{19} = 222.1150605 \quad \text{and} \quad s = 14.9035$$

The test statistic is

$$t = \frac{\bar{x} - \mu}{s/\sqrt{n}} = \frac{89.8547 - 100}{\frac{14.9035}{\sqrt{20}}} = -3.044$$

The critical value of t with $\alpha = .01$ and $n - 1 = 19$ degrees of freedom is $t_{.01} = 2.539$ and the rejection region is $t < -2.539$. The null hypothesis is rejected and we conclude that μ is less than 100 DL.

b The 95% upper one-sided confidence bound, based on $n-1=19$ degrees of freedom, is
$$\bar{x} + t_{.05}\frac{s}{\sqrt{n}} \Rightarrow 89.8547 + 2.539\frac{14.9035}{\sqrt{20}} \Rightarrow \mu < 98.316$$
This confirms the results of part **a** in which we concluded that the mean is less than 100 DL.

10.11 Calculate $\bar{x} = \frac{\sum x_i}{n} = \frac{37.82}{10} = 3.782$

$$s^2 = \frac{\sum x_i^2 - \frac{(\sum x_i)^2}{n}}{n-1} = \frac{143.3308 - \frac{(37.82)^2}{10}}{9} = .03284 \quad \text{and} \quad s = .1812$$

The 95% confidence interval based on $df = 9$ is
$$\bar{x} \pm t_{.025}\frac{s}{\sqrt{n}} \Rightarrow 3.782 \pm 2.262\frac{.1812}{\sqrt{10}} \Rightarrow 3.782 \pm .130$$
or $3.652 < \mu < 3.912$.

10.13 a The hypothesis to be tested is
$$H_0: \mu = 25 \quad \text{versus} \quad H_a: \mu < 25$$
The test statistic is
$$t = \frac{\bar{x} - \mu_0}{s/\sqrt{n}} = \frac{20.3 - 25}{\frac{5}{\sqrt{21}}} = -4.31$$
The critical value of t with $\alpha = .05$ and $n-1 = 20$ degrees of freedom is $t_{.05} = 1.725$ and the rejection region is $t < -1.725$. Since the observed value does fall in the rejection region, H_0 is rejected, and we conclude that pre-treatment mean is less than 25.

b The 95% confidence interval based on $df = 20$ is
$$\bar{x} \pm t_{.025}\frac{s}{\sqrt{n}} \Rightarrow 26.6 \pm 2.086\frac{7.4}{\sqrt{21}} \Rightarrow 26.6 \pm 3.37 \text{ or } 23.23 < \mu < 29.97.$$

c The pre-treatment mean looks considerably smaller than the other two means.

10.15 a The t test of the hypothesis
$$H_0: \mu = 1 \quad \text{versus} \quad H_a: \mu \neq 1$$
is not significant, since the p-value $= .113$ associated with the test statistic
$$t = \frac{\bar{x} - \mu}{s/\sqrt{n}} = \frac{1.05222 - 1}{.16565/\sqrt{27}} = 1.64$$
is greater than .10. There is insufficient evidence to indicate that the mean weight per package is different from one-kilogram.

b In fact, the 95% confidence limits for the average weight per package are
$$\bar{x} \pm t_{.025}\frac{s}{\sqrt{n}} \Rightarrow 1.05222 \pm 2.056\frac{.16565}{\sqrt{27}} \Rightarrow 1.05222 \pm .06554$$
or $.98668 < \mu < 1.11776$. These values agree (except in the last decimal place) with those given in the printout. Remember that you used the rounded values of $\bar{x}$ and s from the printout, causing a small rounding error in the results.

10.17 Refer to Exercise 10.16. If we use the large sample method of Chapter 8, the large sample confidence interval is

$$\bar{x} \pm z_{.025} \frac{s}{\sqrt{n}} \Rightarrow 246.96 \pm 1.96 \frac{46.8244}{\sqrt{50}} \Rightarrow 246.96 \pm 12.98$$

or $233.98 < \mu < 259.94$. The intervals are fairly similar, which is why we choose to approximate the sampling distribution of $\frac{\bar{x} - \mu}{s/\sqrt{n}}$ with a z distribution when $n > 30$.

10.19 **a** $s^2 = \frac{(n_1 - 1)s_1^2 + (n_2 - 1)s_2^2}{n_1 + n_2 - 2} = \frac{9(3.4) + 3(4.9)}{10 + 4 - 2} = 3.775$

b $s^2 = \frac{(n_1 - 1)s_1^2 + (n_2 - 1)s_2^2}{n_1 + n_2 - 2} = \frac{11(18) + 20(23)}{12 + 21 - 2} = 21.2258$

10.21 **a** The hypothesis to be tested is: $H_0 : \mu_1 - \mu_2 = 0$ versus $H_a : \mu_1 - \mu_2 \neq 0$

b The rejection region is two-tailed, based on $df = n_1 + n_2 - 2 = 16 + 13 - 2 = 27$ degrees of freedom. With $\alpha = .01$, from Table 4, the rejection region is $|t| > t_{.005} = 2.771$.

c The pooled estimator of σ^2 is calculated as

$$s^2 = \frac{(n_1 - 1)s_1^2 + (n_2 - 1)s_2^2}{n_1 + n_2 - 2} = \frac{15(4.8) + 12(5.9)}{16 + 13 - 2} = 5.2889$$

and the test statistic is

$$t = \frac{(\bar{x}_1 - \bar{x}_2) - 0}{\sqrt{s^2 \left(\frac{1}{n_1} + \frac{1}{n_2}\right)}} = \frac{34.6 - 32.2}{\sqrt{5.2889 \left(\frac{1}{16} + \frac{1}{13}\right)}} = 2.795$$

d The p-value is

$$p\text{-value} = P(|t| > 2.795) = 2P(t > 2.795), \text{ so that } P(t > 2.795) = \frac{1}{2} p\text{-value}.$$

From Table 4 with $df = 27$, $t = 2.795$ is greater than the largest tabulated value ($t_{.005} = 2.771$). Therefore, the area to the right of $t = 2.795$ must be less than .005 so that

$$\tfrac{1}{2} p\text{-value} < .005 \quad \text{and} \quad p\text{-value} < .01$$

e Comparing the observed $t = 2.795$ to the critical value $t_{.005} = 2.771$ or comparing the p-value ($< .01$) to $\alpha = .01$, H_0 is rejected and we conclude that $\mu_1 \neq \mu_2$.

10.23 **a** If you check the ratio of the two variances using the rule of thumb given in this section you will find:

$$\frac{\text{larger } s^2}{\text{smaller } s^2} = \frac{(4.67)^2}{(4.00)^2} = 1.36$$

which is less than three. Therefore, it is reasonable to assume that the two population variances are equal.

b From the *Minitab* printout, the test statistic is $t = .06$ with p-value $= .95$.

c The value of $s = 4.38$ is labeled "Pooled StDev" in the printout, so that $s^2 = (4.38)^2 = 19.1844$.

d Since the p-value $= .95$ is greater than .10, the results are not significant. There is insufficient evidence to indicate a difference in the two population means.

e   A 95% confidence interval for $(\mu_1 - \mu_2)$ is given as

$$(\bar{x}_1 - \bar{x}_2) \pm t_{.025}\sqrt{s^2\left(\frac{1}{n_1} + \frac{1}{n_2}\right)}$$

$$(29 - 28.86) \pm 2.201\sqrt{19.1844\left(\frac{1}{6} + \frac{1}{7}\right)}$$

$$.14 \pm 5.363 \quad \text{or} \quad -5.223 < (\mu_1 - \mu_2) < 5.503$$

Since the value $\mu_1 - \mu_2 = 0$ falls in the confidence interval, it is possible that the two population means are the same. There insufficient evidence to indicate a difference in the two population means.

10.25   a   The hypothesis to be tested is

$$H_0: \mu_1 - \mu_2 = 0 \quad \text{versus} \quad H_a: \mu_1 - \mu_2 \neq 0$$

From the ***Minitab*** printout, the following information is available:

$$\bar{x}_1 = .896 \qquad s_1^2 = (.400)^2 \qquad n_1 = 14$$
$$\bar{x}_2 = 1.147 \qquad s_2^2 = (.679)^2 \qquad n_2 = 11$$

and the test statistic is

$$t = \frac{(\bar{x}_1 - \bar{x}_2) - 0}{\sqrt{s^2\left(\frac{1}{n_1} + \frac{1}{n_2}\right)}} = -1.16$$

The rejection region is two-tailed, based on $n_1 + n_2 - 2 = 23$ degrees of freedom. With $\alpha = .05$, from Table 4, the rejection region is $|t| > t_{.025} = 2.069$ and H_0 is not rejected. There is not enough evidence to indicate a difference in the population means.

b   It is not necessary to bound the *p*-value using Table 4, since the exact *p*-value is given on the printout as P-Value = .260.

c   If you check the ratio of the two variances using the rule of thumb given in this section you will find:

$$\frac{\text{larger } s^2}{\text{smaller } s^2} = \frac{(.679)^2}{(.400)^2} = 2.88$$

which is less than three. Therefore, it is reasonable to assume that the two population variances are equal.

10.27   a   Check the ratio of the two variances using the rule of thumb given in this section:

$$\frac{\text{larger } s^2}{\text{smaller } s^2} = \frac{2.78095}{.17143} = 16.22$$

which is greater than three. Therefore, it is not reasonable to assume that the two population variances are equal.

b You should use the unpooled t test with Satterthwaite's approximation to the degrees of freedom for testing
$$H_0 : \mu_1 - \mu_2 = 0 \quad \text{versus} \quad H_a : \mu_1 - \mu_2 \neq 0$$
The test statistic is
$$t = \frac{(\bar{x}_1 - \bar{x}_2) - 0}{\sqrt{\frac{s_1^2}{n_1} + \frac{s_2^2}{n_2}}} = \frac{3.73 - 4.8}{\sqrt{\frac{2.78095}{15} + \frac{.17143}{15}}} = -2.412$$
with
$$df = \frac{\left(\frac{s_1^2}{n_1} + \frac{s_2^2}{n_2}\right)^2}{\frac{\left(\frac{s_1^2}{n_1}\right)^2}{n_1 - 1} + \frac{\left(\frac{s_2^2}{n_2}\right)^2}{n_2 - 1}} = \frac{(.185397 + .0114287)^2}{.002455137 + .00000933} = 15.7$$

With $df \approx 15$, the p-value for this test is bounded between .02 and .05 so that H_0 can be rejected at the 5% level of significance. There is evidence of a difference in the mean number of uncontaminated eggplants for the two disinfectants.

10.29 a The *Minitab* stem and leaf plots are shown below. Notice the mounded shapes which justify the assumption of normality.

Stem-and-Leaf Display: Generic, Sunmaid
```
Stem-and-leaf of Generic  N = 14      Stem-and-leaf of Sunmaid  N = 14
Leaf Unit = 0.10                      Leaf Unit = 0.10

    1   24  0                             1   22  0
    4   25  000                           1   23
  (5)   26  00000                         5   24  0000
    5   27  00                            7   25  00
    3   28  000                           7   26
                                          7   27  0
                                          6   28  0000
                                          2   29  0
                                          1   30  0
```

b Use your scientific calculator or the computing formulas to find:
$$\bar{x}_1 = 26.214 \qquad s_1^2 = 1.565934 \qquad s_1 = 1.251$$
$$\bar{x}_2 = 26.143 \qquad s_2^2 = 5.824176 \qquad s_2 = 2.413$$

Since the ratio of the variances is greater than 3, you must use the unpooled t test with Satterthwaite's approximate df.
$$df = \frac{\left(\frac{s_1^2}{n_1} + \frac{s_2^2}{n_2}\right)^2}{\frac{\left(\frac{s_1^2}{n_1}\right)^2}{n_1 - 1} + \frac{\left(\frac{s_2^2}{n_2}\right)^2}{n_2 - 1}} \approx 19$$

For testing $H_0 : \mu_1 - \mu_2 = 0$ versus $H_a : \mu_1 - \mu_2 \neq 0$, the test statistic is
$$t = \frac{(\bar{x}_1 - \bar{x}_2) - 0}{\sqrt{\frac{s_1^2}{n_1} + \frac{s_2^2}{n_2}}} = \frac{26.214 - 26.143}{\sqrt{\frac{1.565934}{14} + \frac{5.824176}{14}}} = .10$$

For a two-tailed test with $df = 19$, the p-value can be bounded using Table 4 so that
$$\frac{1}{2}p\text{-value} > .10 \quad \text{or} \quad p\text{-value} > .20$$
Since the p-value is greater than .10, $H_0 : \mu_1 - \mu_2 = 0$ is not rejected. There is insufficient evidence to indicate that there is a difference in the mean number of raisins per box.

10.31 a If swimmer 2 is faster, his(her) average time should be less than the average time for swimmer 1. Therefore, the hypothesis of interest is
$$H_0 : \mu_1 - \mu_2 = 0 \quad \text{versus} \quad H_a : \mu_1 - \mu_2 > 0$$
and the preliminary calculations are as follows:

Swimmer 1	Swimmer 2
$\sum x_{1i} = 596.46$	$\sum x_{2i} = 596.27$
$\sum x_{1i}^2 = 35576.6976$	$\sum x_{2i}^2 = 35554.1093$
$n_1 = 10$	$n_2 = 10$

Then
$$s^2 = \frac{\sum x_{1i}^2 - \frac{(\sum x_{1i})^2}{n_1} + \sum x_{2i}^2 - \frac{(\sum x_{2i})^2}{n_2}}{n_1 + n_2 - 2}$$
$$= \frac{35576.6976 - \frac{(596.46)^2}{10} + 35554.1093 - \frac{(596.27)^2}{10}}{5+5-2} = .03124722$$

Also, $\bar{x}_1 = \frac{596.46}{10} = 59.646$ and $\bar{x}_2 = \frac{596.27}{10} = 59.627$

The test statistic is
$$t = \frac{(\bar{x}_1 - \bar{x}_2) - 0}{\sqrt{s^2\left(\frac{1}{n_1} + \frac{1}{n_2}\right)}} = \frac{59.646 - 59.627}{\sqrt{.03124722\left(\frac{1}{10} + \frac{1}{10}\right)}} = 0.24$$

For a one-tailed test with $df = n_1 + n_2 - 2 = 18$, the p-value can be bounded using Table 4 so that p-value $> .10$, and H_0 is not rejected. There is insufficient evidence to indicate that swimmer 2's average time is still faster than the average time for swimmer 1.

Introduction to Probability and Statistics, 2ce

10.33   **a**   The hypothesis of interest is
$$H_0: \mu_1 - \mu_2 = 0 \quad \text{versus} \quad H_a: \mu_1 - \mu_2 \neq 0$$
and the preliminary calculations are as follows:

Lemieux	Hull
$\sum x_{1i} = 599$	$\sum x_{2i} = 615$
$\sum x_{1i}^2 = 33769$	$\sum x_{2i}^2 = 33371$
$\bar{x}_1 = \dfrac{599}{15} = 39.93$	$\bar{x}_2 = \dfrac{615}{15} = 41.0$
$s_1^2 = 703.50$	$s_2^2 = 582.57$
$n_1 = 15$	$n_2 = 15$

Then
$$s^2 = \frac{(n_1-1)s_1^2 + (n_2-1)s_2^2}{n_1 + n_2 - 2}$$
$$= \frac{14(703.50) + 14(582.57)}{15+15-2} = 643.035$$

The test statistic is
$$t = \frac{(\bar{x}_1 - \bar{x}_2) - 0}{\sqrt{s^2\left(\dfrac{1}{n_1} + \dfrac{1}{n_2}\right)}} = \frac{39.93 - 41.0}{\sqrt{643.035\left(\dfrac{1}{15} + \dfrac{1}{15}\right)}} = -.12$$

The degrees of freedom for this test are $df = n_1 + n_2 - 2 = 28$, we estimate the rejection region using a value of t with $df = 28$ and the rejection region is $|t| > 2.048$. The null hypothesis is not rejected; there is insufficient evidence to indicate that there is a difference in the average number of goals scored for the two players.

b   Again, we estimate the value of t with $df = 28$ and the 95% confidence interval for $(\mu_1 - \mu_2)$ is given as

$$(\bar{x}_1 - \bar{x}_2) \pm t_{.025}\sqrt{s^2\left(\dfrac{1}{n_1} + \dfrac{1}{n_2}\right)}$$

$$(39.93 - 41.0) \pm 2.048\sqrt{643.035\left(\dfrac{1}{15} + \dfrac{1}{15}\right)}$$

$$1.07 \pm 18.96 \quad \text{or} \quad -17.89 < (\mu_1 - \mu_2) < 20.03$$

Since the value $\mu_1 - \mu_2 = 0$ is in the interval, it is possible that the two means might be equal. We do not have enough evidence to indicate that there is a difference in the means.

10.35 **a** The test statistic is

$$t = \frac{\bar{d} - \mu_d}{s_d/\sqrt{n}} = \frac{.3 - 0}{\sqrt{\frac{.16}{10}}} = 2.372$$

with $n - 1 = 9$ degrees of freedom. The p-value is then

$$P(|t| > 2.372) = 2P(t > 2.372) \text{ so that } P(t > 2.372) = \frac{1}{2} p\text{-value}$$

Since the value $t = 2.372$ falls between two tabled entries for $df = 9$ ($t_{.025} = 2.262$ and $t_{.01} = 2.821$), you can conclude that

$$.01 < \frac{1}{2} p\text{-value} < .025$$
$$.02 < p\text{-value} < .05$$

Since the p-value is less than $\alpha = .05$, the null hypothesis is rejected and we conclude that there is a difference in the two population means.

b A 95% confidence interval for $\mu_1 - \mu_2 = \mu_d$ is

$$\bar{d} \pm t_{.025} \frac{s_d}{\sqrt{n}} \Rightarrow 3 \pm 2.262 \sqrt{\frac{.16}{10}} \Rightarrow .3 \pm .286 \text{ or } .014 < (\mu_1 - \mu_2) < .586.$$

c Using $s_d^2 = .16$ and B = .1, the inequality to be solved is approximately

$$1.96 \frac{s_d}{\sqrt{n}} \leq .1$$

$$\sqrt{n} \geq \frac{1.96\sqrt{.16}}{.1} = 7.84 \Rightarrow n \geq 61.47 \text{ or } n = 62$$

Since this value of n is greater than 30, the sample size, $n = 62$ pairs, will be valid.

10.37 **a** It is necessary to use a paired-difference test, since the two samples are not random and independent. The hypothesis of interest is

$$H_0: \mu_1 - \mu_2 = 0 \quad \text{or} \quad H_0: \mu_d = 0$$
$$H_a: \mu_1 - \mu_2 \neq 0 \quad \text{or} \quad H_a: \mu_d \neq 0$$

The table of differences, along with the calculation of $\bar{d}$ and s_d^2, is presented below.

d_i	.1	.1	0	.2	−.1	$\sum d_i = .3$
d_i^2	.01	.01	.00	.04	.01	$\sum d_i^2 = .07$

$$\bar{d} = \frac{\sum d_i}{n} = \frac{.3}{5} = .06 \text{ and } s_d^2 = \frac{\sum d_i^2 - \frac{(\sum d_i)^2}{n}}{n-1} = \frac{.07 - \frac{(.3)^2}{5}}{4} = .013$$

The test statistic is

$$t = \frac{\bar{d} - \mu_d}{s_d/\sqrt{n}} = \frac{.06 - 0}{\sqrt{\frac{.013}{5}}} = 1.177$$

with $n - 1 = 4$ degrees of freedom. The rejection region with $\alpha = .05$ is $|t| > t_{.025} = 2.776$, and H_0 is not rejected. We cannot conclude that the means are different.

b The p-value is

$$P(|t| > 1.177) = 2P(t > 1.177) > 2(.10) = .20$$

c   A 95% confidence interval for $\mu_1 - \mu_2 = \mu_d$ is

$$\bar{d} \pm t_{.025} \frac{s_d}{\sqrt{n}} \Rightarrow .06 \pm 2.776 \sqrt{\frac{.013}{5}} \Rightarrow .06 \pm .142$$

or $-.082 < (\mu_1 - \mu_2) < .202$.

d   In order to use the paired-difference test, it is necessary that the n paired observations be randomly selected from normally distributed populations.

10.39   **a**   The hypothesis of interest is $H_0 : \mu_1 - \mu_2 = 0$ versus $H_a : \mu_1 - \mu_2 \neq 0$ for the two independent samples of runners and cyclists before exercise. Since the ratio of the sample variances is greater than 3, the population variances cannot be assumed to be equal and you must use the unpooled t test with Satterthwaite's approximate df. Calculate

$$t = \frac{(\bar{x}_1 - \bar{x}_2) - 0}{\sqrt{\frac{s_1^2}{n_1} + \frac{s_2^2}{n_2}}} = \frac{255.63 - 173.8}{\sqrt{\frac{(115.48)^2}{10} + \frac{(60.69)^2}{10}}} = 1.984$$

which has a t distribution with $df = \dfrac{\left(\dfrac{s_1^2}{n_1} + \dfrac{s_2^2}{n_2}\right)^2}{\dfrac{\left(\dfrac{s_1^2}{n_1}\right)^2}{n_1 - 1} + \dfrac{\left(\dfrac{s_2^2}{n_2}\right)^2}{n_2 - 1}} = 13.619 \approx 13$

A two-tailed rejection region is then $|t| > t_{.025} = 2.160$ and H_0 is not rejected. A 95% confidence interval for $\mu_1 - \mu_2$ is given as

$$(\bar{x}_1 - \bar{x}_2) \pm t_{.025} \sqrt{\frac{s_1^2}{n_1} + \frac{s_2^2}{n_2}}$$

$$(255.63 - 173.8) \pm 2.160 \sqrt{\frac{(115.48)^2}{10} + \frac{(60.69)^2}{10}}$$

$$81.83 \pm 89.11 \quad \text{or} \quad -7.28 < (\mu_1 - \mu_2) < 170.94$$

b   Similar to part **a**. The hypothesis of interest is $H_0 : \mu_1 - \mu_2 = 0$ versus $H_a : \mu_1 - \mu_2 \neq 0$ for the two independent samples of runners and cyclists after exercise. Since the ratio of the sample variances is greater than 3, you must again use the unpooled t test with Satterthwaite's approximate df. Calculate

$$t = \frac{(\bar{x}_1 - \bar{x}_2) - 0}{\sqrt{\frac{s_1^2}{n_1} + \frac{s_2^2}{n_2}}} = \frac{284.75 - 177.1}{\sqrt{\frac{(132.64)^2}{10} + \frac{(64.63)^2}{10}}} = 2.307$$

which has a t distribution with $df = \dfrac{\left(\dfrac{s_1^2}{n_1} + \dfrac{s_2^2}{n_2}\right)^2}{\dfrac{\left(\dfrac{s_1^2}{n_1}\right)^2}{n_1 - 1} + \dfrac{\left(\dfrac{s_2^2}{n_2}\right)^2}{n_2 - 1}} = 13.046 \approx 13$

A two-tailed rejection region is then $|t| > t_{.025} = 2.160$ and H_0 is rejected. A 95% confidence interval for $\mu_1 - \mu_2$ is given as

$$\left(\overline{x}_1 - \overline{x}_2\right) \pm t_{.025}\sqrt{\frac{s_1^2}{n_1} + \frac{s_2^2}{n_2}}$$

$$\left(284.75 - 177.1\right) \pm 2.160\sqrt{\frac{(132.64)^2}{10} + \frac{(64.63)^2}{10}}$$

$$107.65 \pm 100.783 \quad \text{or} \quad 6.867 < \left(\mu_1 - \mu_2\right) < 208.433$$

c To test the difference between runners before and after exercise, you use a paired difference test, and the hypothesis of interest is

$$H_0 : \mu_1 - \mu_2 = 0 \quad \text{or} \quad H_0 : \mu_d = 0$$
$$H_a : \mu_1 - \mu_2 \neq 0 \quad \text{or} \quad H_a : \mu_d \neq 0$$

It is given that $\overline{d} = 29.13$ and $s_d = 21.01$, so that the test statistic is

$$t = \frac{\overline{d} - \mu_d}{s_d/\sqrt{n}} = \frac{29.13 - 0}{\frac{21.01}{\sqrt{10}}} = 4.38$$

The rejection region with $\alpha = .05$ and $n - 1 = 9$ df is $|t| > t_{.025} = 2.262$, and H_0 is rejected. We can conclude that the means are different.

d The difference in mean CPK values for the 10 cyclists before and after exercise uses $\overline{d} = 3.3$ with $s_d = 6.85$. The 95% confidence interval for $\mu_1 - \mu_2 = \mu_d$ is

$$\overline{d} \pm t_{.05}\frac{s_d}{\sqrt{n}} \Rightarrow 3.3 \pm 2.262\frac{6.85}{\sqrt{10}} \Rightarrow 3.3 \pm 4.90$$

or $-1.6 < \left(\mu_1 - \mu_2\right) < 8.2$. Since the interval contains the value $\mu_1 - \mu_2 = 0$, we cannot conclude that there is a significant difference between the means.

10.41 a Each subject was presented with both signs in random order. If his reaction time in general is high, both responses will be high; if his reaction time in general is low, both responses will be low. The large variability from subject to subject will mask the variability due to the difference in sign types. The paired-difference design will eliminate the subject to subject variability.

b The hypothesis of interest is

$$H_0 : \mu_1 - \mu_2 = 0 \quad \text{or} \quad H_0 : \mu_d = 0$$
$$H_a : \mu_1 - \mu_2 \neq 0 \quad \text{or} \quad H_a : \mu_d \neq 0$$

The table of differences, along with the calculation of $\overline{d}$ and s_d^2, is presented below.

Driver	1	2	3	4	5	6	7	8	9	10	Totals
d_i	122	141	97	107	37	56	110	146	104	149	1069

$$\overline{d} = \frac{\sum d_i}{n} = \frac{1069}{10} = 106.9$$

$$s_d^2 = \frac{\sum d_i^2 - \frac{(\sum d_i)^2}{n}}{n - 1} = \frac{126{,}561 - \frac{(1069)^2}{10}}{9} = 1364.98889 \quad \text{and} \quad s_d = 36.9458$$

and the test statistic is

$$t = \frac{\overline{d} - \mu_d}{s_d/\sqrt{n}} = \frac{106.9 - 0}{\frac{36.9458}{\sqrt{10}}} = 9.150$$

Since $t = 9.150$ with $df = n - 1 = 9$ is greater than the tabled value $t_{.005}$,
$$p\text{-value} < 2(.005) = .01$$
for this two tailed test and H_0 is rejected. We cannot conclude that the means are different.

c The 95% confidence interval for $\mu_1 - \mu_2 = \mu_d$ is

$$\bar{d} \pm t_{.025} \frac{s_d}{\sqrt{n}} \Rightarrow 106.9 \pm 2.262 \frac{36.9458}{\sqrt{10}} \Rightarrow 106.9 \pm 26.428 \text{ or } 80.472 < (\mu_1 - \mu_2) < 133.328.$$

10.43 a A paired-difference test is used, since the two samples are not random and independent (at any location, the ground and air temperatures are related). The hypothesis of interest is
$$H_0: \mu_1 - \mu_2 = 0 \quad H_a: \mu_1 - \mu_2 \neq 0$$
The table of differences, along with the calculation of $\bar{d}$ and s_d^2, is presented below.

Location	1	2	3	4	5	Totals
d_i	$-.4$	-2.7	-1.6	-1.7	-1.5	-7.9

$$\bar{d} = \frac{\sum d_i}{n} = \frac{-7.9}{5} = -1.58$$

$$s_d^2 = \frac{\sum d_i^2 - \frac{(\sum d_i)^2}{n}}{n-1} = \frac{15.15 - \frac{(-7.9)^2}{5}}{4} = .667 \quad \text{and} \quad s_d = .8167$$

and the test statistic is

$$t = \frac{\bar{d} - \mu_d}{s_d / \sqrt{n}} = \frac{-1.58 - 0}{\frac{.8167}{\sqrt{5}}} = -4.326$$

A rejection region with $\alpha = .05$ and $df = n - 1 = 4$ is $|t| > t_{.025} = 2.776$, and H_0 is rejected at the 5% level of significance. We conclude that the air-based temperature readings are biased.

b The 95% confidence interval for $\mu_1 - \mu_2 = \mu_d$ is

$$\bar{d} \pm t_{.025} \frac{s_d}{\sqrt{n}} \Rightarrow -1.58 \pm 2.776 \frac{.8167}{\sqrt{5}} \Rightarrow -1.58 \pm 1.014 \text{ or } -2.594 < (\mu_1 - \mu_2) < -.566.$$

c The inequality to be solved is
$$t_{\alpha/2} SE \leq B$$

We need to estimate the difference in mean temperatures between ground-based and air-based sensors to within .2 degrees centigrade with 95% confidence. Since this is a paired experiment, the inequality becomes

$$t_{.025} \frac{s_d}{\sqrt{n}} \leq .2$$

With $s_d = .8167$ and n represents the number of pairs of observations, consider the sample size obtained by replacing $t_{.025}$ by $z_{.025} = 1.96$.

$$1.96 \frac{.8167}{\sqrt{n}} \leq .2$$

$$\sqrt{n} \geq 8.0019 \Rightarrow n = 64.03 \text{ or } n = 65$$

Since the value of n is greater than 30, the use of $z_{\alpha/2}$ for $t_{\alpha/2}$ is justified.

10.45 a Use the *Minitab* printout given in the text below. The hypothesis of interest is
$$H_0 : \mu_A - \mu_B = 0 \qquad H_a : \mu_A - \mu_B > 0$$
and the test statistic is
$$t = \frac{\bar{d} - \mu_d}{s_d/\sqrt{n}} = \frac{1.4875 - 0}{\frac{1.49134}{\sqrt{8}}} = 2.82$$

The *p*-value shown in the printout is *p*-value = .013. Since the *p*-value is less than .05, H_0 is rejected at the 5% level of significance. We conclude that assessor A gives higher assessments than assessor B.

b A 95% lower one-sided confidence bound for $\mu_1 - \mu_2 = \mu_d$ is
$$\bar{d} - t_{.05}\frac{s_d}{\sqrt{n}} \Rightarrow 1.4875 - 1.895\frac{1.49134}{\sqrt{8}} \Rightarrow 1.4875 - .999 \text{ or } (\mu_1 - \mu_2) > .4885.$$

c In order to apply the paired-difference test, the 8 properties must be randomly and independently selected and the assessments must be normally distributed.

d Yes. If the individual assessments are normally distributed, then the mean of four assessments will be normally distributed. Hence, the difference $\bar{x}_A - \bar{x}$ will be normally distributed and the *t* test on the differences is valid as in **c**.

10.47 A paired-difference analysis must be used. The hypothesis of interest is
$$H_0 : \mu_1 - \mu_2 = 0 \quad \text{or} \quad H_0 : \mu_d = 0$$
$$H_a : \mu_1 - \mu_2 > 0 \quad \text{or} \quad H_a : \mu_d > 0$$

The table of differences is presented below. Use your scientific calculator to find $\bar{d}$ and s_d,

| d_i | 3 | 3 | −2 | 1 | −1 | 3 | −1 |

Calculate $\bar{d} = .857$, $s_d = 2.193$, and the test statistic is
$$t = \frac{\bar{d} - \mu_d}{s_d/\sqrt{n}} = \frac{.857 - 0}{\frac{2.193}{\sqrt{7}}} = 1.03$$

Since $t = 1.03$ with $df = n - 1 = 6$ is smaller than the smallest tabled value $t_{.10}$,
$$p\text{-value} > .10$$
for this one-tailed test and H_0 is not rejected. We cannot conclude that the average time outside the office is less when music is piped in.

10.49 For this exercise, $s^2 = .3214$ and $n = 15$.

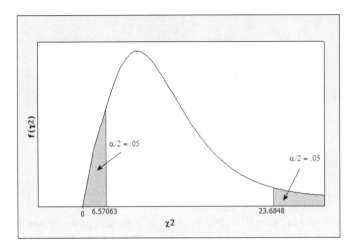

A 90% confidence interval for σ^2 will be

$$\frac{(n-1)s^2}{\chi^2_{\alpha/2}} < \sigma^2 < \frac{(n-1)s^2}{\chi^2_{(1-\alpha/2)}}$$

where $\chi^2_{\alpha/2}$ represents the value of χ^2 such that 5% of the area under the curve (shown in the figure above) lies to its right. Similarly, $\chi^2_{(1-\alpha/2)}$ will be the χ^2 value such that an area .95 lies to its right. Hence, we have located one-half of α in each tail of the distribution. Indexing $\chi^2_{.05}$ and $\chi^2_{.95}$ with $n-1 = 14$ degrees of freedom in Table 5 yields

$$\chi^2_{.05} = 23.6848 \quad \text{and} \quad \chi^2_{.95} = 6.57063$$

and the confidence interval is

$$\frac{14(.3214)}{23.6848} < \sigma^2 < \frac{14(.3214)}{6.57063} \quad \text{or} \quad .190 < \sigma^2 < .685$$

10.51 The hypothesis of interest is

$$H_0 : \sigma = .7 \quad \text{versus} \quad H_a : \sigma > .7$$

or equivalently

$$H_0 : \sigma^2 = .49 \quad \text{versus} \quad H_a : \sigma^2 > .49$$

Calculate

$$s^2 = \frac{\sum x_i^2 - \frac{(\sum x_i)^2}{n}}{n-1} = \frac{36 - \frac{(10)^2}{4}}{3} = 3.6667$$

The test statistic is

$$\chi^2 = \frac{(n-1)s^2}{\sigma_0^2} = \frac{3(3.6667)}{.49} = 22.449$$

The one-tailed rejection region with $\alpha = .05$ and $n-1 = 3$ degrees of freedom is $\chi^2 > \chi^2_{.05} = 7.81$ and H_0 is rejected. There is sufficient evidence to indicate that σ^2 is greater than .49.

10.53 **a** The hypothesis to be tested is
$$H_0 : \mu = 5 \quad H_a : \mu \neq 5$$

Calculate $\bar{x} = \dfrac{\sum x_i}{n} = \dfrac{19.96}{4} = 4.99 \qquad s^2 = \dfrac{\sum x_i^2 - \dfrac{(\sum x_i)^2}{n}}{n-1} = \dfrac{99.6226 - \dfrac{(19.96)^2}{4}}{3} = .0074$

and the test statistic is
$$t = \dfrac{\bar{x} - \mu_0}{s/\sqrt{n}} = \dfrac{4.99 - 5}{\sqrt{\dfrac{.0074}{4}}} = -.232$$

The rejection region with $\alpha = .05$ and $n - 1 = 3$ degrees of freedom is found from Table 4 as $|t| > t_{.025} = 3.182$. Since the observed value of the test statistic does not fall in the rejection region, H_0 is not rejected. There is insufficient evidence to show that the mean differs from 5 mg/cc.

b The manufacturer claims that the range of the potency measurements will equal .2. Since this range is given to equal 6σ, we know that $\sigma \approx .0333$. Then
$$H_0 : \sigma^2 = (.0333)^2 = .0011 \qquad H_a : \sigma^2 > .0011$$
The test statistic is
$$\chi^2 = \dfrac{(n-1)s^2}{\sigma_0^2} = \dfrac{3(.0074)}{.0011} = 20.18$$

and the one-tailed rejection region with $\alpha = .05$ and $n - 1 = 3$ degrees of freedom is
$$\chi^2 > \chi^2_{.05} = 7.81$$

H_0 is rejected; there is sufficient evidence to indicate that the range of the potency will exceed the manufacturer's claim.

10.55 **a** The force transmitted to a wearer, x, is known to be normally distributed with $\mu = 363$ and $\sigma = 18.1$. Hence,
$$P(x > 454) = P\left(z > \dfrac{454 - 363}{18.1}\right) = P(z > 5.03) \approx 0$$

It is highly improbable that any particular helmet will transmit a force in excess of 454 kilograms.

b Since $n = 40$, a large sample test will be used to test
$$H_0 : \mu = 454 \qquad H_a : \mu > 454$$
The test statistic is
$$z = \dfrac{\bar{x} - \mu_0}{s/\sqrt{n}} = \dfrac{374 - 454}{\sqrt{\dfrac{1070}{40}}} = -15.47$$

and the rejection region with $\alpha = .05$ is $z > 1.645$. H_0 is not rejected and we conclude that the mean force transmitted by the helmets does not exceed 454 kilograms. [Note: Here the p-value ≈ 1.00]

10.57 The hypothesis of interest is $\qquad H_0 : \sigma = 150 \qquad H_a : \sigma < 150$
Calculate
$$(n-1)s^2 = \sum x_i^2 - \dfrac{(\sum x_i)^2}{n} = 92,305,600 - \dfrac{(42,812)^2}{20} = 662,232.8$$

and the test statistic is $\chi^2 = \dfrac{(n-1)s^2}{\sigma_0^2} = \dfrac{662,232.8}{150^2} = 29.433$. The one-tailed rejection region with $\alpha = .01$ and $n - 1 = 19$ degrees of freedom is $\chi^2 < \chi^2_{.99} = 7.63273$, and H_0 is not rejected. There is insufficient evidence to indicate that he is meeting his goal.

10.59 Refer to Exercise 10.58. From Table 6, $F_{df_1,df_2} = 2.62$ and $F_{df_2,df_1} \approx 2.76$. The 95% confidence interval for σ_1^2/σ_2^2 is

$$\frac{s_1^2}{s_2^2}\frac{1}{F_{df_1,df_2}} < \frac{\sigma_1^2}{\sigma_2^2} < \frac{s_1^2}{s_2^2}F_{df_2,df_1}$$

$$\frac{55.7}{31.4}\left(\frac{1}{2.62}\right) < \frac{\sigma_1^2}{\sigma_2^2} < \frac{55.7}{31.4}(2.76) \quad \text{or} \quad .667 < \frac{\sigma_1^2}{\sigma_2^2} < 4.896$$

10.61 The hypothesis of interest is $H_0: \sigma_1^2 = \sigma_2^2$ versus $H_a: \sigma_1^2 \neq \sigma_2^2$ and the test statistic is

$$F = \frac{s_1^2}{s_2^2} = \frac{.71^2}{.69^2} = 1.059.$$

The critical values of F for various values of α are given below using $df_1 = 14$ and $df_2 = 14$.

α	.10	.05	.025	.01	.005
F_α	2.02	2.48	2.97	3.70	4.30

Hence, $p\text{-value} = 2P(F > 1.059) > 2(.10) = .20$

Since the p-value is so large, H_0 is not rejected. There is no evidence to indicate that the variances are different.

10.63 Refer to Exercise 10.62. Noting that $F_{df_1,df_2} = F_{df_2,df_1} \approx 1.61$, the 90% confidence interval for σ_1^2/σ_2^2 is

$$\frac{s_1^2}{s_2^2}\frac{1}{F_{df_1,df_2}} < \frac{\sigma_1^2}{\sigma_2^2} < \frac{s_1^2}{s_2^2}F_{df_2,df_1}$$

$$\frac{92,000}{37,000}\left(\frac{1}{1.61}\right) < \frac{\sigma_1^2}{\sigma_2^2} < \frac{92,000}{37,000}(1.61) \quad \text{or} \quad 1.544 < \frac{\sigma_1^2}{\sigma_2^2} < 4.003$$

10.65 For each of the three tests, the hypothesis of interest is

$$H_0: \sigma_1^2 = \sigma_2^2 \quad \text{versus} \quad H_a: \sigma_1^2 \neq \sigma_2^2$$

and the test statistics are

$$F = \frac{s_1^2}{s_2^2} = \frac{3.98^2}{3.92^2} = 1.03 \qquad F = \frac{s_1^2}{s_2^2} = \frac{4.95^2}{3.49^2} = 2.01 \quad \text{and} \quad F = \frac{s_1^2}{s_2^2} = \frac{16.9^2}{4.47^2} = 14.29$$

The critical values of F for various values of α are given below using $df_1 = 9$ and $df_2 = 9$.

α	.10	.05	.025	.01	.005
F_α	2.44	3.18	4.03	5.35	6.54

Hence, for the first two tests,

$$p\text{-value} > 2(.10) = .20$$

while for the last test,

$$p\text{-value} < 2(.005) = .01$$

There is no evidence to indicate that the variances are different for the first two tests, but H_0 is rejected for the third variable. The two-sample t-test with a pooled estimate of σ^2 cannot be used for the third variable.

10.67 A Student's *t* test can be employed to test a hypothesis about a single population mean when the sample has been randomly selected from a normal population. It will work quite satisfactorily for populations which possess mound-shaped frequency distributions resembling the normal distribution.

10.69 Paired observations are used to estimate the difference between two population means in preference to an estimation based on independent random samples selected from the two populations because of the increased information caused by blocking the observations. We expect blocking to create a large reduction in the standard deviation, if differences do exist among the blocks.
Paired observations are not always preferable. The degrees of freedom that are available for estimating σ^2 are less for paired than for unpaired observations. If there were no difference between the blocks, the paired experiment would then be less beneficial.

10.71 The 90% confidence interval is
$$\bar{x} \pm t_{.05} \frac{s}{\sqrt{n}} \Rightarrow 25 \pm 1.746 \frac{7}{\sqrt{17}} \Rightarrow 25 \pm 2.96 \text{ or } 22.04 < \mu < 27.96.$$

10.73 Since it is necessary to determine whether the injected rats drink more water than noninjected rates, the hypothesis to be tested is
$$H_0: \mu = 22.0 \quad H_a: \mu > 22.0$$
and the test statistic is
$$t = \frac{\bar{x} - \mu_0}{s/\sqrt{n}} = \frac{31.0 - 22.0}{\frac{6.2}{\sqrt{17}}} = 5.985.$$

Using the *critical value approach*, the rejection region with $\alpha = .05$ and $n - 1 = 16$ degrees of freedom is located in the upper tail of the *t*-distribution and is found from Table 4 as $t > t_{.05} = 1.746$. Since the observed value of the test statistic falls in the rejection region, H_0 is rejected and we conclude that the injected rats do drink more water than the noninjected rats. The 90% confidence interval is
$$\bar{x} \pm t_{.05} \frac{s}{\sqrt{n}} \Rightarrow 31.0 \pm 1.746 \frac{6.2}{\sqrt{17}} \Rightarrow 31.0 \pm 2.625$$
or $28.375 < \mu < 33.625$.

10.75 The hypothesis of interest is
$$H_0: \sigma_1^2 = \sigma_2^2 \quad \text{versus} \quad H_a: \sigma_1^2 < \sigma_2^2$$
and the test statistic is
$$F = \frac{s_1^2}{s_2^2} = \frac{(28.2)^2}{(15.6)^2} = 3.268.$$

The critical values of *F* for various values of α are given below using $df_1 = 30$ (since there are no tabled values for $df_1 = 29$) and $df_2 = 29$.

α	.10	.05	.025	.01	.005
F_α	1.62	1.85	2.09	2.41	2.66

Hence,
$$p\text{-value} = P(F > 3.268) < .005$$

Since the *p*-value is so small, H_0 is rejected. There is evidence to indicate that increased maintenance of the older system is needed.

10.77 From Exercise 10.76, the best estimate for σ is $s = 21.5$. Then, with $B = 5$, solve for n in the following inequality:

$$t_{.025}\frac{s}{\sqrt{n}} \leq 5$$

which is approximately

$$1.96\frac{21.5}{\sqrt{n}} \leq 5 \Rightarrow \sqrt{n} \geq \frac{1.96(21.5)}{5} = 8.428$$

$$n \geq 71.03 \quad \text{or} \quad n \geq 72$$

Since n is greater than 30, the sample size, $n = 72$, is valid.

10.79 Use the computing formulas or your scientific calculator to calculate

$$\bar{x} = \frac{\sum x_i}{n} = \frac{322.1}{13} = 24.777$$

$$s^2 = \frac{\sum x_i^2 - \frac{(\sum x_i)^2}{n}}{n-1} = \frac{8114.59 - \frac{(322.1)^2}{13}}{12} = 11.1619$$

$s = 3.3409$ and the 95% confidence interval is $\bar{x} \pm t_{.025}\frac{s}{\sqrt{n}} \Rightarrow 24.777 \pm 2.179\frac{3.3409}{\sqrt{13}} \Rightarrow 24.777 \pm 2.019$

or $22.578 < \mu < 26.796$.

10.81 **a** The inequality to be solved is

$$t_{\alpha/2} \times (\text{std error of estimator}) \leq B$$

In this exercise, it is necessary to estimate the difference in means to within 1 minute with 95% confidence. The inequality is then

$$t_{.025}\sqrt{s^2\left(\frac{1}{n_1} + \frac{1}{n_2}\right)} \leq 1$$

We assume that $n_1 = n_2 = n$ and $s^2 = 17.64$ from Exercise 10.76. Consider the sample size obtained by replacing $t_{.025}$ by $z_{.025} = 1.96$.

$$1.96\sqrt{17.64\left(\frac{1}{n} + \frac{1}{n}\right)} \leq 1 \Rightarrow \sqrt{n} \geq 1.96\sqrt{35.28}$$

$$n \geq 135.53 \quad \text{or} \quad n \geq 136$$

Since the value of n is greater than 30, the sample size is valid.

b Consider the inequality from part **a**,

$$t_{\alpha/2}\sqrt{s^2\left(\frac{1}{n_1} + \frac{1}{n_2}\right)} \leq 1.$$

When $n_1 = n_2 = n$, this becomes

$$t_{\alpha/2}\sqrt{s^2\left(\frac{2}{n}\right)} \leq 1 \Rightarrow \sqrt{n} \geq t_{\alpha/2}\sqrt{2s^2}$$

To reduce the sample size necessary to achieve this inequality, we must reduce the quantity

$$t_{\alpha/2}\sqrt{2s^2}$$

If we are willing to lower the level of confidence (or equivalently increase the value of α), the value of $t_{\alpha/2}$ will be smaller. This will decrease the size of n, but at the price of decreased confidence. The only other quantity in the expression which is not fixed is s^2. If this variable can be made smaller, the required sample size will also be reduced. From the definition of s^2, it can be seen that s^2 includes the variability associated with the difference in drugs A and B as well as the variability in absorption rates among the people in the experiment. The experiment could be run a bit differently by using the same people for both drugs. Each person would receive a dose of drug A and drug B. The difference in absorption rates for the two drugs would be observed within each individual. Such a design would eliminate the variability in absorption rates from person to person. The value of s^2 would be reduced and a smaller sample size would be required. The resulting experiment would be analyzed as a paired-difference experiment, as discussed in Section 10.5.

10.83 **a** The range of the first sample is 47 while the range of the second sample is only 16. There is probably a difference in the variances.

  b The hypothesis of interest is
$$H_0 : \sigma_1^2 = \sigma_2^2 \text{ versus } H_a : \sigma_1^2 \neq \sigma_2^2$$

Calculate $s_1^2 = \dfrac{177{,}294 - \dfrac{(838)^2}{4}}{3} = 577.6667 \quad s_2^2 = \dfrac{192{,}394 - \dfrac{(1074)^2}{6}}{5} = 29.6$

and the test statistic is
$$F = \frac{s_1^2}{s_2^2} = \frac{577.6667}{29.6} = 19.516.$$

The critical values with $df_1 = 3$ and $df_2 = 5$ are shown below from Table 6.

α	.10	.05	.025	.01	.005
F_α	3.62	5.41	7.76	12.06	16.53

Hence, $p\text{-value} = 2P(F > 19.516) < 2(.005) = .01$

Since the p-value is smaller than .01, H_0 is rejected at the 1% level of significance. There is a difference in variability.

  c Since the Student's t test requires the assumption of equal variance, it would be inappropriate in this instance. You should use the unpooled t test with Satterthwaite's approximation to the degrees of freedom.

10.85 **a** The leaf measurements probably come from mound-shaped, or approximately normal populations, since their length, width, thickness, and so on can be thought of as being a composite sum of many factors which affect their growth (see the Central Limit Theorem). The values of the sample variances are not very different, and we would not question the assumption of equal variances. Finally, since the plants were all given the same experimental treatment, they can be considered random and independent samples within a treatment group.

  b The hypothesis to be tested is
$$H_0 : \mu_1 - \mu_2 = 0 \qquad H_a : \mu_1 - \mu_2 \neq 0$$

Since the ratio of the variances is less than 3, you can use the pooled t test. The pooled estimator of σ^2 is calculated as
$$s^2 = \frac{(n_1 - 1)s_1^2 + (n_2 - 1)s_2^2}{n_1 + n_2 - 2} = \frac{15(43)^2 + 14(41.7)^2}{29} = 1795.8434$$

and the test statistic is
$$t = \frac{(\bar{x}_1 - \bar{x}_2) - 0}{\sqrt{s^2 \left(\dfrac{1}{n_1} + \dfrac{1}{n_2}\right)}} = \frac{128 - 78.7}{\sqrt{1795.8434 \left(\dfrac{1}{16} + \dfrac{1}{15}\right)}} = 3.237$$

The two-tailed p-value with $df = 29$ can be bounded as
$$p\text{-value} < 2(.005) = .01$$

and H_0 is rejected. There is a difference in the means.

c The hypothesis to be tested is
$$H_0: \mu_1 - \mu_2 = 0 \qquad H_a: \mu_1 - \mu_2 \neq 0$$
Since the ratio of the variances is slightly greater than 3, you should use the pooled t test with the test statistic

$$t = \frac{(\bar{x}_1 - \bar{x}_2) - 0}{\sqrt{\frac{s_1^2}{n_1} + \frac{s_2^2}{n_2}}} = \frac{46.8 - 8.1}{\sqrt{\frac{(2.21)^2}{16} + \frac{(1.26)^2}{15}}} = 60.36$$

The p-value is two-tailed, based on

$$df = \frac{\left(\frac{s_1^2}{n_1} + \frac{s_2^2}{n_2}\right)^2}{\frac{\left(\frac{s_1^2}{n_1}\right)^2}{n_1 - 1} + \frac{\left(\frac{s_2^2}{n_2}\right)^2}{n_2 - 1}} \approx 25$$

degrees of freedom and is bounded as
$$p\text{-value} < 2(.005) = .01$$
The results are highly significant and H_0 is rejected. There is a difference in the means.

10.87 A paired-difference test is used, since the two samples are not random and independent. The hypothesis of interest is
$$H_0: \mu_1 - \mu_2 = 0 \qquad H_a: \mu_1 - \mu_2 > 0$$
and the table of differences, along with the calculation of $\bar{d}$ and s_d^2, is presented below.

Pair	1	2	3	4	Totals
d_i	−1	5	11	7	22

$$\bar{d} = \frac{\sum d_i}{n} = \frac{22}{4} = 5.5$$

$$s_d^2 = \frac{\sum d_i^2 - \frac{(\sum d_i)^2}{n}}{n-1} = \frac{196 - \frac{(22)^2}{4}}{3} = 25 \text{ and } s_d = 5 \text{ and the test statistic is } t = \frac{\bar{d} - \mu_d}{s_d/\sqrt{n}} = \frac{5.5 - 0}{\frac{5}{\sqrt{4}}} = 2.2$$

The one-tailed p-value with $df = 3$ can be bounded between .05 and .10. Since this value is greater than .10, H_0 is not rejected. The results are not significant; there is insufficient evidence to indicate that lack of school experience has a depressing effect on IQ scores.

10.89 A paired-difference analysis is used. The hypothesis of interest is
$$H_0: \mu_1 - \mu_2 = 0 \quad H_a: \mu_1 - \mu_2 \neq 0$$
and the differences are shown below.
$$156, 447, -3, 42$$

Calculate $\bar{d} = \dfrac{\sum d_i}{n} = \dfrac{642}{4} = 160.5$

$s_d = \sqrt{\dfrac{\sum d_i^2 - \dfrac{(\sum d_i)^2}{n}}{n-1}} = \sqrt{\dfrac{225918 - \dfrac{(642)^2}{4}}{3}} = 202.3833$ and the test statistic is

$t = \dfrac{\bar{d} - \mu_d}{s_d/\sqrt{n}} = \dfrac{160.5 - 0}{\dfrac{202.3833}{\sqrt{4}}} = 1.586$

The two-tailed p-value with $df = 3$ is greater than $2(.10) = .20$. Since this value is greater than .05, H_0 is not rejected. The results are not significant; there is insufficient evidence to indicate a difference in the average cost of repair for the Honda Civic and the Hyundai Elantra.

10.91 The object is to determine whether or not there is a difference between the mean responses for the two different stimuli to which the people have been subjected. The samples are independently and randomly selected, and the assumptions necessary for the t test of Section 10.4 are met. The hypothesis to be tested is
$$H_0: \mu_1 - \mu_2 = 0 \quad H_a: \mu_1 - \mu_2 \neq 0$$
and the preliminary calculations are as follows:
$$\bar{x}_1 = \dfrac{15}{8} = 1.875 \quad \text{and} \quad \bar{x}_2 = \dfrac{21}{8} = 2.625$$

$s_1^2 = \dfrac{33 - \dfrac{(15)^2}{8}}{7} = .69643$ and $s_2^2 = \dfrac{61 - \dfrac{(21)^2}{8}}{7} = .83929$

Since the ratio of the variances is less than 3, you can use the pooled t test. The pooled estimator of σ^2 is calculated as
$$s^2 = \dfrac{(n_1-1)s_1^2 + (n_2-1)s_2^2}{n_1 + n_2 - 2} = \dfrac{4.875 + 5.875}{14} = .7679$$

and the test statistic is
$$t = \dfrac{(\bar{x}_1 - \bar{x}_2) - 0}{\sqrt{s^2\left(\dfrac{1}{n_1} + \dfrac{1}{n_2}\right)}} = \dfrac{1.875 - 2.625}{\sqrt{.7679\left(\dfrac{1}{8} + \dfrac{1}{8}\right)}} = -1.712$$

The two-tailed rejection region with $\alpha = .05$ and $df = 14$ is $|t| > t_{.025} = 2.145$, and H_0 is not rejected. There is insufficient evidence to indicate that there is a difference in means.

10.93 Refer to Exercise 10.91 and 10.92. For the unpaired design, the 95% confidence interval for $(\mu_1 - \mu_2)$ is

$$(\bar{x}_1 - \bar{x}_2) \pm t_{.025}\sqrt{s^2\left(\frac{1}{n_1} + \frac{1}{n_2}\right)}$$

$$-.75 \pm 2.145\sqrt{.7679\left(\frac{1}{8} + \frac{1}{8}\right)}$$

$$-.75 \pm .94 \quad \text{or} \quad -1.69 < (\mu_1 - \mu_2) < 0.19$$

while for the paired design, the 95% confidence interval is

$$\bar{d} \pm t_{.025}\frac{s_d}{\sqrt{n}} \Rightarrow -.75 \pm 2.365\frac{.88641}{\sqrt{8}} \Rightarrow -.75 \pm .74$$

or $-1.49 < (\mu_1 - \mu_2) < -.01$. Although the width of the confidence interval has decreased slightly, it does not appear that blocking has increased the amount of information by much.

10.95 a Note that the experiment has been designed so that two cake pans, one containing batter A and one containing batter B, were placed side by side at each of six different locations in an oven. The two samples are therefore not independent, and a paired-difference analysis must be used. To test $H_0: \mu_2 - \mu_1 = 0$ versus $H_a: \mu_2 - \mu_1 \neq 0$, calculate the differences:

$$-.006, .018, .014, .011, .004, .019$$

Then $\bar{d} = \frac{\sum d_i}{n} = \frac{.06}{6} = .01$

$$s_d^2 = \frac{\sum d_i^2 - \frac{(\sum d_i)^2}{n}}{n-1} = \frac{.001054 - \frac{(.06)^2}{6}}{5} = .0000908 \text{ and } s_d = .009529 \text{ and the test statistic is}$$

$$t = \frac{\bar{d} - \mu_d}{s_d/\sqrt{n}} = \frac{.06 - 0}{\frac{.009529}{\sqrt{6}}} = 2.5706$$

For a two-tailed test with $df = 5$, the p-value is bounded as

$$2(.025) < p\text{-value} < 2(.05) \quad \text{or} \quad .05 < p\text{-value} < .10$$

Since the p-value is greater than .10, H_0 is not rejected. There is insufficient evidence to indicate a difference between mean densities for batters A and B.

b The 95% confidence interval is

$$\bar{d} \pm t_{.025}\frac{s_d}{\sqrt{n}} \Rightarrow .01 \pm 2.571\frac{.009529}{\sqrt{6}} \Rightarrow .01 \pm .010 \text{ or } .000 < (\mu_1 - \mu_2) < .020.$$

10.97 It is possible to test the null hypothesis $H_0: \sigma_1^2 = \sigma_2^2$ against any one of three alternative hypotheses:

(1) $H_a: \sigma_1^2 \neq \sigma_2^2$ (2) $H_a: \sigma_1^2 < \sigma_2^2$ (3) $H_a: \sigma_1^2 > \sigma_2^2$

a The first alternative would be preferred by the manager of the dairy. He does not know anything about the variability of the two machines and would wish to detect departures from equality of the type $\sigma_1^2 < \sigma_2^2$ or $\sigma_1^2 > \sigma_2^2$. These alternatives are implied in (1).

b The salesman for company A would prefer that the experimenter select the second alternative. Rejection of the null hypothesis would imply that his machine had smaller variability. Moreover, even if the null hypothesis were not rejected, there would be no evidence to indicate that the variability of the company A machine was greater than the variability of the company B machine.

c The salesman for company B would prefer the third alternative for a similar reason.

10.99 a The manufacturer claims that the range of the random variable x (purity of his product) is no more than 2%. In terms of the standard deviation σ, he is claiming that $\sigma \leq .5$ since

$$\text{Range} \approx 4\sigma = 2 \quad \Rightarrow \quad \sigma = .5$$

Hence, the hypothesis to be tested is

$$H_0: \sigma = .5 \qquad H_a: \sigma > .5 \quad \text{or equivalently}$$
$$H_0: \sigma^2 = .25 \qquad H_a: \sigma^2 > .25$$

Calculate $\quad s^2 = \dfrac{47{,}982.56 - \dfrac{(489.8)^2}{5}}{4} = .438 \quad$ and the test statistic is

$$\chi^2 = \frac{(n-1)s^2}{\sigma_0^2} = \frac{4(.438)}{.25} = 7.088$$

The one-tailed p-value with $n - 1 = 4$ degrees of freedom is

$$p\text{-value} > .10$$

and H_0 is not rejected. There is insufficient evidence to contradict the manufacturer's claim.

b Indexing $\chi^2_{.05}$ and $\chi^2_{.95}$ with $n - 1 = 4$ degrees of freedom in Table 5 yields

$$\chi^2_{.05} = 9.48773 \quad \text{and} \quad \chi^2_{.95} = .710721$$

and the confidence interval is

$$\frac{4(.438)}{9.48773} < \sigma^2 < \frac{4(.438)}{.710721} \quad \text{or} \quad .185 < \sigma^2 < 2.465$$

10.101 A paired-difference analysis is used. To test $H_0: \mu_1 - \mu_2 = 0$ versus $H_a: \mu_1 - \mu_2 > 0$, where μ_2 is the mean reaction time after injection and μ_1 is the mean reaction time before injection, calculate the differences $(x_2 - x_1): 6, 1, 6, 1$

Then $\bar{d} = \dfrac{\sum d_i}{n} = \dfrac{14}{4} = 3.5$

$$s_d^2 = \frac{\sum d_i^2 - \dfrac{(\sum d_i)^2}{n}}{n-1} = \frac{74 - \dfrac{(14)^2}{4}}{3} = 8.33 \text{ and } s_d = 2.88675 \text{ and the test statistic is}$$

$$t = \frac{\bar{d} - \mu_d}{s_d/\sqrt{n}} = \frac{3.5 - 0}{\dfrac{2.88675}{\sqrt{4}}} = 2.425$$

For a one-tailed test with $df = 3$, the rejection region with $\alpha = .05$ is $t > t_{.05} = 2.353$, and H_0 is rejected. We conclude that the drug significantly increases with reaction time.

10.103 The hypothesis to be tested is $\qquad H_0: \mu_1 - \mu_2 = 0 \qquad H_a: \mu_1 - \mu_2 \neq 0$
and the preliminary calculations are as follows:

$$\bar{x}_1 = \frac{.6}{6} = .1 \quad \text{and} \quad \bar{x}_2 = \frac{.83}{6} = .1383$$

$$s_1^2 = \frac{.0624 - \dfrac{(.6)^2}{6}}{5} = .00048 \quad \text{and} \quad s_2^2 = \frac{.1175 - \dfrac{(.83)^2}{6}}{5} = .00053667$$

Since the ratio of the variances is less than 3, you can use the pooled t test. The pooled estimator of σ^2 is calculated as

$$s^2 = \frac{(n_1 - 1)s_1^2 + (n_2 - 1)s_2^2}{n_1 + n_2 - 2} = \frac{.0024 + .00268}{10} = .0005083$$

and the test statistic is

$$t = \frac{(\bar{x}_1 - \bar{x}_2) - 0}{\sqrt{s^2\left(\frac{1}{n_1} + \frac{1}{n_2}\right)}} = \frac{-.0383}{\sqrt{.0005083\left(\frac{1}{6} + \frac{1}{6}\right)}} = -2.945$$

The *p*-value for a one-tailed test with 10 degrees of freedom is bounded as
$$.005 < p\text{-value} < .010$$
Hence, the null hypothesis H_0 is rejected. There is sufficient evidence to indicate that $\mu_1 < \mu_2$.

10.105 The underlying populations are ratings and can only take on the finite number of values, 1, 2, …, 9, 10. Neither population has a normal distribution, but both are discrete. Further, the samples are not independent, since the same person is asked to rank each car design. Hence, two of the assumptions required for the Student's *t* test have been violated.

10.107 A paired-difference test is used. To test $H_0 : \mu_1 - \mu_2 = 0$ versus $H_a : \mu_1 - \mu_2 > 0$, calculate the differences:
$$29, 31, -35, -17, 99, 73, 54$$
The test statistic is given on the *Minitab* printout as

$$t = \frac{\bar{d} - \mu_d}{s_d/\sqrt{n}} = 1.86$$

with *p*-value = .112. Since this value is greater than .10, the results are not significant, and H_0 is not rejected. There is insufficient evidence to indicate a greater mean demand for one of the entrees.

10.109 The *Minitab* printout below shows the summary statistics for the two samples:

Descriptive Statistics: Method 1, Method 2
```
Variable   N    Mean   SE Mean   StDev
Method 1   5   137.00    4.55    10.17
Method 2   5   147.20    3.29     7.36
```

Since the ratio of the two sample variances is less than 3, you can use the pooled *t* test to compare the two methods of measurement, using the remainder of the *Minitab* printout below:

Two-Sample T-Test and CI: Method 1, Method 2
```
Difference = mu (Method 1) - mu (Method 2)
Estimate for difference:  -10.2000
95% CI for difference:  (-23.1506, 2.7506)
T-Test of difference = 0 (vs not =): T-Value = -1.82   P-Value = 0.107   DF = 8
Both use Pooled StDev = 8.8798
```

The test statistic is $t = -1.82$ with *p*-value = .107 and the results are not significant. There is insufficient evidence to declare a difference in the two population means.

10.111 a Calculate $\bar{x} = \frac{\sum x_i}{n} = \frac{68.5}{10} = 6.85$

$$s^2 = \frac{\sum x_i^2 - \frac{(\sum x_i)^2}{n}}{n-1} = \frac{478.375 - \frac{(68.5)^2}{10}}{9} = 1.016667 \quad \text{and} \quad s = 1.0083$$

The 99% confidence interval based on $df = 9$ is

$$\bar{x} \pm t_{.005}\frac{s}{\sqrt{n}} \Rightarrow 6.85 \pm 3.25\frac{1.0083}{\sqrt{10}} \Rightarrow 6.85 \pm 1.036 \text{ or } 5.814 < \mu < 7.886.$$

b The sample must have been randomly selected from a normal population.

10.113 The hypothesis to be tested is $H_0: \mu = 280$ versus $H_a: \mu > 280$

The test statistic is

$$t = \frac{\bar{x} - \mu}{s/\sqrt{n}} = \frac{358 - 280}{\frac{54}{\sqrt{10}}} = 4.57$$

The critical value of t with $\alpha = .01$ and $n - 1 = 9$ degrees of freedom is $t_{.01} = 2.821$ and the rejection region is $t > 2.821$. Since the observed value falls in the rejection region, H_0 is rejected. There is sufficient evidence to indicate that the average number of calories is greater than advertised.

10.115 a A paired difference test is required, since the costs are paired according to the type of drug. To test $H_0: \mu_1 - \mu_2 = 0$ versus $H_a: \mu_1 - \mu_2 \neq 0$, the 9 differences are

111, 201, 45, 14, 175, 105, 288, 94, 79

Calculate $\bar{d} = \frac{\sum d_i}{n} = \frac{1112}{9} = 123.5556$

$$s_d^2 = \frac{194{,}614 - \frac{(1112)^2}{9}}{8} = 7152.527778 \text{ and } s_d = 84.572618$$

The test statistic is

$$t = \frac{\bar{d} - \mu_d}{s_d/\sqrt{n}} = \frac{123.5556 - 0}{\frac{84.572618}{\sqrt{9}}} = 4.38$$

For a two-tailed test with $df = 8$, the rejection region with $\alpha = .01$ is $|t| > t_{.005} = 3.355$, and H_0 is rejected. There is sufficient evidence to indicate that the average cost of prescription drugs in the United States is different from the average cost in Canada.

b The observed value of the test statistic, $t = 4.38$ is larger than $t_{.005} = 3.355$, so that the $\frac{1}{2}(p\text{-value}) < .005$ or $p\text{-value} < .01$. Since $\alpha = .01$, the p-value would cause us to reject H_0, as in part **a**.

10.117 Use the **Student's t Probabilities** applet. Select the proper df using the slider on the right and use either the one- or two-tailed applet, depending on the type of rejection region. Type the positive value of α into the box marked "prob:" and press enter. The answers are shown below.
a $t > 1.8$
b $|t| > 2.37$
c $t < -2.6$

10.119 a Use the **Interpreting Confidence Intervals** applet. Answers will vary from student to student.
b The widths of the ten intervals will not be the same, since the value of s changes with each new sample.
c The student should find that approximately 95% of the intervals in the first applet contain μ.
d. Roughly 99% of the intervals in the second applet contain μ.

10.121 Use the **Small Sample Test of a Population Mean** applet. A screen capture is shown below. The test statistic

$$t = \frac{\bar{x} - \mu}{s/\sqrt{n}} = \frac{47.1 - 48}{\sqrt{\frac{4.7}{20}}} = -1.438$$

has a two-tailed *p*-value of .1782 and H_0 is not rejected. There is insufficient evidence to indicate that μ differs from 48.

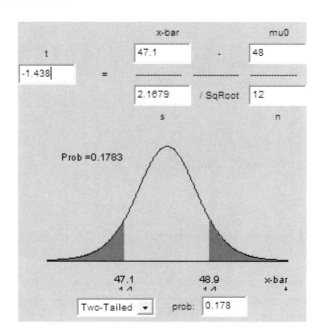

10.123 Use the **Two Sample t Test: Independent Samples** applet. The hypothesis to be tested concerns the differences between mean recovery rates for the two surgical procedures. Let μ_1 be the population mean for Procedure I and μ_2 be the population mean for Procedure II. The hypothesis to be tested is

$$H_0 : \mu_1 - \mu_2 = 0 \text{ versus } H_a : \mu_1 - \mu_2 \neq 0$$

Since the ratio of the variances is less than 3, you can use the pooled *t* test. Enter the appropriate statistics into the applet and you will find that test statistic is

$$t = \frac{(\bar{x}_1 - \bar{x}_2) - 0}{\sqrt{s^2\left(\frac{1}{n_1} + \frac{1}{n_2}\right)}} = -3.33$$

with a two-tailed *p*-value of .0030. Since the *p*-value is very small, H_0 can be rejected for any value of α greater than .003 and the results are judged highly significant. There is sufficient evidence to indicate a difference in the mean recovery rates for the two procedures.

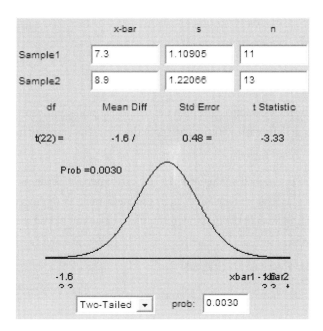

Case Study:
How does Bait Type Affect the Visit of the American Marten in Ontario?

1 A two sampled *t*-test is used to test $H_0: \mu_1 - \mu_2 = 0$ versus $H_a: \mu_1 - \mu_2 < 0$, where μ_1 is the mean number of days until a track box with jam-lard-fish oil mixture bait was visited and μ_2 is the mean number of days until a track box with chicken bait was visited. Using Minitab to perform the two sampled *t*-test, you obtain the following printout.

```
Two-Sample T-Test and CI: Jam, Chicken
Two-sample T for Jam vs Chicken
          N   Mean   StDev   SE Mean
Jam       5   3.80   1.79    0.80
Chicken   5   4.40   1.67    0.75

Difference = mu (Jam) - mu (Chicken)
Estimate for difference:  -0.60
95% upper bound for difference:  1.48
T-Test of difference = 0 (vs <): T-Value = -0.55   P-Value = 0.300   DF = 7
```

The test statistic of $t = -0.55$ with *p*-value = .300 indicates that the results are not significant. There is not enough evidence to indicate that Jam-lard-fish oil mixture attracts the martens more efficiently than chicken does.

2 A two sampled *t*-test is used to test $H_0: \mu_1 - \mu_2 = 0$ versus $H_a: \mu_1 - \mu_2 < 0$, where μ_1 is the mean number of days until a track box with peanut butter bait was visited and μ_2 is the mean number of days until a track box with control was visited. Using Minitab to perform the two sampled *t*-test, you obtain the following printout.

```
Two-Sample T-Test and CI: Peanut, Control
Two-sample T for Peanut vs Control
          N   Mean   StDev   SE Mean
Peanut    5   5.00   1.41    0.63
Control   5   5.60   1.14    0.51

Difference = mu (Peanut) - mu (Control)
Estimate for difference:  -0.600
95% upper bound for difference:  0.939
T-Test of difference = 0 (vs <): T-Value = -0.74   P-Value = 0.242   DF = 7
```

The test statistic of $t = -0.74$ with *p*-value = .242 indicates that the results are not significant. There is not enough evidence to indicate that American martens have a better taste for peanut butter.

3 From the printout below, the 95% confidence interval for the difference in days taken for the marten to visit a chicken track box and a peanut butter track box is
$$-2.917 < \mu_1 - \mu_2 < 1.717$$

Two-Sample T-Test and CI: Chicken, Peanut
```
Two-sample T for Chicken vs Peanut
          N    Mean   StDev   SE Mean
Chicken   5    4.40   1.67    0.75
Peanut    5    5.00   1.41    0.63

Difference = mu (Chicken) - mu (Peanut)
Estimate for difference:  -0.600
95% CI for difference:  (-2.917, 1.717)
T-Test of difference = 0 (vs not =): T-Value = -0.61  P-Value = 0.560  DF = 7
```

Based on this interval, which contains the value $\mu_1 - \mu_2 = 0$, there is no evidence to indicate the significant difference.

4 Based on the analysis of these variables, no bait seems significantly more effective than the other.

Project 10: Watch your Sugar Level!

a To test whether the blood glucose level is higher than 7.0 mmol/L we have the following null and alternative hypotheses

H_0: µ = 7.0

H_a: µ > 7.0

Since we have a small sample size ($n=7$), we use a *t*-test statistic to test our hypotheses. Specifically, we have

$$\begin{aligned} t_{test} &= \frac{\bar{x} - \mu}{\frac{s}{\sqrt{n}}} \\ &= \frac{7.014286 - 7}{\frac{0.4913538}{\sqrt{7}}} \\ &= 0.07692308 \end{aligned}$$

where

$$\begin{aligned} \bar{x} &= \frac{1}{n}\sum_{i=1}^{n} x_i \\ &= 7.014286 \\ s^2 &= \frac{1}{n-1}\sum_{i=1}^{n}(x_i - \bar{x})^2 \\ &= 0.2414286 \end{aligned}$$

Our degrees of freedom for this test are n-1=7-1=6. Using Table 4 in Appendix I, we find that the approximate *p*-value for this test is P(t>0.07692308)>0.10. At the α=0.05 level, we would fail to reject the null hypothesis. There is insufficient evidence to suggest that the blood glucose level is higher than 7.0mmol/L. The test fails to support the alternative hypothesis.

A 95% confidence interval is calculated as

$$\bar{x} \pm t_{\frac{\alpha}{2}, n-1} \frac{s}{\sqrt{n}}$$

$$7.014286 \pm 2.447 \frac{0.4913538}{\sqrt{7}}$$

Which gives us an interval of (6.559859, 7.468712). Since our interval covers the null hypothesized value of 7.0mmol/L, we do not have sufficient evidence to reject the null.

b To test whether the blood glucose level is higher than 7.0 mmol/L we have the following null and alternative hypotheses

H_0: $\sigma^2 = 0.8$

H_a: $\sigma^2 < 0.8$

To test the null hypothesis, we calculate a χ^2 test statistic. Specifically

$$\chi^2_{test} = \frac{(n-1)s^2}{\sigma_0^2}$$
$$= \frac{6 \cdot 0.2414286}{0.8}$$
$$= 1.810714$$

The test is rejected if our test statistic exceeds χ^2 on n-1=6 degrees of freedom (with α=0.05). Thus, we reject the null hypothesis in favour of the alternative if our test statistic exceeds 12.59159. Since our test statistic is less than the critical value, we do not reject the null hypothesis. There is insufficient evidence to suggest the variance is less than 0.8.

The p-value is determined using Table 5 in Appendix I. On 6 degrees of freedom, the *p*-value would be $P(\chi^2 > 1.810714) > 0.10$.

A confidence interval for the population variance is

$$\frac{(n-1)s^2}{\chi^2_{\frac{\alpha}{2},n-1}} < \sigma^2 < \frac{(n-1)s^2}{\chi^2_{1-\frac{\alpha}{2},n-1}}$$
$$\frac{6 \cdot 0.2414286}{12.59159} < \sigma^2 < \frac{6 \cdot 0.2414286}{1.635383}$$
$$0.1150428 < \sigma^2 < 0.885769$$

c To compare the treatments (placebo vs. treatment), we can assume that the population variances are the same. This is the case since the ratio of the sample variances is less than 3 (i.e., 0.21/0.17<3). The null and alternative hypothesis are

H_0: $\mu_1 = \mu_2$ or $\mu_1 - \mu_2 = 0$

H_a: $\mu_1 > \mu_2$ or $\mu_1 - \mu_2 > 0$

Here μ_1 represents the population mean for the placebo group, and μ_2 represents the population mean for the treatment group. To test the null hypothesis, we calculate an independent samples *t*-test statistic. Since we have assumed equal variances, we calculate the pooled estimate of variance. Specifically, we have

$$s_p^2 = \frac{(n_1-1)s_1^2 + (n_2-1)s_2^2}{n_1 + n_2 - 2}$$
$$= \frac{9 \cdot 0.21 + 9 \cdot 0.17}{10 + 10 - 2}$$
$$= 0.19$$

Introduction to Probability and Statistics, 2ce

And thus our test statistic is

$$t_{test} = \frac{\bar{x}_1 - \bar{x}_2}{\sqrt{s_p^2 \left(\frac{1}{n_1} + \frac{1}{n_2}\right)}}$$

$$= \frac{6.5 - 5.4}{\sqrt{0.19 \left(\frac{1}{10} + \frac{1}{10}\right)}}$$

$$= 5.642881$$

The critical *t* value on $n_1+n_2-2=18$ degrees of freedom (with α=0.05) is 1.734064. Our decision rule is to reject the null hypothesis if our test statistic exceeds the critical value. Since our test statistic 5.642881>1.734064, we reject the null hypothesis. There is sufficient evidence to suggest that the strict diet plan lowers the blood glucose level.

The approximate *p*-value for this test is obtained from Table 4 of Appendix I. With 18 degrees of freedom we have P(t>5.642881) < 0.005.

Since our p-value < 0.005 is less than α=0.10, we would reject the null hypothesis. That is, the blood glucose levels following treatment are significantly lower than the blood glucose levels for the placebo group.

A 95% confidence interval for the difference in the two means is

$$\bar{x}_1 - \bar{x}_2 \ \pm \ t_{\frac{\alpha}{2}, n_1+n_2-2} \sqrt{s_p^2 \left(\frac{1}{n_1} + \frac{1}{n_2}\right)}$$

$$1.1 \ \pm \ 2.101 \sqrt{0.19 \left(\frac{1}{10} + \frac{1}{10}\right)}$$

which gives (0.6904549, 1.5095451). Since the interval does not cover zero, we can say there is a significant difference between the two means. Further, since the interval for the difference between the placebo and treatment groups is positive, there is significant evidence to suggest that the placebo group has a higher blood glucose reading than the treatment group.

d   To test the equality of the variances we perform an F test. Our hypotheses are

$H_0: \sigma^2_1 = \sigma^2_2$

$H_a: \sigma^2_1 > \sigma^2_2$

Where σ21 represents the population variance for the placebo group, and σ22 represents the population variance for the treatment group. Our F test statistic is

$$F_{test} = \frac{s_1^2}{s_2^2}$$

$$= \frac{0.21}{0.17}$$

$$= 1.235294$$

The decision to reject the null hypothesis will be made if the test statistic exceeds the critical F value on n1-1=9 and n2-1=9 degrees of freedom. Using Table 6 from Appendix I, our critical F value with α=0.10 is 2.44. Since our test statistic is less than 2.440340, we do not reject the null hypothesis. There is insufficient evidence to suggest that the variances are different.

A 90% confidence interval for the ratio of the two population variances is

$$\left(\frac{s_2^1}{s_2^2}\right)\frac{1}{F_{\frac{\alpha}{2},df1,df2}} < \frac{\sigma_1^2}{\sigma_2^2} < \left(\frac{s_2^1}{s_2^2}\right)F_{\frac{\alpha}{2},df2,df1}$$

$$\frac{1.235294}{3.18} < \frac{\sigma_1^2}{\sigma_2^2} < 1.235294 \cdot 3.18$$

$$0.388 < \frac{\sigma_1^2}{\sigma_2^2} < 3.927$$

Since the interval covers 1, we have insufficient evidence to suggest that the ratio of the variances is 1. Hence, we can assume that the variances are not significantly different.

e The samples are not independent since the readings made after medication will depend on the pre treatment results. With this in mind, performing a paired-difference t-test would be valid (assuming the data comes from a normal distribution).
The null and alternative hypotheses are thus based on the differences before and after medication. Specifically, we have

H_0: $\mu_d = 0$

H_a: $\mu_d > 0$, where $\mu_d = \mu_{before} - \mu_{after}$

$$1.2, 1.8, 0.4, 3.0, 2.2, 3.2, 4.0, 2.3, 2.9, 2.4$$

for each of the adults respectively. The sample mean and sample variance are

$$\begin{aligned}
\bar{x}_d &= \frac{1}{n}\sum_{i=1}^{n} x_i \\
&= \frac{1}{10}(1.2 + 1.8 + \ldots + 2.4) \\
&= 2.34 \\
s_d^2 &= \frac{1}{n-1}\sum_{i=1}^{n}(x_i - \bar{x}_d)^2 \\
&= \frac{1}{9}\left((1.2 - 2.34)^2 + (1.8 - 2.34)^2 + \ldots + (2.4 - 2.34)^2\right) \\
&= 1.069333
\end{aligned}$$

where $n=10$, the number of differences.

The test statistic is

$$\begin{aligned}
t_{test} &= \frac{\bar{x}_d - \mu_d}{\frac{s_d}{\sqrt{n}}} \\
&= \frac{2.34}{\frac{1.034086}{\sqrt{10}}} \\
&= 7.155817
\end{aligned}$$

Given the one sided alternative, we find the critical t value from Table 4 in Appendix I. With n-1=9 degrees of freedom, and α=0.05, we will reject the null hypothesis if our test statistic is greater than 1.833.

Since our test statistic is greater than the critical value, we reject the null hypothesis in favour of the alternative. There is sufficient evidence to suggest that the blood glucose levels are lower after treatment with medication.

The approximate p-value can be determined using Table 4 in Appendix I. In this case, with 9 degrees of freedom, we obtain a p-value = $P(t>7.155817) < 0.005$.

A 90% confidence interval for the differences in blood glucose level between the two groups is

$$\bar{x}_d \pm t_{\frac{\alpha}{2}, n-1} \frac{s_d}{\sqrt{n}}$$

$$2.34 \pm 1.833 \frac{1.034086}{\sqrt{10}}$$

which gives an interval of (1.740560, 2.939440). Since this interval does not cover zero, there is evidence to suggest that there is a significant difference in the average blood glucose levels between the two groups. This is consistent with the findings of the hypothesis test.

Chapter 11: The Analysis of Variance

11.1 In comparing 6 populations, there are $k-1$ degrees of freedom for treatments and $n = 6(10) = 60$. The ANOVA table is shown below.

Source	df
Treatments	5
Error	54
Total	59

11.5 a Refer to Exercise 11.4. The given sums of squares are inserted and missing entries found by subtraction. The mean squares are found as $MS = SS/df$.

Source	df	SS	MS	F
Treatments	3	339.8	113.267	16.98
Error	20	133.4	6.67	
Total	23	473.2		

b The F statistic, $F = MST/MSE$, has $df_1 = 3$ and $df_2 = 20$ degrees of freedom.

c With $\alpha = .05$ and degrees of freedom from **b**, H_0 is rejected if $F > F_{.05} = 3.10$.

d Since $F = 16.98$ falls in the rejection region, the null hypothesis is rejected. There is a difference among the means.

e The critical values of F with $df_1 = 3$ and $df_2 = 20$ (Table 6) for bounding the p-value for this one-tailed test are shown below.

α	.10	.05	.025	.01	.005
F_α	2.38	3.10	3.86	4.94	5.82

Since the observed value $F = 16.98$ is greater than $F_{.005}$, p-value $< .005$ and H_0 is rejected as in part **d**.

11.7 The following preliminary calculations are necessary:
$$T_1 = 14 \quad T_2 = 19 \quad T_3 = 5 \quad G = 38$$

a $CM = \dfrac{\left(\sum x_{ij}\right)^2}{n} = \dfrac{(38)^2}{14} = 103.142857$

Total $SS = \sum x_{ij}^2 - CM = 3^2 + 2^2 + \cdots + 2^2 + 1^2 - CM = 130 - 103.142857 = 26.8571$

b $SST = \sum \dfrac{T_i^2}{n_i} - CM = \dfrac{14^2}{5} + \dfrac{19^2}{5} + \dfrac{5^2}{4} - CM = 117.65 - 103.142857 = 14.5071$

and $MST = \dfrac{SST}{k-1} = \dfrac{14.5071}{2} = 7.2536$

c By subtraction, $SSE = $ Total $SS - SST = 26.8571 - 14.5071 = 12.3500$ and the degrees of freedom, by subtraction, are $13 - 2 = 11$. Then

$$MSE = \dfrac{SSE}{11} = \dfrac{12.3500}{11} = 1.1227$$

d The information obtained in parts **a-c** is consolidated in an ANOVA table.

Source	df	SS	MS
Treatments	2	14.5071	7.2536
Error	11	12.3500	1.1227
Total	13	26.8571	

e The hypothesis to be tested is
$$H_0 : \mu_1 = \mu_2 = \mu_3 \quad \text{versus} \quad H_a : \text{at least one pair of means are different}$$

f The rejection region for the test statistic $F = \dfrac{\text{MST}}{\text{MSE}} = \dfrac{7.2536}{1.1227} = 6.46$ is based on an F-distribution with 2 and 11 degrees of freedom. The critical values of F for bounding the p-value for this one-tailed test are shown below.

α	.10	.05	.025	.01	.005
F_α	2.86	3.98	5.26	7.21	8.91

Since the observed value $F = 6.46$ is between $F_{.01}$ and $F_{.025}$, $.01 < p$-value $< .025$ and H_0 is rejected at the 5% level of significance. There is a difference among the means.

11.9 **a** The 90% confidence interval for μ_1 is
$$\bar{x}_1 \pm t_{.05}\sqrt{\frac{\text{MSE}}{n_1}} \Rightarrow 2.8 \pm 1.796\sqrt{\frac{1.1227}{5}} \Rightarrow 2.8 \pm .85$$

or $1.95 < \mu_1 < 3.65$.

b The 90% confidence interval for $\mu_1 - \mu_3$ is
$$(\bar{x}_1 - \bar{x}_3) \pm t_{.05}\sqrt{\text{MSE}\left(\frac{1}{n_1} + \frac{1}{n_3}\right)}$$
$$(2.8 - 1.25) \pm 1.796\sqrt{1.1227\left(\frac{1}{5} + \frac{1}{4}\right)}$$
$$1.55 \pm 1.28 \quad \text{or} \quad .27 < \mu_1 - \mu_3 < 2.83$$

11.11 **a** The 95% confidence interval for μ_A is
$$\bar{x}_A \pm t_{.025}\sqrt{\frac{\text{MSE}}{n_A}} \Rightarrow 76 \pm 2.306\sqrt{\frac{62.333}{5}} \Rightarrow 76 \pm 8.142$$

or $67.86 < \mu_A < 84.14$.

b The 95% confidence interval for μ_B is
$$\bar{x}_B \pm t_{.025}\sqrt{\frac{\text{MSE}}{n_B}} \Rightarrow 66.33 \pm 2.306\sqrt{\frac{62.333}{3}} \Rightarrow 66.33 \pm 10.51$$

or $55.82 < \mu_B < 76.84$.

c The 95% confidence interval for $\mu_A - \mu_B$ is
$$(\bar{x}_A - \bar{x}_B) \pm t_{.025}\sqrt{\text{MSE}\left(\frac{1}{n_A} + \frac{1}{n_B}\right)}$$
$$(76 - 66.33) \pm 2.306\sqrt{62.333\left(\frac{1}{5} + \frac{1}{3}\right)}$$
$$9.667 \pm 13.296 \quad \text{or} \quad -3.629 < \mu_A - \mu_B < 22.963$$

d Note that these three confidence intervals cannot be jointly valid because all three employ the same value of $s = \sqrt{\text{MSE}}$ and are dependent.

11.13 **a** We would be reasonably confident that the data satisfied the normality assumption because each measurement represents the average of 10 continuous measurements. The Central Limit Theorem assures us that this mean will be approximately normally distributed.

b We have a completely randomized design with four treatments, each containing 6 measurements. The analysis of variance table is given in the *Minitab* printout. The F test is

$$F = \frac{\text{MST}}{\text{MSE}} = \frac{6.580}{.115} = 57.38$$

with p-value = .000 (in the column marked "P"). Since the p-value is very small (less than .01), H_0 is rejected. There is a significant difference in the mean leaf length among the four locations with $P < .01$ or even $P < .001$.

c The hypothesis to be tested is $H_0: \mu_1 = \mu_4$ versus $H_a: \mu_1 \neq \mu_4$ and the test statistic is

$$t = \frac{\bar{x}_1 - \bar{x}_4}{\sqrt{\text{MSE}\left(\dfrac{1}{n_1} + \dfrac{1}{n_4}\right)}} = \frac{6.0167 - 3.65}{\sqrt{.115\left(\dfrac{1}{6} + \dfrac{1}{6}\right)}} = 12.09$$

The p-value with $df = 20$ is $2P(t > 12.09)$ is bounded (using Table 4) as

$$p\text{-value} < 2(.005) = .01$$

and the null hypothesis is rejected. We conclude that there is a difference between the means.

d The 99% confidence interval for $\mu_1 - \mu_4$ is

$$(\bar{x}_1 - \bar{x}_4) \pm t_{.005}\sqrt{\text{MSE}\left(\dfrac{1}{n_1} + \dfrac{1}{n_4}\right)}$$

$$(6.0167 - 3.65) \pm 2.845\sqrt{.115\left(\dfrac{1}{6} + \dfrac{1}{6}\right)}$$

$$2.367 \pm .557 \quad \text{or} \quad 1.810 < \mu_1 - \mu_4 < 2.924$$

e When conducting the t tests, remember that the stated confidence coefficients are based on random sampling. If you looked at the data and only compared the largest and smallest sample means, the randomness assumption would be disturbed.

11.15 The design is completely randomized with 3 treatments and 5 replications per treatment. The *Minitab* printout on the next page shows the analysis of variance for this experiment.

One-way ANOVA: Calcium versus Method
```
Source   DF         SS         MS       F      P
Method    2  0.0000041  0.0000021   16.38  0.000
Error    12  0.0000015  0.0000001
Total    14  0.0000056

S = 0.0003545   R-Sq = 73.19%   R-Sq(adj) = 68.72%

Individual 95% CIs For Mean Based on
                            Pooled StDev
Level  N     Mean      StDev  -+---------+---------+---------+--------
1      5  0.027620  0.000421                   (------*------)
2      5  0.026780  0.000396  (------*------)
3      5  0.028040  0.000207                              (------*------)
                              -+---------+---------+---------+--------
                            0.02650    0.02700    0.02750    0.02800

Pooled StDev = 0.000354
```

The test statistic, $F = 16.38$ with p-value $= .000$ indicates the results are highly significant; there is a difference in the mean calcium contents for the three methods. All assumptions appear to have been satisfied.

11.17 **a** The design is a completely randomized design (four independent samples).
   b The following preliminary calculations are necessary:

$$T_1 = 1211 \quad T_2 = 1074 \quad T_3 = 1158 \quad T_4 = 1243 \quad G = 4686$$

$$\text{CM} = \frac{\left(\sum x_{ij}\right)^2}{n} = \frac{(4686)^2}{20} = 1,097,929.8 \quad \text{Total SS} = \sum x_{ij}^2 - \text{CM} = 1,101,862 - \text{CM} = 3932.2$$

$$\text{SST} = \sum \frac{T_i^2}{n_i} - \text{CM} = \frac{1211^2}{5} + \frac{1074^2}{5} + \frac{1158^2}{5} + \frac{1243^2}{5} - \text{CM} = 3272.2$$

Calculate $\text{MS} = \text{SS}/df$ and consolidate the information in an ANOVA table.

Source	df	SS	MS
Treatments	3	3272.2	1090.7333
Error	16	660	41.25
Total	19	3932.2	

   c The hypothesis to be tested is
$$H_0 : \mu_1 = \mu_2 = \mu_3 = \mu_4 \quad \text{versus} \quad H_a : \text{at least one pair of means are different}$$
and the F test to detect a difference in average prices is
$$F = \frac{\text{MST}}{\text{MSE}} = 26.44.$$

The rejection region with $\alpha = .05$ and 3 and 16 df is approximately $F > 3.24$ and H_0 is rejected. There is enough evidence to indicate a difference in the average prices for the four provinces.

11.19 Sample means must be independent and based upon samples of equal size.

11.21 **a** $\omega = q_{.05}(4,12)\dfrac{s}{\sqrt{5}} = 4.20\dfrac{s}{\sqrt{5}} = 1.878s$

   b $\omega = q_{.01}(6,12)\dfrac{s}{\sqrt{8}} = 6.10\dfrac{s}{\sqrt{8}} = 2.1567s$

11.23 With $k = 4$, $df = 20$, $n_t = 6$,
$$\omega = q_{.01}(4,20)\sqrt{\frac{\text{MSE}}{n_t}} = 5.02\sqrt{\frac{.115}{6}} = .69$$

The ranked means are shown below.

$$\underline{\begin{array}{cccc} 6.0167 & 5.65 & 5.35 & 3.65 \\ \overline{x}_1 & \overline{x}_2 & \overline{x}_3 & \overline{x}_4 \end{array}}$$

11.25 The design is completely randomized with 3 treatments and 5 replications per treatment. The Minitab printout below shows the analysis of variance for this experiment.

```
One-way ANOVA: mg/dl versus Lab
Source  DF     SS    MS     F      P
Lab      2   42.6  21.3  0.60  0.562
Error   12  422.5  35.2
Total   14  465.0

S = 5.933   R-Sq = 9.15%   R-Sq(adj) = 0.00%

                           Individual 95% CIs For Mean Based on
                           Pooled StDev
Level   N    Mean   StDev  --+---------+---------+---------+-------
1       5  108.86    7.47                       (-------------*--------------)
2       5  105.04    6.01  (--------------*-------------)
3       5  105.60    3.70     (-------------*------------)
                           --+---------+---------+---------+-------
                           100.0     104.0     108.0     112.0
Pooled StDev = 5.93

Tukey 95% Simultaneous Confidence Intervals
All Pairwise Comparisons among Levels of Lab
Individual confidence level = 97.94%

Lab = 1 subtracted from:
Lab    Lower   Center   Upper    +---------+---------+---------+---------
2    -13.824   -3.820   6.184    (--------------*-------------)
3    -13.264   -3.260   6.744     (-------------*--------------)
                                 +---------+---------+---------+---------
                               -14.0      -7.0      0.0       7.0

Lab = 2 subtracted from:
Lab    Lower   Center   Upper    +---------+---------+---------+---------
3     -9.444    0.560  10.564             (--------------*-------------)
                                 +---------+---------+---------+---------
                               -14.0      -7.0      0.0       7.0
```

a The analysis of variance F test for $H_0 : \mu_1 = \mu_2 = \mu_3$ is $F = .60$ with p-value $= .562$. The results are not significant and H_0 is not rejected. There is insufficient evidence to indicate a difference in the treatment means.

b Since the treatment means are not significantly different, there is no need to use Tukey's test to search for the pairwise differences. Notice that all three intervals generated by *Minitab* contain zero, indicating that the pairs cannot be judged different.

11.27 **a** The following preliminary calculations are necessary:

$$T_1 = 2835 \quad T_2 = 3300 \quad T_3 = 2724 \quad G = 8859$$

$$CM = \frac{(\sum x_{ij})^2}{n} = \frac{(8859)^2}{15} = 5232125.4 \quad \text{Total SS} = \sum x_{ij}^2 - CM = 5,295,693 - CM = 63567.6$$

$$SST = \sum \frac{T_i^2}{n_i} - CM = \frac{2835^2}{5} + \frac{3300^2}{5} + \frac{2724^2}{5} - CM = 37354.8$$

Calculate $MS = SS/df$ and consolidate the information in an ANOVA table.

Source	df	SS	MS
Treatments	2	37354.8	18677.4
Error	12	26212.8	2184.4
Total	14	63567.6	

c The hypothesis to be tested is
$$H_0: \mu_1 = \mu_2 = \mu_3 \quad \text{versus} \quad H_a: \text{at least one pair of means are different}$$
and the F test to detect a difference in average scores is
$$F = \frac{MST}{MSE} = 8.55.$$
The rejection region with $\alpha = .05$ and 2 and 12 df is approximately $F > 3.89$ and H_0 is rejected. There is evidence of a difference in the average scores for the three graduate programs.

b The 95% confidence interval for $\mu_1 - \mu_2$ is
$$(\bar{x}_1 - \bar{x}_2) \pm t_{.025}\sqrt{MSE\left(\frac{1}{n_1} + \frac{1}{n_2}\right)}$$
$$\left(\frac{2835}{5} - \frac{3300}{5}\right) \pm 2.179\sqrt{2184.4\left(\frac{1}{5} + \frac{1}{5}\right)}$$
$$-93 \pm 64.41 \quad \text{or} \quad -157.41 < \mu_1 - \mu_2 < -28.59$$

c With $k = 3$, $df = 12$, $n_t = 5$,
$$\omega = q_{.05}(3,12)\frac{\sqrt{MSE}}{\sqrt{n_t}} = 3.77\sqrt{\frac{2184.4}{5}} = 78.80$$
The ranked means are shown below.

544.8	567	660
$\bar{x}_3$	$\bar{x}_1$	$\bar{x}_2$

There is no significant difference between programs 1 and 3, but programs 1 and 2, 2, and 3 are different from each other.

11.29 Refer to Exercise 11.28. The given sums of squares are inserted and missing entries found by subtraction. The mean squares are found as $MS = SS/df$.

Source	df	SS	MS	F
Treatments	2	11.4	5.70	4.01
Blocks	5	17.1	3.42	2.41
Error	10	14.2	1.42	
Total	17	42.7		

11.31 The 95% confidence interval for $\mu_A - \mu_B$ is then
$$(\bar{x}_A - \bar{x}_B) \pm t_{.025}\sqrt{MSE\left(\frac{2}{b}\right)}$$
$$(21.9 - 24.2) \pm 2.228\sqrt{1.42\left(\frac{2}{6}\right)}$$
$$-2.3 \pm 1.533 \quad \text{or} \quad -3.833 < \mu_A - \mu_B < -.767$$

11.33 Use *Minitab* to obtain an ANOVA printout, or use the following calculations:

$$CM = \frac{\left(\sum x_{ij}\right)^2}{n} = \frac{(113)^2}{12} = 1064.08333$$

Total SS $= \sum x_{ij}^2 - CM = 6^2 + 10^2 + \cdots + 14^2 - CM = 1213 - CM = 148.91667$

$$SST = \sum \frac{T_j^2}{3} - CM = \frac{22^2 + 34^2 + 27^2 + 30^2}{3} - CM = 25.58333$$

$$SSB = \sum \frac{B_i^2}{4} - CM = \frac{33^2 + 25^2 + 55^2}{4} - CM = 120.66667 \text{ and}$$

SSE = Total SS − SST − SSB = 2.6667

Calculate MS = SS/df and consolidate the information in an ANOVA table.

Source	df	SS	MS	F
Treatments	3	25.5833	8.5278	19.19
Blocks	2	120.6667	60.3333	135.75
Error	6	2.6667	0.4444	
Total	11	148.9167		

a To test the difference among treatment means, the test statistic is

$$F = \frac{MST}{MSE} = \frac{8.528}{.4444} = 19.19$$

and the rejection region with $\alpha = .05$ and 3 and 6 df is $F > 4.76$. There is a significant difference among the treatment means.

b To test the difference among block means, the test statistic is

$$F = \frac{MSB}{MSE} = \frac{60.3333}{.4444} = 135.75$$

and the rejection region with $\alpha = .05$ and 2 and 6 df is $F > 5.14$. There is a significant difference among the block means.

c With $k = 4$, $df = 6$, $n_t = 3$,

$$\omega = q_{.01}(4,6)\sqrt{\frac{MSE}{n_t}} = 7.03\sqrt{\frac{.4444}{3}} = 2.71$$

The ranked means are shown below.

7.33	9.00	10.00	11.33
$\bar{x}_1$	$\bar{x}_3$	$\bar{x}_4$	$\bar{x}_2$

d The 95% confidence interval is

$$(\bar{x}_A - \bar{x}_B) \pm t_{.025}\sqrt{MSE\left(\frac{2}{b}\right)}$$

$$(7.333 - 11.333) \pm 2.447\sqrt{.4444\left(\frac{2}{3}\right)}$$

$$-4 \pm 1.332 \quad \text{or} \quad -5.332 < \mu_A - \mu_B < -2.668$$

e Since there is a significant difference among the block means, blocking has been effective. The variation due to block differences can be isolated using the randomized block design.

11.35 **a** By subtraction, the degrees of freedom for blocks is $b - 1 = 34 - 28 = 6$. Hence, there are $b = 7$ blocks.
  b There are always $b = 7$ observations in a treatment total.
  c There are $k = 4 + 1 = 5$ observations in a block total.
  d

Source	df	SS	MS	F
Treatments	4	14.2	3.55	9.68
Blocks	6	18.9	3.15	8.59
Error	24	8.8	0.3667	
Total	34	41.9		

  e To test the difference among treatment means, the test statistic is
$$F = \frac{\text{MST}}{\text{MSE}} = \frac{3.55}{.3667} = 9.68$$
and the rejection region with $\alpha = .05$ and 4 and 24 df is $F > 2.78$. There is a significant difference among the treatment means.

  f To test the difference among block means, the test statistic is
$$F = \frac{\text{MSB}}{\text{MSE}} = \frac{3.15}{.3667} = 8.59$$
and the rejection region with $\alpha = .05$ and 6 and 24 df is $F > 2.51$. There is a significant difference among the block means.

11.37 Similar to previous exercises. The *Minitab* printout for this randomized block experiment is shown below.

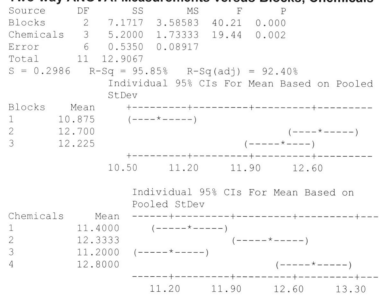

Both the treatment and block means are significantly different. Since the four chemicals represent the treatments in this experiment, Tukey's test can be used to determine where the differences lie:
$$\omega = q_{.05}(4,6)\frac{\sqrt{\text{MSE}}}{\sqrt{n_t}} = 4.90\sqrt{\frac{.08917}{3}} = .845$$

The ranked means are shown below.   11.20   11.40   12.33   12.80
                                     $\overline{x}_3$ $\overline{x}_1$ $\overline{x}_2$ $\overline{x}_4$

The chemical falls into two significantly different groups – A and C versus B and D.

11.39 The factor of interest is "soil preparation", and the blocking factor is "locations". A randomized block design is used and the analysis of variance table can be obtained using the computer printout.

a The F statistic to detect a difference due to soil preparations is
$$F = \frac{\text{MST}}{\text{MSE}} = 10.06$$
with p-value $= .012$. The null hypothesis can be rejected at the 5% level of significance; there is a significant difference among the treatment means.

b The F statistic to detect a difference due to locations is
$$F = \frac{\text{MSB}}{\text{MSE}} = 10.88$$
with p-value $= .008$. The null hypothesis can be rejected at the 1% level of significance; there is a highly significant difference among the block means.

c Tukey's test can be used to determine where the differences lie:
$$\omega = q_{.01}(3,6)\frac{\sqrt{\text{MSE}}}{\sqrt{n_t}} = 6.33\sqrt{\frac{1.8889}{4}} = 4.35$$

The ranked means are shown below.

$$\begin{array}{ccc} 12.0 & 12.5 & 16.0 \\ \overline{x}_3 & \overline{x}_1 & \overline{x}_2 \end{array}$$

Preparations 2 and 3 are the only two treatments that can be declared significantly different.

d The 95% confidence interval is
$$(\overline{x}_B - \overline{x}_A) \pm t_{.025}\sqrt{\text{MSE}\left(\frac{2}{b}\right)}$$
$$(16.5 - 12.5) \pm 2.447\sqrt{1.89\left(\frac{2}{4}\right)}$$
$$3.5 \pm 2.38 \quad \text{or} \quad 1.12 < \mu_B - \mu_A < 5.88$$

11.41 A randomized block design has been used with "estimators" as treatments and "construction job" as the block factor. The analysis of variance table is found in the *Minitab* printout below..

Two-way ANOVA: Cost versus Estimator, Job
```
Source      DF      SS        MS       F      P
Estimator    2   10.8617   5.4308   7.20   0.025
Job          3   37.6073  12.5358  16.61   0.003
Error        6    4.5283   0.7547
Total       11   52.9973
S = 0.8687    R-Sq = 91.46%    R-Sq(adj) = 84.34%

                     Individual 95% CIs For Mean Based on
                     Pooled StDev
Estimator    Mean    -------+---------+---------+---------+--
A          32.6125   (--------*--------)
B          34.8875                            (--------*--------)
C          34.1875                    (--------*--------)
                     -------+---------+---------+---------+--
                         32.4      33.6      34.8      36.0
```

Both treatments and blocks are significant. The treatment means can be further compared using Tukey's test with

$$\omega = q_{.05}(3,6)\frac{\sqrt{MSE}}{\sqrt{n_t}} = 4.34\sqrt{\frac{.7547}{4}} = 1.885$$

The ranked means are shown below.

$$\begin{array}{ccc} 32.6125 & 34.1875 & 34.8875 \\ \overline{x}_A & \overline{x}_C & \overline{x}_B \end{array}$$

Estimators A and B show a significant difference in average costs.

11.43 **a** The complete ANOVA table is shown below. Since factor A is run at 3 levels, it must have 2 *df*. Other entries are found by similar reasoning.

Source	df	SS	MS	F
A	2	5.3	2.6500	1.30
B	3	9.1	3.0333	1.49
A × B	6	4.8	0.8000	0.39
Error	12	24.5	2.0417	
Total	23	43.7		

b The test statistic is $F = \text{MS}(AB)/\text{MSE} = 0.39$ and the rejection region is $F > 3.00$. Hence, H_0 is not rejected. There is insufficient evidence to indicate interaction between A and B.

c The test statistic for testing factor A is $F = 1.30$ with $F_{.05} = 3.89$. The test statistic for factor B is $F = 1.49$ with $F_{.05} = 3.49$. Neither A nor B are significant.

11.45 **a** The nine treatment (cell) totals needed for calculation are shown in the table.

	Factor A			
Factor B	*1*	*2*	*3*	*Total*
1	12	16	10	38
2	15	25	17	57
3	25	17	27	69
Total	52	58	54	164

$$\text{CM} = \frac{164^2}{18} = 1492.2222 \qquad \text{Total SS} = 1662 - \text{CM} = 167.7778$$

$$\text{SSA} = \frac{52^2 + 58^2 + 54^2}{6} - \text{CM} = 3.1111 \qquad \text{SSB} = \frac{38^2 + 57^2 + 69^2}{6} - \text{CM} = 81.4444$$

$$\text{SS}(AB) = \frac{12^2 + 16^2 + \cdots + 27^2}{2} - \text{SSA} - \text{SSB} - \text{CM} = 62.2222$$

Source	df	SS	MS	F
A	2	3.1111	1.5556	.67
B	2	81.4444	40.7222	2.02
A × B	4	62.2222	15.5556	6.67
Error	9	21.0000	2.3333	
Total	17	167.7778		

b-c The test statistic is $F = \text{MS}(AB)/\text{MSE} = 6.67$ and the rejection region is $F > 3.63$. There is evidence of a significant interaction. That is, the effect of factor A depends upon the level of factor B at which A is measured.

d Since $F = 6.67$ lies between $F_{.01}$ and $F_{.005}$, $.005 < p\text{-value} < .01$.

e Since the interaction is significant, the differences in the four factor-level combinations should be explored individually, using an interaction plot such as the one generated by *Minitab* below.

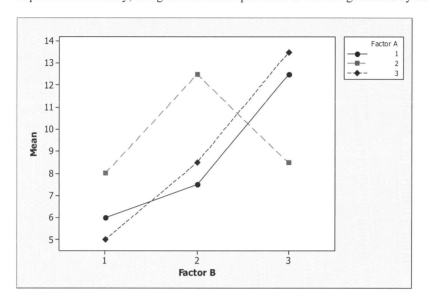

Look at the differences between the three levels of factor A when factor B changes from level 1 to level 2. Levels 2 and 3 behave very similarly while level 1 behaves quite differently. When factor B changes from level 2 to level 3, levels 1 and 3 of factor A behave similarly, and level 2 behaves differently.

11.47 Use the computing formulas given in this section or a computer software package to generate the ANVOA table for this 2×3 factorial experiment. The following printout was generated using *Minitab*.

Two-way ANOVA: Percent Gain versus Markup, Location
```
Source         DF        SS        MS        F       P
Markup          2     835.17   417.583    11.87   0.008
Location        1     280.33   280.333     7.97   0.030
Interaction     2      85.17    42.583     1.21   0.362
Error           6     211.00    35.167
Total          11    1411.67
S = 5.930    R-Sq = 85.05%    R-Sq(adj) = 72.60%
```

a From the printout, $F = 1.21$ with $p\text{-value} = .362$. Hence, at the $\alpha = .05$ level, H_0 is not rejected. There is insufficient evidence to indicate interaction.

b Since no interaction is found, the effects of A and B can be tested individually. Both A and B are significant.

c The interaction plot generated by *Minitab* is shown on the next page. Notice that the lines, although not exactly parallel, do not indicate a significant difference in the behavior of the mean responses for the two different locations.

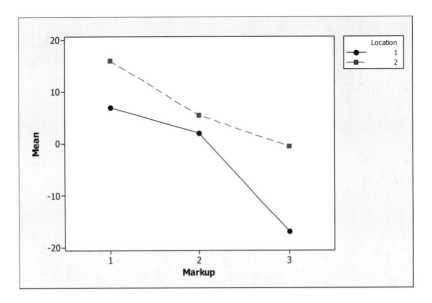

d The 95% confidence interval is

$$(\bar{x}_{31} - \bar{x}_{32}) \pm t_{.025}\sqrt{MSE\left(\frac{2}{r}\right)}$$

$$(-17 + .5) \pm 2.447\sqrt{35.167\left(\frac{2}{2}\right)}$$

$$-16.5 \pm 14.51 \quad \text{or} \quad -31.01 < \mu_{31} - \mu_{32} < -1.99$$

11.49 Answers will vary from student to student. There is no significant interaction, nor is the main effect for cities significant. There is a significant difference in the average cost per mile based on the distance traveled, with the cost per mile decreasing as the distance increases. Perhaps a straight line may model the costs as a function of time.

11.51 **a** The experimental units are the supervisors.
b The two factors are the training method (trained or untrained) and the situation (standard or emergency).
c There are two levels of each factor.
d There are $2 \times 2 = 4$ treatments.
e The design is a 2×2 factorial experiment, with 4 replications per treatment.

11.53 The design is completely randomized with five treatments, containing four, seven, six, five and five measurements, respectively. The analysis of variance table can be found using the computer printout or the following calculations:

$$CM = \frac{\left(\sum x_{ij}\right)^2}{n} = \frac{(20.6)^2}{27} = 15.717$$

Total SS $= \sum x_{ij}^2 - CM = 17.500 - CM = 1.783$

$$SST = \sum \frac{T_i^2}{n_i} - CM = \frac{(2.5)^2}{4} + \frac{(4.7)^2}{7} + \cdots + \frac{(2.4)^2}{5} - CM = 1.212$$

SSE = Total SS $-$ SST $= .571$

a The F test is $F = 11.67$ with p-value $= .000$. The results are highly significant, and H_0 is rejected. There is a difference in mean reaction times due to the five stimuli.

b The hypothesis to be tested is $H_0: \mu_A = \mu_D$ versus $H_a: \mu_A \neq \mu_D$ and the test statistic is

$$t = \frac{\bar{x}_A - \bar{x}_D}{\sqrt{MSE\left(\frac{1}{n_A} + \frac{1}{n_D}\right)}} = \frac{.625 - .920}{\sqrt{.026\left(\frac{1}{4} + \frac{1}{5}\right)}} = -2.73$$

The rejection region with $\alpha = .05$ and 22 degrees of freedom is $|t| > t_{.025} = 2.074$ and the null hypothesis is rejected. We conclude that there is a difference between the means.

11.55 The residuals in the upper tail of the normal probability plot are smaller than expected, but overall, there is not a problem with normality. The spreads of the residuals when plotted against the fitted values is relatively constant.

11.57 A completely randomized design has been used. The analysis of variance table can be found using a computer program or the following calculations:

$$CM = \frac{\left(\sum x_{ij}\right)^2}{n} = \frac{(1161)^2}{40} = 33,698.025$$

Total SS = $\sum x_{ij}^2 - CM = 34,701 - CM = 1002.975$

$$SST = \sum \frac{T_i^2}{n_i} - CM = \frac{309^2}{10} + \frac{275^2}{10} + \frac{295^2}{10} + \frac{282^2}{10} - CM = 67.475$$

Calculate MS = SS/df and consolidate the information in an ANOVA table.

Source	df	SS	MS
Treatments	3	67.475	22.4917
Error	36	935.500	25.9861
Total	39	1002.975	

The Minitab computer printout is shown below.

One-way ANOVA: 10-19, 20-39, 40-59, 60-69
```
Source   DF      SS     MS     F      P
Factor    3    67.5   22.5   0.87   0.468
Error    36   935.5   26.0
Total    39  1003.0
S = 5.098    R-Sq = 6.73%    R-Sq(adj) = 0.00%

                                  Individual 95% CIs For Mean Based on
                                  Pooled StDev
Level    N    Mean   StDev   ---+---------+---------+---------+------
10-19   10  30.900   5.195                        (------------*------------)
        20-39  10  27.500   4.882        (------------*------------)
40-59   10  29.500   4.696               (------------*------------)
60-69   10  28.200   5.574          (------------*------------)
                                  ---+---------+---------+---------+------
                                  25.0       27.5      30.0      32.5

Pooled StDev = 5.098
```

a The F test for treatments is

$$F = \frac{MST}{MSE} = .87$$

with p-value = .468 and H_0 is not rejected. There is no evidence to suggest a difference among the four groups.

b The 90% confidence interval for $\mu_1 - \mu_4$ is

$$(\bar{x}_1 - \bar{x}_4) \pm t_{.05}\sqrt{\text{MSE}\left(\frac{1}{n_1} + \frac{1}{n_4}\right)}$$

$$(30.9 - 28.2) \pm 1.645\sqrt{25.9861\left(\frac{2}{10}\right)}$$

$$2.7 \pm 3.750 \quad \text{or} \quad -1.050 < \mu_1 - \mu_4 < 6.450$$

c The 90% confidence interval for μ_2 is

$$\bar{x}_2 \pm t_{.05}\sqrt{\frac{\text{MSE}}{n_2}} \Rightarrow 27.5 \pm 1.645\sqrt{\frac{25.9861}{10}} \Rightarrow 27.5 \pm 2.652$$

or $24.848 < \mu_2 < 30.152$.

d With B = 2, $\sigma^2 \approx \text{MSE}$ and $t \approx 2$, the necessary inequality is

$$2\sqrt{\frac{\text{MSE}}{n}} \leq 2 \Rightarrow \sqrt{n} \geq \sqrt{\text{MSE}} = \sqrt{25.9861} \quad \text{or} \quad n \geq 25.9861$$

Samples of size $n = 26$ will be required. In this case, the degrees of freedom associated with MSE will be $4n - 4 = 100$, which is large enough that $t \approx 2$ is a valid approximation.

11.59 This is similar to Exercise 11.42.
a-b There are $4 \times 2 = 8$ treatments and $4 \times 2 \times r = 8r$ total observations.
c The sources of variation and associated degrees of freedom are given below.

Source	df
A	3
B	1
A × B	3
Error	8r – 8
Total	8r – 1

11.61 Refer to Exercise 11.63. The 95% confidence interval is

$$(\bar{x}_1 - \bar{x}_2) \pm t_{.025}\sqrt{\text{MSE}\left(\frac{1}{n_1} + \frac{1}{n_2}\right)}$$

$$(3.7 - 1.4) \pm 2.064\sqrt{.175\left(\frac{2}{15}\right)}$$

$$2.3 \pm .315 \quad \text{or} \quad 1.985 < \mu_1 - \mu_2 < 2.615$$

11.63 The completely randomized design has been used. The analysis of variance table can be obtained using a computer program or the computing formulas.

One-way ANOVA: A, B, C, D
```
Source  DF    SS       MS      F      P
Factor   3  0.4649  0.1550  5.20  0.011
Error   16  0.4768  0.0298
Total   19  0.9417

S = 0.1726   R-Sq = 49.37%   R-Sq(adj) = 39.87%

                          Individual 95% CIs For Mean Based on
                          Pooled StDev
Level  N   Mean    StDev   -+---------+---------+---------+--------
A      5  1.5680  0.1366   (-------*--------)
B      5  1.7720  0.2160              (--------*-------)
C      5  1.5460  0.1592   (-------*-------)
D      5  1.9160  0.1689                     (-------*-------)
                          -+---------+---------+---------+--------
                         1.40      1.60      1.80      2.00
Pooled StDev = 0.1726
```

a To test the difference in treatment means, use $F = \dfrac{\text{MST}}{\text{MSE}} = 5.20$ with p-value $= .011$. H_0 is rejected at the 5% level of significance; there is evidence to suggest a difference in mean discharge for the four plants.

b The hypothesis to be tested is $H_0: \mu_A = 1.5$ versus $H_a: \mu_A > 1.5$ and the test statistic is

$$t = \frac{\bar{x}_A - \mu_A}{\sqrt{\dfrac{\text{MSE}}{n_A}}} = \frac{1.568 - 1.5}{\sqrt{\dfrac{.0298}{5}}} = .88$$

The rejection region with $\alpha = .05$ and 16 df is $t > t_{.05} = 1.746$ and the null hypothesis is not rejected. We cannot conclude that the limit is exceeded at plant A.

c The 95% confidence interval for $\mu_A - \mu_D$ is

$$(\bar{x}_A - \bar{x}_D) \pm t_{.025}\sqrt{\text{MSE}\left(\frac{1}{n_A} + \frac{1}{n_D}\right)}$$

$$(1.568 - 1.916) \pm 2.12\sqrt{.0298\left(\frac{2}{5}\right)}$$

$$-.348 \pm .231 \quad \text{or} \quad -.579 < \mu_A - \mu_D < -.117$$

11.65 Answers will vary from student to student. The students should mention the significance of both block and treatment effects. There appear to be no violations of the normality and common variance assumptions. Since the treatment means were significantly different, Tukey's test is used to explore the differences with

$$\omega = q_{.05}(5,20)\frac{\sqrt{\text{MSE}}}{\sqrt{n_t}} = 4.23\sqrt{\frac{1.9165}{6}} = 2.39$$

The ranked means are shown below.

E	B	A	C	D
31.20	32.28	34.35	36.30	36.78

11.67 **a** This is a 2×3 factorial design with $r = 5$ replications.

b The *Minitab* analysis of variance is shown below.

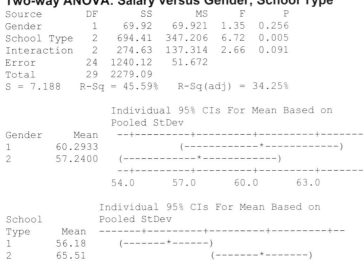

```
Two-way ANOVA: Salary versus Gender, School Type
Source         DF      SS         MS        F       P
Gender          1    69.92     69.921    1.35   0.256
School Type     2   694.41    347.206    6.72   0.005
Interaction     2   274.63    137.314    2.66   0.091
Error          24  1240.12     51.672
Total          29  2279.09
S = 7.188    R-Sq = 45.59%    R-Sq(adj) = 34.25%
```

The F-test for interaction is $F = 2.66$ with a p-value of .091. There is no evidence of significant interaction. The F-test for gender is $F = 1.35$ with a p-value of .256 and the F-test for school type is $F = 6.72$ with a p-value of .005. There is sufficient evidence to indicate a difference in the average salaries due faculty type, but not due to gender.

c The 95% confidence interval for $\mu_M - \mu_F$ is

$$\left(\bar{x}_M - \bar{x}_F\right) \pm t_{.025}\sqrt{\text{MSE}\left(\frac{1}{n_F} + \frac{1}{n_M}\right)}$$

$$(60.2933 - 57.24) \pm 2.064\sqrt{51.672\left(\frac{2}{15}\right)}$$

$$3.053 \pm 5.418 \quad \text{or} \quad -2.365 < \mu_M - \mu_F < 8.471$$

Since the value $\mu_M - \mu_F = 0$ falls in the interval, there is not enough evidence to indicate a difference in the average salaries for males and females.

d Tukey's test is used to explore the differences due to school type with

$$\omega = q_{.01}(3, 24)\frac{\sqrt{\text{MSE}}}{\sqrt{n_t}} \approx 4.55\sqrt{\frac{51.672}{10}} = 10.34$$

The ranked means are shown below.

Nursing	Arts	Science
54.61	56.18	65.51

There is a difference in average salary between Nursing and Science faculties, but not between the other two pairs.

11.69 **a** The experiment is run in a randomized block design, with telephone companies as treatments and cities as blocks.

b Use the computing formulas in Section 11.8 or the *Minitab* printout below.

Two-way ANOVA: Score versus City, Carrier
```
Source    DF       SS        MS       F        P
City       3   55.688   18.5625    3.88    0.049
Carrier    3  285.688   95.2292   19.90    0.000
Error      9   43.063    4.7847
Total     15  384.438
S = 2.187   R-Sq = 88.80%   R-Sq(adj) = 81.33%
```

c The F test for treatments (carriers) has a test statistic $F = 19.90$ with p-value = .000. The null hypothesis is rejected and we conclude that there is a significant difference in the average satisfaction scores for the four carriers.

d The F test for blocks (cities) has a test statistic $F = 3.88$ with p-value = .049. The null hypothesis is rejected and we conclude that there is a significant difference in the average satisfaction scores for the four cities.

11.71 **a** The experiment is a 2×3 factorial experiment, with two factors (rank and gender). There are $r = 10$ replications per factor-level combination.

b Use the computing formulas in Section 11.10 or the *Minitab* printout below.

Two-way ANOVA: Salary versus Gender, Rank
```
Source        DF          SS             MS          F        P
Gender         1   1.18409E+08    1.18409E+08    17.25    0.000
Rank           2   7.29444E+10    3.64722E+10  5314.50    0.000
Interaction    2   3.39075E+07    1.69537E+07     2.47    0.094
Error         54   3.70589E+08    6.86277E+06
Total         59   7.34673E+10
S = 2620   R-Sq = 99.50%   R-Sq(adj) = 99.45%

           Individual 95% CIs For Mean Based on
              Pooled StDev
Rank    Mean   -----+---------+---------+---------+---
1    12313.6   (*
2    93564.3                                        *)
3    75733.0                              *)
               -----+---------+---------+---------+---
               25000      50000     75000    100000
```

c The F test for Interaction has a test statistic $F = 2.47$ with p-value = .094. The null hypothesis is not rejected and we conclude that there is no significant interaction between rank and gender.

d The F test for rank has a test statistic $F = 5314.50$ with p-value = .000, and the F test for gender has a test statistic $F = 17.25$ with p-value = .000. Both factors are highly significant. We conclude that there is a difference in average salary due to both gender and rank.

e The interaction plot is shown below. Notice the differences in salary due to both rank and gender.

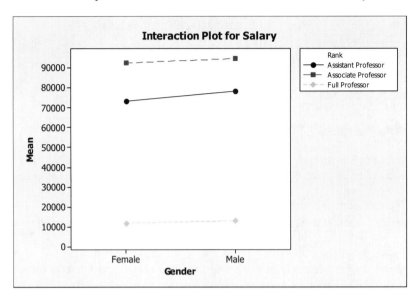

Using Tukey's test is used to explore the differences with
$$\omega = q_{.05}(3,54)\frac{\sqrt{\text{MSE}}}{\sqrt{n_t}} \approx 3.44\sqrt{\frac{6862770}{20}} = 2015.084$$

The ranked means are shown below. All three of the ranks have significantly different average salaries.

Assistant Professor	Associate Professor	Full Professor
75733.0	93564.3	12313.6

Case Study:
"A Fine Mess"

1 The design is a two-way classification, with type of ticket as the treatment and cities as blocks.

2 The Minitab printout for the randomized block design is shown below.

```
Two-way ANOVA: Fine versus City, Type
Source   DF       SS       MS       F       P
City     10    3255.5   325.55    1.77   0.134
Type      2   10321.3  5160.64   27.99   0.000
Error    20    3687.4   184.37
Total    32   17264.2

S = 13.58   R-Sq = 78.64%   R-Sq(adj) = 65.83%

                Individual 95% CIs For Mean Based on
                Pooled StDev
Type   Mean   ----+---------+---------+---------+-----
1     17.3636  (-----*----)
2     57.2727                                (-----*-----)
3     22.7273      (-----*-----)
              ----+---------+---------+---------+-----
                  15        30        45        60
```

Students should notice the significant difference in average ticket prices for the three types of tickets, but not from city to city. It does not appear that blocking has been effective. To explore the differences between the three types of tickets, use Tukey's procedure with

$$\omega = q_{.05}(3,24)\frac{\sqrt{MSE}}{\sqrt{n_t}} = 3.53\sqrt{\frac{184.37}{11}} = 14.45$$

The ranked means are shown below.

Expired Meter	Fire Route	No Parking Zone
17.36	57.27	22.73

The expired meter and no parking zone ticket amounts are not significantly different, but the amount for Fire route parking appears to be significantly higher than the other two types.

3 Answers will vary, but should summarize the above results.

Project 11: Hard to Shake: *Globe and Mail* Series Exposes the Pervasive Health Risks Associated with Canada's Excessive Salt Consumption

Part 1

i. The experimental design is called a Completely Randomized Design. There are 4 treatments (Provinces), each with 5 observations.

ii. To complete the ANOVA table, we need to determine several quantities (provided in the table below)

	PQ	ON	AB	BC	
	2.50	2.90	2.80	2.60	
	3.10	2.70	3.20	2.70	
	2.60	3.20	3.20	2.90	
	2.80	3.00	2.90	2.90	
	2.40	2.20	3.30	3.10	
$T_i = \sum x_i$	13.40	14.00	15.40	14.20	57.00
$\sum x_i^2$	36.22	39.78	47.62	40.48	164.10
mean	2.68	2.80	3.08	2.84	2.85

Our ANOVA table will have k-1=4-1=3 degrees of freedom for treatment, N-1=20-1=19 degrees of freedom total, and thus 19-3=16 degrees of freedom for error.

The total sums of squares (Total Sum of Squares) is

$$\text{Total SS} = \sum_{i=1}^{k}\sum_{j=1}^{n_i} x_{ij}^2 - \frac{1}{n}\left(\sum_{i=1}^{k}\sum_{j=1}^{n_i} x_{ij}\right)^2$$

$$= 164.10 - \frac{1}{20}(57)^2$$

$$= 1.65$$

The sums of squares treatment (SST) is

$$\text{SST} = \sum_{i=1}^{k} \frac{T_i^2}{n_i} - \frac{1}{n}\left(\sum_{i=1}^{k}\sum_{j=1}^{n_i} x_{ij}\right)^2$$

$$= \frac{1}{5}\left(13.4^2 + 14.0^2 + 15.4^2 + 14.2^2\right) - \frac{57^2}{20}$$

$$= 0.422$$

The sums of squares error (SSE) can be calculated using the formula SSE=Total SS-SST=1.65-0.422=1.2880.

Our ANOVA table is thus

Source	Df	SS	MS	F
Treatment	3	0.422	0.422/3=0.14067	0.14067/0.07675=1.83
Error	16	1.288	1.288/16=0.07675	
Total	19	1.650		

iii. To test the null hypothesis of no differences in the true mean amount of sodium for the four provinces with α=0.05, we compare the F test value for treatment (1.83) to the critical value on 3 and 16 degrees of freedom. The critical value is obtained from Table 6 in Appendix I. The F critical value with α=0.05 on 3 and 16 degrees of freedom is 3.24. Since the F test statistic is smaller than our critical value, we do not reject the null hypothesis. There is insufficient evidence to suggest that the mean amount of sodium for the four provinces is different.

The approximate p-value can be obtained from Table 6 in Appendix I. With 3 and 16 degrees of freedom, we have p-value=p(F>1.83)>0.10.

iv. A 90% confidence interval to compare Quebec and Alberta is

$$\bar{x}_Q - \bar{x}_A \pm t_{\frac{\alpha}{2}, n-k}\sqrt{MSE}\sqrt{\frac{1}{n_Q} + \frac{1}{n_A}}$$

$$2.68 - 3.08 \pm 1.746\sqrt{0.07675}\sqrt{\frac{1}{5} + \frac{1}{5}}$$

which gives (-0.706, -0.094). We use the root of the mean square error as an estimate for σ. Since the interval does not cover zero, there is sufficient evidence to suggest that the means are different at the 90% level.

v. If we perform a two-sample independent *t*-test on the Quebec and Alberta data, we obtain

$$t_{test} = \frac{\bar{x}_A - \bar{x}_Q}{\sqrt{MSE}\sqrt{\frac{1}{n_Q} + \frac{1}{n_A}}}$$

$$= \frac{2.68 - 3.08}{\sqrt{0.07675}\sqrt{\frac{2}{5}}}$$

$$= 2.282921$$

Our null hypothesis is that there is no difference between the two means. The alternative is two sided, suggesting there is a difference between the two means. We again use the root of the mean square error (MSE) as an estimate for σ as it represents a pooled estimate of the standard deviation. We compare this value to a critical t value with α/2=0.025 and degrees of freedom (associated with MSE) of 16. The critical value is 2.120. Since our test statistic is greater than the critical value, we reject the null hypothesis. There is sufficient evidence to suggest that the two means are different.

vi. A 90% confidence interval for the mean salt intake in Ontario is

$$\bar{x}_O \pm t_{\frac{\alpha}{2}, n-k}\sqrt{MSE}\sqrt{\frac{1}{n_O}}$$

$$2.80 \pm 1.746\sqrt{0.07675}\sqrt{\frac{1}{5}}$$

which gives (2.584, 3.016).

vii. If we were to rank the means simply on magnitude, they would be ordered from smallest to largest as
PQ < ON < BC < AB

If we perform ordered pair-wise comparisons for each of the means (as we did in part v) we obtain the following t-test statistics (see table below). These are compared to a critical t value with 16 degrees of freedom. Assuming α=0.05, our critical t value is 2.120. Hence, we reject the null hypothesis (of equivalence between means) in favour of the two-sided alternative, if our test statistic is greater than 2.120, or less than -2.120. We can also compare the differences with tcrit×s×√(1/n1+1/n2)=0.3714. Differences found to be larger than 0.3714 would be considered significant (i.e., would result in rejecting the null hypothesis).

Comparison	Difference	t-test	Comment
ON-PQ	0.12	0.6848762	Do not reject the null
BC-ON	0.04	0.2282921	Do not reject the null
AB-BC	0.24	1.369752	Do not reject the null

Thus we find no differences between the ordered pair-wise comparison of means.

viii. To use Tukey's method, we require the yardstick. In our particular example, we have k=4 treatments each with n=5 observations, and an MSE=0.07675 with 16 degrees of freedom. We also find our critical q value of 4.05 with α=0.05 from Tables 11a in Appendix I. Thus our yardstick is

$$\begin{aligned} \omega &= q_\alpha(k, df)\left(\frac{s}{\sqrt{n_t}}\right) \\ &= q_{0.05}(4, 16)\left(\frac{s}{\sqrt{n_t}}\right) \\ &= 4.05\sqrt{\frac{0.07675}{5}} \\ &= 0.5017752 \end{aligned}$$

Our pairwise comparisons are

Comparison	|Difference|	Comment
PQ-ON	0.12	Not significantly different
PQ-AB	0.40	Not significantly different
PQ-BC	0.16	Not significantly different
ON-AB	0.28	Not significantly different
ON-BC	0.04	Not significantly different
AB-BC	0.24	Not significantly different

Thus, none of the means are significantly different than the others.

Part 2

i The experimental design is called a Randomized Complete Block Design. There are 6 treatments (Strategies), each with 4 observations over 4 blocks (Provinces).

ii. The blocks are the provinces.

iii. The treatments are the strategies for lowering salt intake.

iv. Blocking is necessary in this particular problem since the experimental units (i.e., the provinces) are not necessarily homogenous. As a result, the provinces may represent a source of variation that needs to be accounted for.

v. The ANOVA for the RCBD is (obtained using the R statistical package)

Source	Df	SS	MS	F	p-value
Treatment	5	0.60708	0.12142	5.4981	0.004516
Block	3	0.48125	0.16042	7.2642	0.003099
Error	15	0.33125	0.02208		
Total	23	1.41958			

vi. It is necessary to treat provinces as block in order to account for the variability observed within each of the provinces.

vii. The p-value for treatment effects (i.e., the strategies) is 0.004516<α=0.05. This suggests that there is sufficient evidence to reject the null hypothesis (that the treatment means are equivalent) in favour of the alternative (at least one of the treatment means differs from the rest). There is a significant different among the five strategies in reducing added unnecessary salt.

viii. The p-value for block effects (i.e., the provinces) is 0.003099<α=0.05. This suggests that there is sufficient evidence to reject the null hypothesis (that the block effects are equivalent) in favour of the alternative (at least one of the block effects differs from the rest). Thus, there is a significant different among the provinces' salt intake at the 0.05 level.

ix. To calculate a 95% confidence interval for strategies 1 and 6, we have

$$\bar{T}_1 - \bar{T}_6 \pm t_{\frac{\alpha}{2}}\sqrt{s^2\left(\frac{1}{b} + \frac{1}{b}\right)}$$

$$2.625 - 2.200 \pm 2.131\sqrt{0.02208\left(\frac{2}{4}\right)}$$

which gives (0.2011, 0.6489). This interval indicates a significant difference between strategies 1 and 6.

x. It does appear that the RCBD was justified. That is, the RCBD ANOVA has indicated that both strategies and blocks (i.e., provinces) are significant. In this particular case, the blocks explain a significant portion of the total variability in the observed data. By including blocks, we have improved the power of the experiment.

Part 3

i The experimental design is called a Factorial Design. There are 6 levels of the first treatment (Strategies) and 4 levels of the second treatment (Education Level). Further, 3 observations are recorded for each of the strategy and education level combinations.

ii. The ANOVA for the Factorial Design (using the R statistical package)

Source	Df	SS	MS	F	p-value
A (Education)	3	3.6450	1.2150	24.7819	7.839x10^{-10}
B (Strategy)	5	0.9583	0.1917	3.9093	0.004714
AB	15	1.7783	0.1186	2.4181	0.010530
Error	48	2.3533	0.0490		
Total	71	1.41958			

The p-value associated with strategy is 0.004714. Given α=0.05, we will reject the null hypothesis of equivalence between the means for each of the strategies. That is, there is sufficient evidence to suggest that at least one of the strategy means is different than the others.

iii. The F-test for interaction between education level and strategy is 2.4649. On 15 and 48 degrees of freedom, this F-test has a p-value of 0.010530. At the α=0.05 level, there is sufficient evidence to suggest that the interaction is significant. The data provides sufficient evidence to suggest the existence of an interaction between education level and strategy.

iv. Since interactions are significant, the simple effects of strategy and education level should be discussed. The means for each of the education level and strategy combinations are summarized below.

	Less than High School	High School	University/College	Postgraduate
Strategy 1	2.80	2.80	2.50	2.40
Strategy 2	2.40	2.53	2.60	2.30
Strategy 3	3.00	2.67	2.43	2.27
Strategy 4	2.77	2.53	2.53	2.20
Strategy 5	2.83	2.53	2.37	2.37
Strategy 6	3.03	2.27	2.10	1.67

Practically speaking, to obtain the largest drop in salt intake, researchers should consider the education level of individuals before implementing a particular strategy. For individuals with an education level less than high school, strategy 6 had the highest impact. The strategies with the highest impact (i.e., the largest mean reduction in salt intake) for the remaining education levels (high school, university and postgraduate respectively) are strategies 1, 2 and 1 respectively. Of course, the significant differences between treatment combinations has not been considered here.

v. At the 10% level of significance, we can reject the null hypothesis that there are no differences among the means of the education levels of the subjects. Our F-test statistic on 3 and 48 degrees of freedom is 7.839×10^{-10} which is less than 0.10. Hence, there is sufficient evidence to suggest that at least one of the education levels differs from the rest.

vi. A 95% confidence interval for the difference in mean between high school and postgraduate education levels for Strategy 6 is

$$\bar{x}_h - \bar{x}_p \quad \pm \quad t_{\frac{\alpha}{2}, ab(r-1)} \sqrt{MSE \left(\frac{1}{n_h} + \frac{1}{n_p} \right)}$$

$$2.27 - 1.67 \quad \pm \quad 2.011 \sqrt{0.0490 \left(\frac{1}{3} + \frac{1}{3} \right)}$$

which is equivalent to (0.2365, 0.9635).

Chapter 12: Linear Regression and Correlation

12.1 The line corresponding to the equation $y = 2x + 1$ can be graphed by locating the y values corresponding to $x = 0, 1$, and 2.

$$\text{When } x = 0, y = 2(0) + 1 = 1$$
$$\text{When } x = 1, y = 2(1) + 1 = 3$$
$$\text{When } x = 2, y = 2(2) + 1 = 5$$

The graph is shown below.

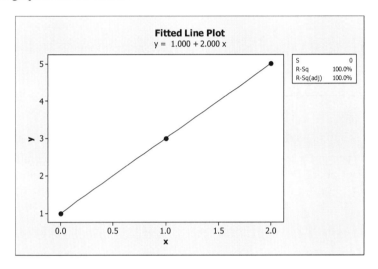

Note that the equation is in the form $y = \alpha + \beta x$.

Thus, the slope of the line is $\beta = 2$ and the y-intercept is $\alpha = 1$.

12.3 If $\alpha = 3$ and $\beta = -1$, the straight line is $y = 3 - x$.

12.5 A deterministic mathematical model is a model in which the value of a response y is exactly predicted from values of the variables that affect the response. On the other hand, a probabilistic mathematical model is one that contains random elements with specific probability distributions. The value of the response y in this model is not exactly determined.

12.7 a The equations for calculating the quantities a and b are found in Section 12.2 of the text and involve the preliminary calculations:

$$\sum x_i = 21 \qquad \sum y_i = 24.3 \qquad \sum x_i y_i = 75.3$$
$$\sum x_i^2 = 91 \qquad \sum y_i^2 = 103.99 \qquad n = 6$$

Then $S_{xy} = \sum x_i y_i - \dfrac{(\sum x_i)(\sum y_i)}{n} = 75.3 - \dfrac{21(24.3)}{6} = 75.3 - 85.05 = -9.75$

$S_{xx} = \sum x_i^2 - \dfrac{(\sum x_i)^2}{n} = 91 - \dfrac{21^2}{6} = 17.5$

$b = \dfrac{S_{xy}}{S_{xx}} = \dfrac{-9.75}{17.5} = -0.55714$ and $a = \bar{y} - b\bar{x} = \dfrac{24.3}{6} - (-0.557)\left(\dfrac{21}{6}\right) = 6$

and the least squares line is $\hat{y} = a + bx = 6 - 0.557x$.

b The graph of the least squares line and the six data points are shown below.

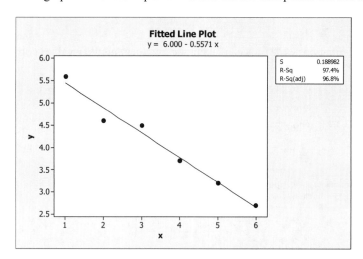

c When $x = 3.5$, the value for y can be predicted using the least squares line as
$$\hat{y} = 6.00 - 0.55714(3.5) = 4.05$$

d Using the additivity properties for the sums of sums of squares and degrees of freedom for an analysis of variance, and the fact that $MS = SS/df$, the completed ANOVA table is shown below.

```
Analysis of Variance
Source      DF      SS        MS
Regression  1       5.4321    5.4321
Error       4       0.1429    0.0357
Total       5       5.5750
```

12.9 a Calculate

$$\sum x_i = 850.8 \qquad \sum y_i = 3.755 \qquad \sum x_i y_i = 443.7727$$
$$\sum x_i^2 = 101,495.78 \qquad \sum y_i^2 = 1.941467 \qquad n = 9$$

Then $S_{xy} = \sum x_i y_i - \dfrac{(\sum x_i)(\sum y_i)}{n} = 88.80003333$

$S_{xx} = \sum x_i^2 - \dfrac{(\sum x_i)^2}{n} = 21,066.82$

$S_{yy} = \sum y_i^2 - \dfrac{(\sum y_i)^2}{n} = 0.3747976$

b-c $b = \dfrac{S_{xy}}{S_{xx}} = \dfrac{88.800033}{21066.82} = 0.00421516$ and $a = \bar{y} - b\bar{x} = 0.41722 - 0.00421516(94.5333) = 0.187$

and the least squares line is $\hat{y} = a + bx = 0.187 + 0.0042x$.

The graph of the least squares line and the nine data points are shown below. The assumption of linear Relationship appears to be appropriate.

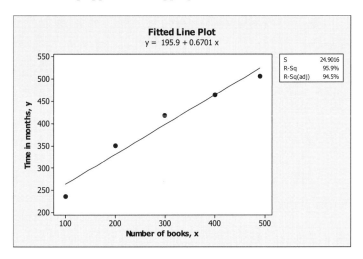

d When $x = 100$, the value for y can be predicted using the least squares line as
$$\hat{y} = 0.187 + 0.0042(100) = 0.44$$

e Calculate $SSR = \dfrac{(S_{xy})^2}{S_{xx}} = \dfrac{(88.80003333)^2}{21,066.82} = .3743064$ and

$SSE = \text{Total SS} - SSR = S_{yy} - \dfrac{(S_{xy})^2}{S_{xx}} = .3747976 - .374306417 = .00049118$

The ANOVA table with 1 df for regression and $n - 2$ df for error is shown on below. Remember that the mean squares are calculated as $MS = SS/df$.

Source	df	SS	MS
Regression	1	.3743064	.374306
Error	7	.0004912	.0000070
Total	8	.3747976	

12.11 a The Temperature Anomaly is the dependent or response variable, y, and Year is the independent or explanatory variable, x.

b The scatterplot is shown below. Notice the linear pattern in the points.

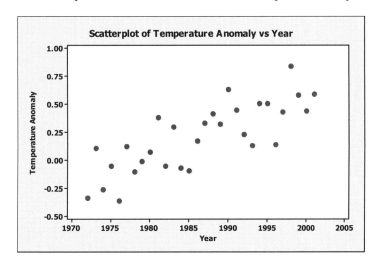

Introduction to Probability and Statistics, 2ce

c Calculate

$$\sum x_i = 59595.0 \qquad \sum y_i = 6.513 \qquad \sum x_i y_i = 12999$$
$$\sum x_i^2 = 118387715.0 \qquad \sum y_i^2 = 4.0928 \qquad n = 30$$

Then $S_{xy} = \sum x_i y_i - \dfrac{(\sum x_i)(\sum y_i)}{n} = 60.9255$

$S_{xx} = \sum x_i^2 - \dfrac{(\sum x_i)^2}{n} = 2247.5$

$S_{yy} = \sum y_i^2 - \dfrac{(\sum y_i)^2}{n} = 2.6788277$

Then $b = \dfrac{S_{xy}}{S_{xx}} = \dfrac{60.9255}{2247.5} = .027$ and $a = \bar{y} - b\bar{x} = .2171 - (.027)(1986.5) = -53.42$

and the least squares line is $\hat{y} = a + bx = -53.42 - .027x$.

d The fitted line and the plotted points are shown below.

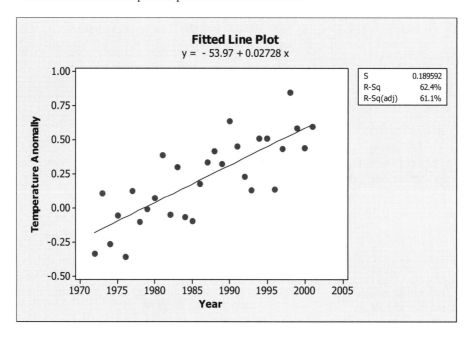

12.13 **a** The scatterplot generated by *Minitab* is shown below. The assumption of linearity is reasonable.

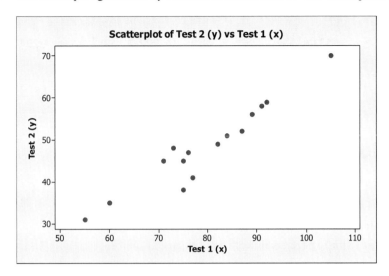

b Calculate

$$\sum x_i = 1192 \qquad \sum y_i = 725 \qquad \sum x_i y_i = 59,324$$
$$\sum x_i^2 = 96,990 \qquad \sum y_i^2 = 36,461 \qquad n = 15$$

Then $S_{xy} = \sum x_i y_i - \dfrac{(\sum x_i)(\sum y_i)}{n} = 1710.6667$

$S_{xx} = \sum x_i^2 - \dfrac{(\sum x_i)^2}{n} = 2265.7333$

$S_{yy} = \sum y_i^2 - \dfrac{(\sum y_i)^2}{n} = 1419.3333$

$b = \dfrac{S_{xy}}{S_{xx}} = \dfrac{1710.6667}{2265.7333} = .75502$ and $a = \bar{y} - b\bar{x} = 48.3333 - (0.75502)(79.4667) = -11.665$

(using full accuracy) and the least squares line is $\hat{y} = a + bx = -11.665 + 0.755x$.

c When $x = 85$, the value for y can be predicted using the least squares line as
$$\hat{y} = a + bx = -11.665 + .755(85) = 52.51.$$

12.15 **a** The scatterplot is shown below. There is a positive linear relationship between armspan and height.

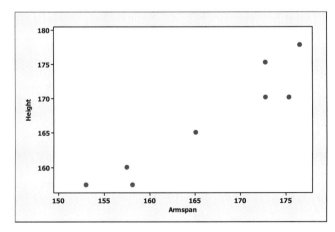

b If armspan and height are roughly equal, the slope of the regression line should be approximately equal to 1.

c Calculate

$$n = 8; \sum x_i = 1330.9; \sum y_i = 1333.6; \sum x_i^2 = 222001.79; \sum y_i^2 = 222749.52; \sum x_i y_i = 222341.87.$$

Then $S_{xy} = \sum x_i y_i - \dfrac{(\sum x_i)(\sum y_i)}{n} = 480.84$

$S_{xx} = \sum x_i^2 - \dfrac{(\sum x_i)^2}{n} = 589.93875$

$S_{yy} = \sum y_i^2 - \dfrac{(\sum y_i)^2}{n} = 438.4$

$b = \dfrac{S_{xy}}{S_{xx}} = \dfrac{480.84}{589.93875} = .815068$ and $a = \bar{y} - b\bar{x} = 166.70 - (0.815068)(166.3625) = 31.103$

and the least squares line is $\hat{y} = a + bx = 31.103 + 0.815x$.

The slope is quite close to the expected value of slope parameter $\beta = 1$.

d When $x = 157.5$, the value for y can be predicted using the least squares line as

$$\hat{y} = a + bx = 31.103 + .815(157.5) = 159.4655$$

12.17 a The hypothesis to be tested is $H_0 : \beta = 0$ versus $H_a : \beta \neq 0$
and the test statistic is a Student's t, calculated as

$$t = \dfrac{b - \beta_0}{\sqrt{MSE/S_{xx}}} = \dfrac{1.2 - 0}{\sqrt{0.533/10}} = 5.20$$

The critical value of t is based on $n - 2 = 3$ degrees of freedom and the rejection region for $\alpha = 0.05$ is $|t| > t_{.025} = 3.182$. Since the observed value of t falls in the rejection region, we reject H_0 and conclude that $\beta \neq 0$. That is, x is useful in the prediction of y.

b From the ANOVA table in Exercise 12.6, calculate

$$F = \dfrac{MSR}{MSE} = \dfrac{14.4}{0.5333} = 27.00$$

which is the square of the t statistic from part **a**: $t^2 = (5.20)^2 = 27.0$.

c The critical value of t from part **a** was $t_{.025} = 3.182$, while the critical value of F from part **b** with $df_1 = 1$ and $df_2 = 3$ is $F_{.05} = 10.13$. Notice that the relationship between the two critical values is

$$F = 10.13 = (3.182)^2 = t^2$$

12.19 a The hypothesis to be tested is $H_0 : \beta = 0$ versus $H_a : \beta \neq 0$
and the test statistic is

$$F = \dfrac{MSR}{MSE} = \dfrac{5.4321}{0.0357} = 152.10$$

with p-value $= 0.000$. Since the p-value is less than $\alpha = 0.01$, the null hypothesis is rejected. There is evidence to indicate that y and x are linearly related.

b Use the formula for r^2 given in this section:

$$r^2 = \dfrac{SSR}{Total\ SS} = \dfrac{5.4321}{5.5750} = 0.974.$$

The coefficient of determination measures the proportion of the total variation in y that is accounted for using the independent variable x. That is, the total variation in y is reduced by 97.4% by using $\hat{y} = a + bx$ rather than $\bar{y}$ to predict the response y.

12.21 a The temperature in (°F) and (°C)

temperature (°F)	temperature (°C)
88.6	31.444444
71.6	22.000000
93.3	34.055556
84.3	29.055556
80.6	27.000000
75.2	24.000000
69.7	20.944444
82.0	27.777778
69.4	20.777778
83.3	28.500000
79.6	26.444444
82.6	28.111111
80.6	27.000000
83.5	28.611111
76.3	24.611111

b The dependent variable (to be predicted) is y = number of chirps per second and the independent variable is x = temperature (°C).

c Preliminary calculations:

$$\sum x_i = 400.333333 \qquad \sum y_i = 249.8 \qquad \sum x_i y_i = 6741.038889$$

$$\sum x_i^2 = 10878.845679 \qquad \sum y_i^2 = 4200.56 \quad n = 15$$

Then $S_{xy} = \sum x_i y_i - \dfrac{(\sum x_i)(\sum y_i)}{n} = 74.1544501066664$

$S_{xx} = \sum x_i^2 - \dfrac{(\sum x_i)^2}{n} = 194.393844940741$

$b = \dfrac{S_{xy}}{S_{xx}} = 0.38147$ and $a = \bar{y} - b\bar{x} = 16.6533333 - 0.38147(26.68888887) = 6.472$

and the least squares line is $\hat{y} = a + bx = 6.472 + 0.38147x$.

d The plot is shown below. The line appears to fit well through the 15 data points.

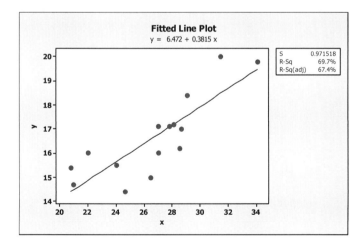

e Calculate Total SS $= S_{yy} = \sum y_i^2 - \frac{(\sum y_i)^2}{n} = 4{,}200.56 - \frac{(249.8)^2}{15} = 40.5573$.

Then SSE $= S_{yy} - \frac{(S_{xy})^2}{S_{xx}} = 40.5573 - \frac{(74.15445)^2}{194.3938} = 12.27$ and MSE $= \frac{SSE}{n-2} = \frac{12.27}{13} = .94385$.

The hypothesis to be tested is $H_0 : \beta = 0$ versus $H_a : \beta \neq 0$ and the test statistic is

$$t = \frac{b - \beta_0}{\sqrt{MSE/S_{xx}}} = \frac{0.38147 - 0}{\sqrt{.94385/194.3938}} = 5.47$$

The critical value of t is based on $n - 2 = 13$ degrees of freedom and the rejection region for $\alpha = 0.05$ is $|t| > t_{.025} = 2.160$, and H_0 is rejected. There is evidence at the 5% level to indicate that x and y are linearly related. That is, the regression model $y = \alpha + \beta x + \varepsilon$ is useful in predicting number of chirps y.

12.23 a From the **Minitab** printout, the test of $H_0 : \beta = 0$ versus $H_a : \beta \neq 0$ is performed using one of two test statistics:

$t = 3.79$ or $F = 14.37$ with p-value $= 0.005$

Since the p-value is smaller than $\alpha = 0.01$, H_0 is rejected, and the results are declared highly significant. There is evidence to indicate that x and y are linearly related.

b If a person is deprived of sleep for as much as 48 hours, their number of errors will probably become extremely high. The relationship will not remain linear, but will become curvilinear.

c From the printout, R-Sq = 64.2%. That is, 64.2% of the total variation in the experiment can be explained by the independent variable x. The total variation in y is reduced by 64.2% by using $\hat{y} = a + bx$ rather than $\bar{y}$ to predict the response y. This is a relatively strong relationship.

d The population variance σ^2 is estimated using $s^2 = MSE = 5.025$.

e The 95% confidence interval for the slope β is

$$b \pm t_{\alpha/2} \sqrt{MSE/S_{xx}} \Rightarrow 0.475 \pm 2.306(0.1253) \Rightarrow 0.475 \pm 0.289$$

or $0.186 < \beta < 0.764$.

12.25 a The scatterplot generated by **Minitab** is shown below. The assumption of linearity is reasonable.

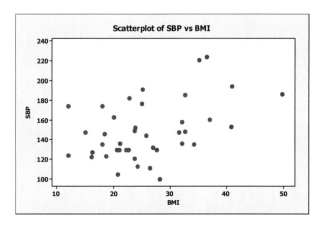

b Using the **Minitab** printout, the equation of the regression line is $y = 109 + 1.55x$.

c The hypothesis to be tested is $H_0: \beta = 0$ versus $H_a: \beta \neq 0$
and the test statistic is a Student's t, calculated as
$$t = \frac{b - \beta_0}{\sqrt{MSE/S_{xx}}} = 3.01$$
with p-value $= .005$. Since the p-value is less than $\alpha = .01$, we reject H_0 and conclude that $\beta \neq 0$. That is, SBP measurement and BMI are linearly related.

d The 99% confidence interval for the slope β is
$$b \pm t_{\alpha/2}\sqrt{MSE/S_{xx}} \Rightarrow 1.55 \pm 2.576(0.5128) \Rightarrow 1.55 \pm 1.3210$$
or $0.229 < \beta < 2.871$.

12.27 Refer to Exercise 12.15.
a The hypothesis to be tested is $H_0: \beta = 0$ versus $H_a: \beta \neq 0$
and the test statistic is a Student's t, calculated as
$$t = \frac{b - \beta_0}{\sqrt{MSE/S_{xx}}} = \frac{.815}{\sqrt{\frac{7.7472}{589.93875}}} = 7.11$$
The rejection region, with $\alpha = .05$, is $|t| > t_{.025,6} = 2.447$ and we reject H_0. That is, there is a linear relationship between armspan and height.

b The 95% confidence interval for the slope β is
$$b \pm t_{\alpha/2}\sqrt{MSE/S_{xx}} \Rightarrow .815 \pm 2.447\sqrt{\frac{7.7472}{589.93875}} \Rightarrow .815 \pm .2804$$
or $.5346 < \beta < 1.0954$.

c Since the value $\beta = 1$ is in the confidence interval, da Vinci's supposition is confirmed by the confidence interval in part **b**.

12.29 Use a plot of residuals versus fits. The plot should appear as a random scatter of points, free of any patterns.

12.31 Although there is one data point in each graph that appears somewhat unusual, there is no reason to doubt the validity of the regression assumptions.

12.33 **a** If you look carefully, there appears to be a slight curve to the five points.
b The fit of the regression line, measured as $r^2 = 0.959$ indicates that 95.9% of the overall variation can be explained by the straight line model.
c When we look at the residuals there is a strong curvilinear pattern that has not been explained by the straight line model. The relationship between time in months and number of books appears to be curvilinear.

12.35 **a** Since $r^2 = .713$, we know that there is a fairly strong positive relationship between price and screen size.
b The residual plot appears to have a curvilinear pattern to it; in fact, it looks like a cubic relationship.

c The scatterplot is shown below. You can see the curvilinear (possibly cubic) relationship in the data points.

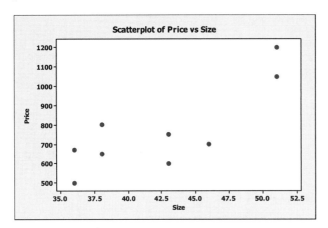

12.37 a From Exercise 12.7, the least squares equation is $\hat{y} = a + bx = 6 - 0.557x$. When $x = 2$,
$$\hat{y} = a + bx = 6 - 0.557(2) = 4.886$$
which is the first line in the section of **Minitab** output labeled "Predicted values for New Observations". The 95% confidence interval for $E(y)$ appears in the fourth column as $4.6006 < E(y) < 5.1708$.

b From the fifth column of the first line in the "Predicted values" section, the 95% prediction interval is $4.2886 < y < 5.4829$.

c The value $\hat{y} = 1.5429$ corresponds to
$$x = \frac{y-a}{b} = \frac{1.5429 - 6}{-0.557} = 8$$
This value is far from the average value of x, $\bar{x} = 3.5$, and outside the experimental region $1 \le x \le 6$. This *extrapolation* can cause a problem with inaccurate predictions.

12.39 a Although very slight, the student might notice a slight curvature to the data points.
b The fit of the linear model is very good, assuming that this is *indeed* the correct model for this data set.
c The normal probability plot follows the correct pattern for the assumption of normality. However, the residuals show the pattern of a quadratic curve, indicating that a quadratic rather than a linear model may have been the correct model for this data.

12.41 Use the preliminary calculations from Exercise 12.16 and 12.24.
a The point estimator for $E(y)$ when $x = -1$ is
$$\hat{y} = 2 - .875(-1) = 2.875$$
and the 99% confidence interval is
$$\hat{y} \pm t_{.005} \sqrt{\text{MSE}\left(\frac{1}{n} + \frac{(x_0 - \bar{x})^2}{S_{xx}}\right)}$$
$$2.875 \pm 5.841 \sqrt{(.08333)\left(\frac{1}{5} + \frac{(-1-0)^2}{16}\right)}$$
$$2.875 \pm .864$$
or $2.011 < E(y) < 3.739$.

b The point estimator for y when $x = 1$ is
$$\hat{y} = 2 - .875(1) = 1.125$$
and the 99% prediction interval is
$$\hat{y} \pm t_{.005}\sqrt{\text{MSE}\left(1 + \frac{1}{n} + \frac{(x_0 - \bar{x})^2}{S_{xx}}\right)}$$
$$1.125 \pm 5.841\sqrt{(.08333)\left(1 + \frac{1}{5} + \frac{(1-0)^2}{16}\right)}$$
$$1.125 \pm 1.895$$
or $-.77 < y < 3.02$.

c The width will be the narrowest when $x_p = \bar{x} = 0$.

12.43 Refer to Exercise 12.42 and calculate
$$\text{SSE} = S_{yy} - \frac{(S_{xy})^2}{S_{xx}} = .2032 - \frac{.7708^2}{2.9325} = .0006 \text{ and } \text{MSE} = \frac{\text{SSE}}{n-2} = \frac{.0006}{5} = .00012.$$

a The point estimator for $E(y)$ when $x = 5.8$ is
$$\hat{y} = 1.769 + .263(5.8) = 3.2944$$
and the 95% confidence interval is
$$\hat{y} \pm t_{.025}\sqrt{\text{MSE}\left(\frac{1}{n} + \frac{(x_0 - \bar{x})^2}{S_{xx}}\right)}$$
$$3.2944 \pm 2.571\sqrt{.00012\left(\frac{1}{7} + \frac{(5.8 - 5.749)^2}{2.9325}\right)}$$
$$3.2944 \pm .0107$$
or $3.2837 < E(y) < 3.3051$.

b The point estimator for y when $x = 5.8$ is still
$$\hat{y} = 1.769 + .263(5.8) = 3.2944$$
and the 95% prediction interval is
$$\hat{y} \pm t_{.025}\sqrt{\text{MSE}\left(1 + \frac{1}{n} + \frac{(x_0 - \bar{x})^2}{S_{xx}}\right)}$$
$$3.2944 \pm 2.571\sqrt{.00012\left(1 + \frac{1}{7} + \frac{(5.8 - 5.749)^2}{2.9325}\right)}$$
$$3.2944 \pm .03012$$
or $3.2643 < y < 3.3245$.

c This would not be advisable, since you are trying to estimate outside the range of experimentation.

12.45 If the value of r is positive, then the least squares line slopes upward to the right. Similarly, if the value of r is negative, the line slopes downward to the right. The coefficient of correlation r will be zero only when b is zero (see Section 12.8 of the text). Moreover, the least squares equation when $b = 0$ is given by $\hat{y} = a$. The variable x has no effect on the value of y and there is no linear correlation between x and y. Finally, r will equal ± 1 only when SSE = 0; that is, all points fall exactly on the fitted line.

12.47 a Refer to the figure given below. The sample correlation coefficient will be positive.

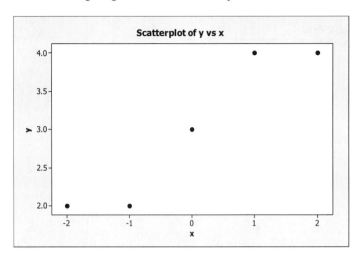

b Calculate $S_{xy} = \sum x_i y_i - \dfrac{(\sum x_i)(\sum y_i)}{n} = 6 - \dfrac{0(15)}{5} = 6$

$S_{xx} = \sum x_i^2 - \dfrac{(\sum x_i)^2}{n} = 10 - \dfrac{0^2}{5} = 10$

$S_{yy} = \sum y_i^2 - \dfrac{(\sum y_i)^2}{n} = 49 - \dfrac{15^2}{5} = 4$

Then $r = \dfrac{S_{xy}}{\sqrt{S_{xx} S_{yy}}} = \dfrac{6}{\sqrt{40}} = 0.9487$ and $r^2 = (0.9487)^2 = 0.9000$. Approximately 90% of the total sum of squares of deviations was reduced by using the least squares equation instead of $\bar{y}$ as a predictor of y.

12.49 The data from Exercise 12.48 are reused here, except that the y observations are reordered. The only calculation that has changed from the previous exercise is

$$S_{xy} = \sum x_i y_i - \dfrac{(\sum x_i)(\sum y_i)}{n} = 100 - \dfrac{21(22)}{6} = 23$$

a Refer to the figure below. The sample correlation coefficient will be positive.

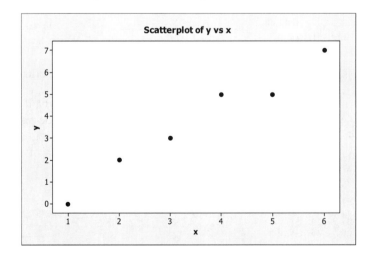

b Calculate $r = \dfrac{S_{xy}}{\sqrt{S_{xx}S_{yy}}} = \dfrac{23}{\sqrt{17.5(31.3333)}} = 0.982$.

c We first calculate the coefficient of determination:
$$r^2 = (0.982)^2 = 0.9647$$

This value implies that the sum of squares of deviations is reduced by 96.47% using the linear model $\hat{y} = a + bx$ instead of $\bar{y}$ to predict values of y. Since the value of r is near 1, a strong positive linear association between the two variables is implied. Note that this value of r is the negative of the value calculated for r in Exercise 12.48.

12.51 When the pre-test score x is high, the post-test score y should also be high. There should be a positive correlation.

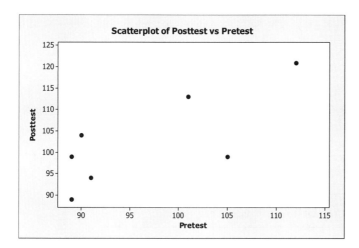

Calculate $S_{xy} = \sum x_i y_i - \dfrac{(\sum x_i)(\sum y_i)}{n} = 70,006 - \dfrac{677(719)}{7} = 468.42857$

$S_{xx} = \sum x_i^2 - \dfrac{(\sum x_i)^2}{n} = 65,993 - \dfrac{677^2}{7} = 517.42857$

$S_{yy} = \sum y_i^2 - \dfrac{(\sum y_i)^2}{n} = 74,585 - \dfrac{719^2}{7} = 733.42857$

Then $r = \dfrac{S_{xy}}{\sqrt{S_{xx}S_{yy}}} = \dfrac{468.42857}{\sqrt{517.42857(733.42857)}} = 0.760$.

The test of hypothesis is $H_0: \rho = 0$ versus $H_a: \rho > 0$ and the test statistic is

$$t = \dfrac{r\sqrt{n-2}}{\sqrt{1-r^2}} = \dfrac{0.760\sqrt{5}}{\sqrt{1-(0.760)^2}} = 2.615$$

The rejection region for $\alpha = 0.05$ is $t > t_{.05} = 2.015$ and H_0 is rejected. There is sufficient evidence to indicate positive correlation.

12.53 The hypothesis of interest is $H_0: \rho = 0$ versus $H_a: \rho \neq 0$
and the test statistic is

$$t = \frac{r\sqrt{n-2}}{\sqrt{1-r^2}} = \frac{0.36\sqrt{67}}{\sqrt{1-(0.36)^2}} = 3.158$$

The p-value can be bounded using Table 4 as
$$p\text{-value} = 2P(t > 3.158) < 2(0.005) = 0.01$$

and H_0 can be rejected at the 1% level of significance. The results are declared highly significant; there is evidence of correlation between x and y.

12.55 **a** Since neither of the two variables, amount of sodium or number of calories, is controlled, the methods of correlation rather than linear regression analysis should be used.

b Use a computer program, your scientific calculator or the computing formulas given in the text to calculate the correlation coefficient r. The Minitab printout for this data set is shown below.

Correlations: Sodium, Calories
```
Pearson correlation of Sodium and Calories = 0.981
P-Value = 0.003
```

There is evidence of a highly significant correlation, since the p-value is so small. The correlation is positive.

12.57 **a** The scatterplot generated by *Minitab* is shown below. There appears to be a strong positive relationship between average run rate and number of completed overs.

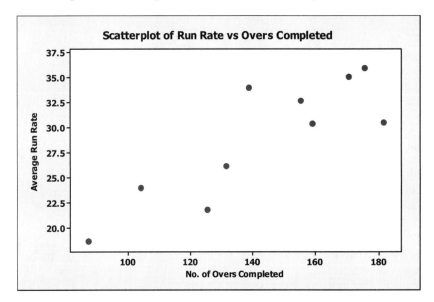

b Calculate $S_{xy} = \sum x_i y_i - \dfrac{(\sum x_i)(\sum y_i)}{n} = 42798.274 - \dfrac{1428.30(289.59)}{10} = 1436.1343$

$S_{xx} = \sum x_i^2 - \dfrac{(\sum x_i)^2}{n} = 212749.65 - \dfrac{1428.3^2}{10} = 8745.561$

$S_{yy} = \sum y_i^2 - \dfrac{(\sum y_i)^2}{n} = 8709.67 - \dfrac{289.59^2}{10} = 323.43319$

Then $r = \dfrac{S_{xy}}{\sqrt{S_{xx} S_{yy}}} = \dfrac{1436.1343}{\sqrt{8745.561(323.43319)}} = 0.854$.

The hypothesis to be tested is $H_0 : \rho = 0$ versus $H_a : \rho > 0$
and the test statistic is

$$t = \dfrac{r\sqrt{n-2}}{\sqrt{1-r^2}} = \dfrac{0.854\sqrt{8}}{\sqrt{1-(0.854)^2}} = 4.643$$

The rejection region with $\alpha = .05$ is $t > t_{.05} = 1.860$ and H_0 is rejected. There is sufficient evidence to indicate a positive correlation between the two variables.

c Answers will vary.

12.59 a The data are plotted with the least squares line below. There appears to be a linear relationship.

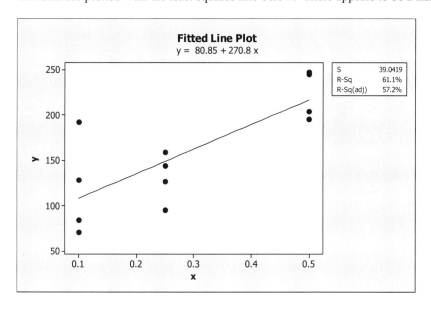

b Using the printout, the least squares line is $\hat{y} = a + bx = 80.85 + 270.82x$.

c To test $H_0 : \beta = 0, H_a : \beta \neq 0$, the test statistic is $t = 3.96$ with p-value $= 0.003$. Since this p-value is less than 0.01, we reject H_0. There is sufficient evidence to indicate that the independent variable x does help in predicting values of the dependent variable y.

e When $x = 0.20$, $\hat{y} = 80.85 + 270.82(0.20) = 135.0$. The 90% confidence interval is shown at the bottom of the printout as $(112.1, 157.9)$ or $112.1 < E(y) < 157.9$.

12.61 Answers will vary. The *Minitab* output for this linear regression problem is shown below.

Regression Analysis: y versus x
```
The regression equation is
y = 21.9 + 15.0 x

Predictor      Coef    SE Coef       T       P
Constant     21.867      3.502    6.24   0.000
x           14.9667     0.9530   15.70   0.000

S = 3.69098    R-Sq = 96.1%    R-Sq(adj) = 95.7%

Analysis of Variance

Source            DF       SS       MS        F       P
Regression         1   3360.0   3360.0   246.64   0.000
Residual Error    10    136.2     13.6
Total             11   3496.2
```

Correlations: x, y
```
Pearson correlation of x and y = 0.980
P-Value = 0.000
```

a   The correlation coefficient is $r = 0.980$.

b   The coefficient of determination is $r^2 = 0.961$ (or 96.1%).

c   The least squares line is $\hat{y} = 21.867 + 14.9667x$.

d   We wish to estimate the mean percentage of kill for an application of 4 pounds of nematicide per acre. Since the percent kill *y* is actually a binomial percentage, the variance of *y* will change depending on the value of *p*, the proportion of nematodes killed for a particular application rate. The residual plot versus the fitted values shows this phenomenon as a "football-shaped" pattern. The normal probability plot also shows some deviation from normality in the tails of the plot. A transformation may be needed to assure that the regression assumptions are satisfied.

12.63 **a-b** Answers will vary. The *Minitab* output is shown below, along with two diagnostic plots. These plots give no indications of any violation of assumptions. The printout indicates a significant linear regression ($t = 6.82$; p-value ≈ 0) with the regression line given as $\hat{y} = -54.0 + 0.0273x$.

Regression Analysis: Temperature Anomaly versus Year
```
The regression equation is
Temperature Anomaly = - 54.0 + 0.0273 Year

Predictor        Coef    SE Coef       T       P
Constant      -53.972      7.944   -6.79   0.000
Year         0.027279   0.003999    6.82   0.000

S = 0.189592    R-Sq = 62.4%    R-Sq(adj) = 61.1%

Analysis of Variance

Source            DF       SS       MS       F       P
Regression         1   1.6724   1.6724   46.53   0.000
Residual Error    28   1.0065   0.0359
Total             29   2.6789
```

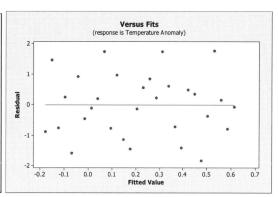

12.65 a Stiffness is inherent material quality, hence the two kinds of stiffness should be positively correlated. Use a computer program, your scientific calculator or the computing formulas given in the text to calculate the correlation coefficient r.

$$S_{xy} = \sum x_i y_i - \frac{(\sum x_i)(\sum y_i)}{n} = 1,233,987 - \frac{5028(2856)}{12} = 37,323$$

$$S_{xx} = \sum x_i^2 - \frac{(\sum x_i)^2}{n} = 2,212,178 - \frac{5028^2}{12} = 105,446$$

$$S_{yy} = \sum y_i^2 - \frac{(\sum y_i)^2}{n} = 723,882 - \frac{2856^2}{12} = 44,154$$

Then $r = \dfrac{S_{xy}}{\sqrt{S_{xx}S_{yy}}} = \dfrac{37,323}{\sqrt{105,446(44,154)}} = 0.5470$.

The test of hypothesis is $H_0: \rho = 0$ versus $H_a: \rho > 0$ and the test statistic is

$$t = \frac{r\sqrt{n-2}}{\sqrt{1-r^2}} = \frac{0.5470\sqrt{10}}{\sqrt{1-(0.5470)^2}} = 2.066$$

with p-value $= P(t > 2.066)$ bounded as

$$0.05 < p\text{-value} < 0.10$$

If the experimenter is willing to tolerate a p-value this large, then H_0 can be rejected. Otherwise, you would declare the results not significant; there is insufficient evidence to indicate that bending stiffness and twisting stiffness are positively correlated.

b $r^2 = (0.5470)^2 = 0.2992$ so that 29.9% of the total variation in y can be explained by the independent variable x.

12.67 The relationship between y = penetrability and x = number of days is apparently nonlinear, as seen by the strong curvilinear pattern in the residual plot. The regression analysis discussed in this chapter is not appropriate; we will discuss the appropriate model in Chapter 13.

Introduction to Probability and Statistics, 2ce

12.69 **a** The plot is shown below. Notice that the relationship is fairly weak.

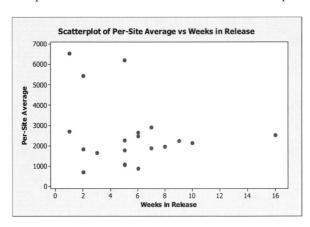

Regression Analysis: y versus x
```
The regression equation is
y = 3091 - 98 x

Predictor     Coef   SE Coef      T      P
Constant    3091.4     701.0   4.41  0.000
x            -97.7     107.2  -0.91  0.374
S = 1657.23   R-Sq = 4.4%   R-Sq(adj) = 0.0%

Analysis of Variance
Source          DF        SS        MS      F      P
Regression       1   2282896   2282896   0.83  0.374
Residual Error  18  49435210   2746401
Total           19  51718106

Unusual Observations
Obs     x       y    Fit   SE Fit   Residual   St Resid
  1   1.0    6551   2994      613       3557      2.31R
 10  16.0    2538   1527     1180       1011      0.87 X
 13   5.0    6219   2603      375       3616      2.24R
```

b From the printout, $r^2 = 0.044$. Only about 4% of the overall variation in y is explained by using the linear model.

c From the printout, the regression equation is $\hat{y} = 3091.4 - 97.7x$ and the regression is not significant ($t = -0.91$ with p-value $= 0.374$).

d Since the regression is not significant, it is not appropriate to use the regression line for estimation or prediction.

12.71 The error will be a maximum for the values of x at the extremes of the experimental region.

12.73

a The calculations shown below are done using the computing formulas. An appropriate computer program will provide identical results to within rounding error.

$$\sum x_i = 150 \qquad \sum y_i = 91 \qquad \sum x_i y_i = 986$$
$$\sum x_i^2 = 2750 \qquad \sum y_i^2 = 1120.04 \qquad n = 10$$

Then $S_{xy} = \sum x_i y_i - \dfrac{(\sum x_i)(\sum y_i)}{n} = 986 - \dfrac{150(91)}{10} = -379$

$S_{xx} = \sum x_i^2 - \dfrac{(\sum x_i)^2}{n} = 2750 - \dfrac{150^2}{10} = 500$

$S_{yy} = \sum y_i^2 - \dfrac{(\sum y_i)^2}{n} = 1120.04 - \dfrac{91^2}{10} = 291.94$

a $b = \dfrac{S_{xy}}{S_{xx}} = \dfrac{-379}{500} = -.758$ and $a = \bar{y} - b\bar{x} = 9.1 - (-.758)(15) = 20.47$

and the least squares line is $\hat{y} = a + bx = 20.47 - .758x$.

b Since Total SS $= S_{yy} = 291.94$ and SSR $= \dfrac{(S_{xy})^2}{S_{xx}} = \dfrac{(-379)^2}{500} = 287.282$

Then SSE = Total SS − SSR $= S_{yy} - \dfrac{(S_{xy})^2}{S_{xx}} = 4.658$

The ANOVA table with 1 df for regression and n − 2 df for error is shown below. Remember that the mean squares are calculated as MS = SS/df.

Source	df	SS	MS
Regression	1	287.282	287.282
Error	8	4.658	.58225
Total	9	291.940	

c To test $H_0 : \beta = 0, H_a : \beta \neq 0$, the test statistic is

$$t = \dfrac{b - \beta_0}{s/\sqrt{S_{xx}}} = \dfrac{-.758}{\sqrt{.58225/500}} = -22.21$$

The rejection region for $\alpha = 0.05$ is $|t| > t_{.025} = 2.306$ and we reject H_0. There is sufficient evidence to indicate that x and y are linearly related.

d The 95% confidence interval for the slope β is

$$b \pm t_{\alpha/2}\sqrt{\text{MSE}/S_{xx}} \Rightarrow -.758 \pm 2.896\sqrt{.58225/500} \Rightarrow -.758 \pm .099$$

or $-.857 < \beta < -.659$.

e When $x = 14$, the estimate of expected freshness $E(y)$ is $\hat{y} = 20.47 - .758(14) = 9.858$ and the 95% confidence interval is

$$\hat{y} \pm t_{.025}\sqrt{\text{MSE}\left(\dfrac{1}{n} + \dfrac{(x_p - \bar{x})^2}{S_{xx}}\right)}$$

$$9.858 \pm 2.306\sqrt{.58225\left(\dfrac{1}{10} + \dfrac{(14-15)^2}{500}\right)}$$

$$9.858 \pm .562$$

or $9.296 < E(y) < 10.420$.

f Calculate $r^2 = \dfrac{SSR}{\text{Total SS}} = \dfrac{287.282}{291.94} = 0.984$

The total variation has been reduced by 98.4%% by using the linear model.

12.75 Answers will vary from student to student.

12.76-12.77 Use the **How a Line Works** applet. The line $y = 0.5x + 3$ has a slope of 0.5 and a y-intercept of 3, while the line $y = -0.5x + 3$ has a slope of -0.5 and a y-intercept of 3. The second line slopes downward at the same rate as the first line slopes upward. They both cross the y axis at the same point.

12.79 Student should follow the directions given in the applet.

Case Study:
Are Foreign Companies "Buying Up the Canadian Economy"?

1 A plot of the data using the years 1975 to 1985 is shown below. The relationship appears to be linear.

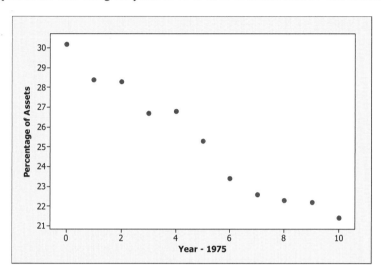

2 A *Minitab* program was run to provide a simple regression analysis for the years 1975 to 1985. The printout is shown below.

Regression Analysis: y versus x
```
The regression equation is
y = 29.7 - 0.895 x

Predictor        Coef    SE Coef        T       P
Constant      29.7091     0.3509    84.67   0.000
x            -0.89455    0.05931   -15.08   0.000

S = 0.622020   R-Sq = 96.2%   R-Sq(adj) = 95.8%
Analysis of Variance

Source           DF       SS        MS        F       P
Regression        1   88.023    88.023   227.50   0.000
Residual Error    9    3.482     0.387
Total            10   91.505

Predicted Values for New Observations
New
Obs    Fit   SE Fit       95% CI              95% PI
 1   5.556    1.318   (2.574, 8.538)    (2.259, 8.854)XX
 2   4.662    1.377   (1.547, 7.777)    (1.244, 8.080)XX
 3   3.767    1.436   (0.520, 7.015)    (0.228, 7.307)XX

XX denotes a point that is an extreme outlier in the predictors.
Values of Predictors for New Observations

New
Obs    x
 1   27.0
 2   28.0
 3   29.0
```

Introduction to Probability and Statistics, 2ce

3   From the printout the test statistic for testing $H_0: \beta = 0$, $H_a: \beta \neq 0$, is $t = -15.08$. Since the observed significance level is p-value = 0.000, there is a strong linear relationship between x and y.

4-5   The prediction intervals for the three years of interest are shown on the printout. Notice that the predictions are not accurate, illustrating the dangers of *extrapolation* – predicting outside of the experimental region.

6   When the data for 1986 – 2004 are added to the database, the following computer output results.

Regression Analysis: y versus x
```
The regression equation is
y = 24.5 + 0.0478 x

Predictor        Coef   SE Coef       T       P
Constant      24.5034    0.8941   27.40   0.000
x             0.04781   0.05295    0.90   0.374

S = 2.51021    R-Sq = 2.8%    R-Sq(adj) = 0.0%
Analysis of Variance
Source            DF        SS       MS      F       P
Regression         1     5.137    5.137   0.82   0.374
Residual Error    28   176.433    6.301
Total             29   181.570
```

Notice that the slope of the line has changed from negative to positive and that error measured by SSE = 176.433 has increased.

7   A quadratic or a cubic model, which will be discussed in Chapter 13, might be more appropriate. You can see the cubic pattern in the plot of the data points from 1975 – 2004 and in the residual plot for the analysis in part **6**, shown below.

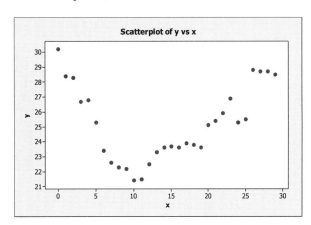
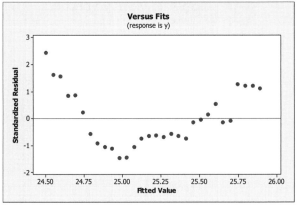

Project 12: Aspen Mixedwood Forests in Alberta

1 A scatterplot for the forest data is below.

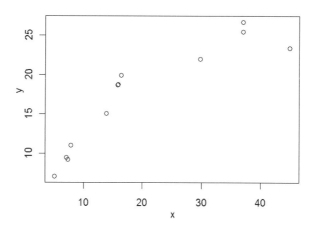

2 From the scatterplot, there is evidence of a positive relationship between DBH and canopy height. The relationship is somewhat curvilinear, but a linear fit would probably be adequate.

3 The appropriate hypothesis test is found on page 564 of the text, where it describes the test statistic as $t = r\sqrt{\frac{n-2}{1-r^2}}$. Now, $r = \frac{S_{xy}}{\sqrt{S_{xx}S_{yy}}} = \frac{920.5808}{\sqrt{(2077.009)(500.8492)}} = 0.902587$, and $t = 0.902587\sqrt{\frac{12-2}{1-0.902587^2}} = 6.63$. Clearly, since this test statistic is so large, at any reasonable level of significance α, the null hypothesis that $\rho = 0$ would be strongly rejected. Therefore yes, there is a significant positive correlation between x and y.

4 Instead of *Minitab*, we will use *R* to perform the analysis. The results are:

```
Coefficients:
             Estimate    Std. Error   t value    Pr(>|t|)
(Intercept)  8.43997     1.59248      5.30       0.000348
x            0.44322     0.06685      6.63       5.85e-05
Residual standard error: 3.047 on 10 degrees of freedom
Multiple R-squared: 0.8147,    Adjusted R-squared: 0.7961
F-statistic: 43.96 on 1 and 10 DF, p-value: 5.854e-05

Analysis of Variance Table

Response: y
           Df  Sum Sq  Mean Sq  F value   Pr(>F)
x           1  408.02   408.02   43.956   5.854e-05
Residuals  10   92.83     9.28
```

5 The equation of the regression line is $\hat{y} = 8.43997 + 0.44322x$.

6 This is represented by the slope of the line, which is 0.44322.

7 Since the r^2 value is fairly high at 0.8147, we can conclude that the strength of the linear relationship between x and y is strong.

8 The coefficient of determination is the same thing as r^2, which equals 0.8147. It can be found in the *R* output above. It can also be interpreted as found in question 9 below.

9 The percentage of total variation in y that can be explained by the regression is $0.8147 = r^2$. This is the classical interpretation of r^2.

10 The text interprets the phrase, "a linear regression model is useful" when the slope is significantly different from 0. Mimicking Example 12.2, the null hypothesis is that $\beta = 0$ versus $\beta \neq 0$. The test statistic is $t = \frac{b-0}{\sqrt{MSE/S_{xx}}} = \frac{0.44322}{\sqrt{(3.047)^2/2077.009}} = 6.629$. At this value, the p-value is essentially 0, and so, since the p-value is less than the 5% significance level, we can reject the null hypothesis, and conclude that the regression is "useful", as the text puts it.

11 From the computer output found in Question 4 above, the value of the F-statistic is 43.956. This can also be calculated by hand using $F = \frac{MSR}{MSE}$, where $MSR = \frac{(S_{xy})^2}{S_{xx}} = \frac{(920.5808)^2}{2077.009} = 408.023754$ and $MSE = \left(S_{yy} - \frac{(S_{xy})^2}{S_{xx}}\right)/(n-2) = \left(500.8492 - \frac{(920.5808)^2}{2077.009}\right)/(12-2) = 9.28254$, implying that $F = \frac{408.023754}{9.28254} = 43.956$. The p-value is also given in the output (5.854 x 10^{-5}), which is clearly less than $\alpha = 5\%$. Thus, by the F-test we can again reject the null hypothesis that $\beta = 0$.

12 The critical value at $\alpha = 2.5\%$ for the t-test is 2.228139 (based on 10 degrees of freedom). The critical value for the F-test at $\alpha = 5\%$ is 4.964603 (based on 1 and 10 degrees of freedom). The relationship between them is as follows: $2.228139^2 = 4.964603$.

13 Plugging $x = 42$ into the regression equation yields, $\hat{y} = 8.43997 + 0.44322(42) = 27.05521$ as the predicted tree canopy height.

14 As a general rule, the linear regression should not be used to make predictions outside of the range of values used to make the line, since we have no evidence that the linear relationship will persist in these areas. In this case, a DBH of 52 is very close to the largest observed value of 45, and so it might be ok in this instance, although technically not allowed.

15 A 95% confidence interval for β is $b \pm t_{\alpha/2}\sqrt{\frac{s^2}{S_{xx}}}$, where the notation is from the text. The value of s is given in the computer output above ($s = 3.047$) and S_{xx} was computed in Question 3 as $S_{xx} = 2077.009$. We have $(12-2)$ degrees of freedom for this question. And so, we obtain $0.44322 \pm 2.228\sqrt{\frac{3.047^2}{2077.009}} = (0.29426, 0.59218)$.

16 The estimate is simply b, or 0.44322. A 99% confidence interval is the same procedure as in the previous question, but $t_{\alpha/2} = 3.169$ instead. Thus, the 99% interval is $0.44322 \pm 3.169\sqrt{\frac{3.047^2}{2077.009}} = (0.23135, 0.65509)$. This interval is wider than the one in Question 15, as it should since it is a 99% interval.

17 The best estimate for σ^2 is s^2, which from the output (Question 4 above) is $s^2 = (3.047)^2 = 9.284$.

18 The residuals are equal to the difference between y and $\hat{y}$ as can be seen below:

y	$\hat{y}$	residual
7.10	10.70	-3.60
11.00	11.94	-0.94
9.50	11.59	-2.09
9.20	11.72	-2.52
18.70	15.44	3.26
19.90	15.71	4.19
18.80	15.49	3.31
15.10	14.60	0.50
25.50	24.84	0.66
22.00	21.65	0.35
23.40	28.38	-4.98
26.70	24.84	1.86

19 A plot of the residuals versus the fitted values is shown below. It does not really appear that the constant variance assumption has been violated, as there is no increasing or decreasing pattern in the size of the residuals to be found in the plot. It is hard to make strong conclusions for such a small number of points, but everything looks ok, other than the fact that lower fitted values tend to have negative residuals, which may be a slight cause for concern.

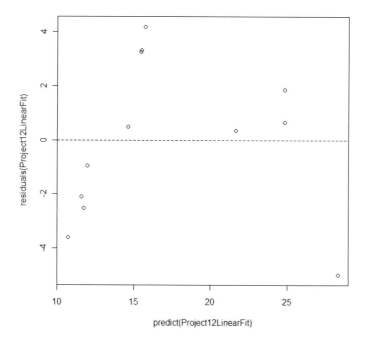

20 The following plot is a frequency histogram of the residuals. No, it does not appear that the errors are normally distributed, as the histogram is not mound-shaped.

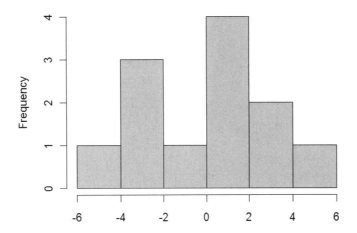

21 The method needed to do this is not covered in Chapter 12.

22 As none of the standardized residuals are greater than 2 in absolute value, we can safely say that there are no major outliers, although point #11 is somewhat suspect.

23 A normal probability plot of the residuals is below.

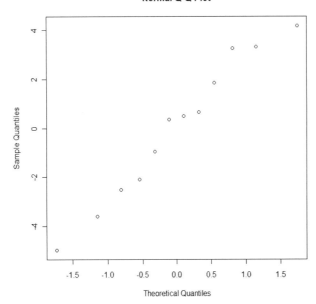

The pattern is linear, indicating that the residuals follow a normal distribution, as they should.

24 A 95% confidence interval for the mean of y when DBH $= x_0 = 10$ can be found using the formula on page 557 of the text: $\hat{y} \pm t_{\alpha/2}\sqrt{MSE\left[\frac{1}{n} + \frac{(x_0-\bar{x})^2}{S_{xx}}\right]} = 12.87217 \pm 2.228\sqrt{(3.047)^2\left[\frac{1}{12} + \frac{(10-19.85833)^2}{2077.009}\right]}$
$= 12.87217 \pm 0.42392 = (12.448, 13.296)$.

25 A 95% prediction interval for the mean of y when DBH $= x_0 = 10$ can be found using the formula on page 557 of the text: $\hat{y} \pm t_{\alpha/2}\sqrt{MSE\left[1 + \frac{1}{n} + \frac{(x_0-\bar{x})^2}{S_{xx}}\right]} = 12.87217 \pm 2.228\sqrt{(3.047)^2\left[1 + \frac{1}{12} + \frac{(10-19.85833)^2}{2077.009}\right]}$
$= 12.87217 \pm 7.21690 = (5.655, 20.089)$.

26 The confidence interval is narrower. As it states on page 557 of the text, "the prediction interval is wider because of the extra variability in predicting the actual value of the response y."

Chapter 13: Multiple Regression Analysis

13.1   **a**   When $x_2 = 2$, $E(y) = 3 + x_1 - 2(2) = x_1 - 1$.
When $x_2 = 1$, $E(y) = 3 + x_1 - 2(1) = x_1 + 1$.
When $x_2 = 0$, $E(y) = 3 + x_1 - 2(0) = x_1 + 3$.
These three straight lines are graphed below.

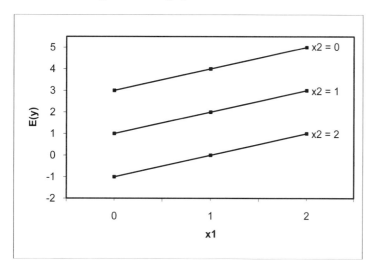

   b   Notice that the lines are parallel (they have the same slope).

13.3   **a**   The hypothesis to be tested is $H_0: \beta_1 = \beta_2 = \beta_3 = 0$ H_a: at least one β_i differs from zero
and the test statistic is
$$F = \frac{\text{MSR}}{\text{MSE}} = 57.44$$
which has an F distribution with $df_1 = k = 3$ and $df_2 = n - k - 1 = 15 - 3 - 1 = 11$. The rejection region for $\alpha = .05$, which is found in the upper tail of the F distribution, is $F > 3.59$ and H_0 is rejected. There is evidence that the model contributes information for the prediction of y.

   b   Use the fact that
$$F = \frac{R^2/k}{(1-R^2)/[n-(k+1)]} = 57.44$$
Solving for R^2 you find
$$\frac{R^2/3}{(1-R^2)/11} = 57.44$$
$$.33R^2 = 5.2218 - 5.2218R^2$$
$$R^2 = \frac{5.2218}{5.5551} = .94$$

If $R^2 = .94$, the total sum of squares of deviations of the y-values about their mean has been reduced by 94% by using the linear model to predict y.

13.5   **a**   The model is quadratic.

b   Since $R^2 = .815$, the sum of squares of deviations is reduced by 81.5% using the quadratic model rather than $\bar{y}$ to predict y.

c   The hypothesis to be tested is $H_0: \beta_1 = \beta_2 = 0$ H_a: at least one β_i differs from zero and the test statistic is

$$F = \frac{\text{MSR}}{\text{MSE}} = 37.37$$

which has an F distribution with $df_1 = k = 2$ and $df_2 = n - k - 1 = 20 - 2 - 1 = 17$. The p-value given in the printout is P = .000 and H_0 is rejected. There is evidence that the model contributes information for the prediction of y.

13.7   **a**   Refer to Exercise 13.5. When $x = 0$, the estimate of $E(y)$ is

$$\hat{y} = 10.5638 + 4.4366(0) - .64754(0)^2 = 10.5638.$$

b   Since $E(y) = \beta_0 + \beta_1 x + \beta_2 x^2$, when $x = 0$, $E(y) = \beta_0$. A test of $E(y$ given $x = 0) = 0$ is equivalent to a test of $H_0: \beta_0 = 0$ $H_a: \beta_0 \neq 0$
The individual t-test is

$$t = \frac{b_0}{SE(b_0)} = \frac{10.5638}{.6951} = 15.20$$

with p-value = .000 and H_0 is rejected. The mean value of y differs from zero when $x = 0$.

13.9   **a**   Rate of increase is measured by the slope of a line tangent to the curve; this line is given by an equation obtained as dy/dx, the derivative of y with respect to x. In particular,

$$\frac{dy}{dx} = \frac{d}{dx}\left(\beta_0 + \beta_1 x + \beta_2 x^2\right) = \beta_1 + 2\beta_2 x$$

which has slope $2\beta_2$. If β_2 is negative, then the rate of increase is decreasing. Hence, the hypothesis of interest is $H_0: \beta_2 = 0,$ $H_a: \beta_2 < 0$

b   The individual t-test is $t = -8.11$ as in Exercise 13.8b. However, the test is one-tailed, which means that the p-value is half of the amount given in the printout. That is, p-value $= \frac{1}{2}(.000) = .000$. Hence, H_0 is again rejected. There is evidence to indicate a decreasing rate of increase.

13.11   Refer to Exercise 13.10.
a   From the printout, SSR = 234.96 and Total SS = $S_{yy} = 236.02$. Then

$$R^2 = \frac{\text{SSR}}{\text{Total SS}} = \frac{234.96}{236.02} = .9955 \text{ which agrees with the printout.}$$

b   Calculate $R^2(\text{adj}) = \left(1 - \frac{\text{MSE}}{\text{Total SS}/(n-1)}\right)100\% = \left(1 - \frac{.35}{236.02/5}\right)100\% = 99.3\%$

The value of R^2(adj) can be used to compare two or more regression models using different numbers of independent predictor variables. Since the value of $R^2(\text{adj}) = 99.3\%$ is just slightly larger than the value of $R^2(\text{adj}) = 95.7\%$ for the linear model, the quadratic model fits just slightly better.

13.13   **a**   The values of R^2(adj) should be used to compare several different regression models. For the seven possible models given in the printout, the largest value of R^2(adj) is 29.7% which occurs when x_2, x_3, and x_4 are included in the model. This agrees with the decision made in Exercise 13.12b.
b   Even using the best of all possible subsets of these four predictor variables, the model does not fit well. The experimenter may want to look for some other possible predictor variables for the taste score.

13.15 **a** The *Minitab* printout fitting the model to the data is shown on the next page. The least squares line is
$$\hat{y} = -8.177 + 0.292x_1 + 4.434x_2$$

Regression Analysis: y versus x1, x2
```
The regression equation is
y = - 8.18 + 0.292 x1 + 4.43 x2

Predictor      Coef    SE Coef       T      P
Constant     -8.177      4.206   -1.94  0.093
x1           0.2921     0.1357    2.15  0.068
x2           4.4343     0.8002    5.54  0.001
S = 3.30335    R-Sq = 82.3%    R-Sq(adj) = 77.2%

Analysis of Variance
Source          DF        SS       MS       F      P
Regression       2    355.22   177.61   16.28  0.002
Residual Error   7     76.38    10.91
Total            9    431.60

Source  DF   Seq SS
x1       1    20.16
x2       1   335.05
```

b The F test for the overall utility of the model is $F = 16.28$ with $P = .002$. The results are highly significant; the model contributes significant information for the prediction of y.

c To test the effect of advertising expenditure, the hypothesis of interest is $H_0 : \beta_2 = 0$, $H_a : \beta_2 \neq 0$ and the test statistic is $t = 5.54$ with p-value $= .001$. Since $\alpha = .01$, H_0 is rejected. We conclude that advertising expenditure contributes significant information for the prediction of y, given that capital investment is already in the model.

d From the *Minitab* printout, R-Sq = 82.3%, which means that 82.3% of the total variation can be explained by the quadratic model. The model is very effective.

13.17 **a** Quantitative
 b Quantitative
 c Qualitative ($x_1 = 1$ if B; 0 otherwise $x_2 = 1$ if C; 0 otherwise)
 d Quantitative
 e Qualitative ($x_1 = 1$ if night shift; 0 otherwise)

13.19 **a** The variable x_2 must be the quantitative variable, since it appears as a quadratic term in the model. Qualitative variables appear only with exponent 1, although they may appear as the coefficient of another quantitative variable with exponent 2 or greater.
 b When $x_1 = 0$, $\hat{y} = 12.6 + 3.9x_2^2$ while when $x_1 = 1$,
$$\hat{y} = 12.6 + .54(1) - 1.2x_2 + 3.9x_2^2$$
$$= 13.14 - 1.2x_2 + 3.9x_2^2$$

c The following graph shows the two parabolas.

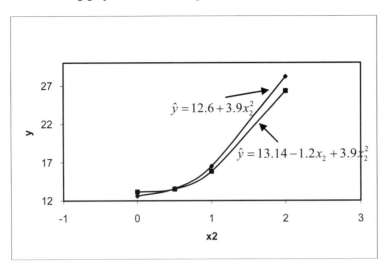

13.21 a The model involves two variables and an interaction between the two:
$$E(y) = \beta_0 + \beta_1 x_1 + \beta_2 x_2 + \beta_3 x_1 x_2$$
where $x_1 = 0$ if cotton; 1 if cucumber and x_2 = temperature.

b The Minitab regression printout is shown below.

Regression Analysis: y versus x1, x2, x1x2
```
The regression equation is
y = 10.9 - 12.0 x1 + 1.13 x2 + 0.97 x1x2

Predictor       Coef   SE Coef       T      P
Constant       10.93     27.11    0.40  0.691
x1            -12.03     37.32   -0.32  0.750
x2             1.128     1.119    1.01  0.324
x1x2           0.969     1.531    0.63  0.533

S = 10.5737    R-Sq = 37.8%    R-Sq(adj) = 29.3%

Analysis of Variance

Source           DF       SS      MS      F      P
Regression        3   1492.8   497.6   4.45  0.014
Residual Error   22   2459.7   111.8
Total            25   3952.5

Source   DF   Seq SS
x1        1    928.6
x2        1    519.4
x1x2      1     44.8
```

c Look first at the interaction effect. The interaction term is not significant ($t = .63$ with $P = .533$). That is, there is insufficient evidence to indicate that the effect of temperature on the number of eggs is different depending on the type of plant.

d Since the interaction term is not significant, it could be removed and the data refit using the model
$$E(y) = \beta_0 + \beta_1 x_1 + \beta_2 x_2$$
The *Minitab* printout for the regression analysis with interaction removed is shown below.

Regression Analysis: y versus x1, x2

```
The regression equation is
y = - 1.5 + 11.4 x1 + 1.65 x2

Predictor     Coef    SE Coef       T       P
Constant     -1.54     18.39    -0.08   0.934
x1          11.442     4.113     2.78   0.011
x2          1.6456    0.7535     2.18   0.039

S = 10.4350    R-Sq = 36.6%    R-Sq(adj) = 31.1%

Analysis of Variance

Source            DF      SS      MS       F       P
Regression         2  1448.0   724.0    6.65   0.005
Residual Error    23  2504.5   108.9
Total             25  3952.5
```

Notice that both variables are significant, and that the overall model contributes significant information for the prediction of y. However, since $R^2 = 36.6\%$, there is still much variation which has not been accounted for. The model without interaction is better, but still does not fit as well as it might. Perhaps there are other variables that the experimenter should explore.

e Answers will vary.

13.23 The basic response equation for a specific type of bonding compound would be
$$E(y) = \beta_0 + \beta_1 x_1 + \beta_2 x_1^2$$
Since the qualitative variable "bonding compound" is at two levels, one dummy variable is needed to incorporate this variable into the model. Define the dummy variable x_2 as follows:

$x_2 = 1$ if bonding compound 2

$\quad\, = 0$ otherwise

The expanded model is now written as
$$E(y) = \beta_0 + \beta_1 x_1 + \beta_2 x_1^2 + \beta_3 x_2 + \beta_4 x_1 x_2 + \beta_5 x_1^2 x_2$$

13.25 **a** From the printout, the prediction equation is $\hat{y} = 8.585 + 3.8208x - 0.21663x^2$.

b R^2 is labeled "R-sq" or $R^2 = .944$. Hence 94.4% of the total variation is accounted for by using x and x^2 in the model.

c The hypothesis of interest is $H_0 : \beta_1 = \beta_2 = 0$ H_a: at least one β_i differs from zero
and the test statistic is $F = 33.44$ with p-value $= .003$. Hence, H_0 is rejected, and we conclude that the model contributes significant information for the prediction of y.

d The hypothesis of interest is $H_0 : \beta_2 = 0$ $H_a : \beta_2 \neq 0$
and the test statistic is $t = -4.93$ with p-value $= .008$. Hence, H_0 is rejected, and we conclude that the quadratic model provides a better fit to the data than a simple linear model.

e The pattern of the diagnostic plots does not indicate any obvious violation of the regression assumptions.

13.27 The *Minitab* printout for the data is shown below.

Regression Analysis: y versus x1, x2, x3, x1x2, x1x3

```
The regression equation is
y = 4.10 + 1.04 x1 + 3.53 x2 + 4.76 x3 - 0.00430 x1x2 - 0.00080 x1x3

Predictor        Coef     SE Coef       T       P
Constant       4.1000      0.3860   10.62   0.000
x1             1.0400      0.1164    8.94   0.000
x2             3.5300      0.5459    6.47   0.000
x3             4.7600      0.5459    8.72   0.000
x1x2        -0.004300    0.001646   -2.61   0.028
x1x3        -0.000800    0.001646   -0.49   0.639
S = 0.368028   R-Sq = 98.4%   R-Sq(adj) = 97.5%

Analysis of Variance
Source            DF       SS       MS        F       P
Regression         5   74.830   14.966   110.50   0.000
Residual Error     9    1.219    0.135
Total             14   76.049
```

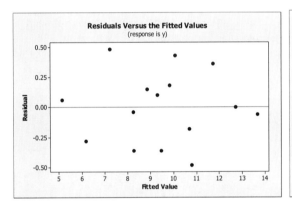
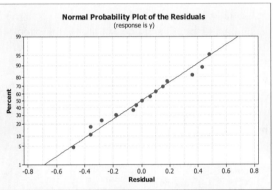

a The model fits very well, with an overall $F = 110.50$ (P = .000) and $R^2 = .984$. The diagnostic plots indicate no violations of the regression assumptions.

b The parameter estimates are found in the column marked "Coef" and the prediction equation is
$$\hat{y} = 4.10 + 1.04x_1 + 3.53x_2 + 4.76x_3 - 0.43x_1x_2 - 0.08x_1x_3$$

Using the dummy variables defined in Exercise 13.26, the coefficients can be combined to give the three lines that are graphed in the figure below.

Men: $\hat{y} = 4.10 + 1.04x_1$

Children: $\hat{y} = 7.63 + 0.61x_1$

Women: $\hat{y} = 8.86 + 0.96x_1$

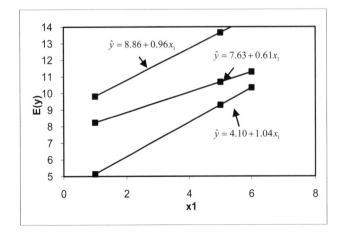

c The hypothesis of interest is $H_0: \beta_4 = 0 \quad H_a: \beta_4 \neq 0$
and the test statistic is $t = -2.61$ with $P = .028$. Since this value is less than .05, the results are significant at the 5% level of significance and H_0 is rejected. There is a difference in the slopes.

d The hypothesis of interest is $H_0: \beta_4 = \beta_5 = 0 \quad H_a:$ at least one β_i differs from zero for $i = 4, 5$
Using the methods of Section 13.5 and the *Minitab* printout above, $SSE_2 = 1.219$ with 9 degrees of freedom, while the printout below, fit using the reduced model gives $SSE_1 = 2.265$ with 11 degrees of freedom.

Regression Analysis (reduced model): y versus x1, x2, x3

```
The regression equation is
y = 4.61 + 0.870 x1 + 2.24 x2 + 4.52 x3

Predictor       Coef   SE Coef       T       P
Constant      4.6100    0.3209   14.37   0.000
x1           0.87000   0.08285   10.50   0.000
x2            2.2400    0.2870    7.81   0.000
x3            4.5200    0.2870   15.75   0.000

S = 0.453772   R-Sq = 97.0%    R-Sq(adj) = 96.2%

Analysis of Variance
Source          DF         SS        MS        F       P
Regression       3     73.784    24.595   119.44   0.000
Residual Error  11      2.265     0.206
Total           14     76.049
```

Hence, the degrees of freedom associated with $SSE_1 - SSE_2 = 1.046$ is $11 - 9 = 2$. The test statistic is

$$F = \frac{(SSE_1 - SSE_2)/2}{SSE_2/9} = \frac{1.046/2}{.1354} = 3.86$$

The rejection region with $\alpha = .05$ is $F > F_{.05} = 4.26$ (with 2 and 9 *df*) and H_0 is not rejected. The interaction terms in the model are not significant. The experimenter should consider eliminating these terms from the model.

e Answers will vary.

13.29 a The model is $y = \beta_0 + \beta_1 x_1 + \beta_2 x_2 + \beta_3 x_1^2 + \beta_4 x_1 x_2 + \beta_5 x_1^2 x_2 + \varepsilon$ and the Minitab printout is shown below.

Regression Analysis: y versus x1, x2, x1sq, x1x2, x1sqx2

```
The regression equation is
y = 4.5 + 6.39 x1 - 50.9 x2 + 0.132 x1sq + 17.1 x1x2 - 0.502 x1sqx2

Predictor       Coef   SE Coef       T       P
Constant        4.51     42.24    0.11   0.916
x1             6.394     5.777    1.11   0.275
x2           -50.85     56.21    -0.90   0.371
x1sq          0.1318    0.1687    0.78   0.439
x1x2          17.064     7.101    2.40   0.021
x1sqx2       -0.5025    0.1992   -2.52   0.016
S = 71.6891    R-Sq = 76.8%    R-Sq(adj) = 73.8%

Analysis of Variance
Source          DF         SS        MS        F       P
Regression       5     664164    132833    25.85   0.000
Residual Error  39     200434      5139
Total           44     864598
```

b The fitted prediction model uses the coefficients given in the column marked "Coef" in the printout:
$$\hat{y} = 4.51 + 6.394x_1 - 50.85x_2 + 17.064x_1x_2 + .1318x_1^2 - .5025x_1^2x_2$$
The F test for the model's utility is $F = 25.85$ with P = .000 and $R^2 = .768$. The model fits quite well.

c If the dolphin is female, $x_2 = 0$ and the prediction equation becomes $\hat{y} = 4.51 + 6.394x_1 + .1318x_1^2$

d If the dolphin is male, $x_2 = 1$ and the prediction equation becomes $\hat{y} = -46.34 + 23.458x_1 - .3707x_1^2$

e The hypothesis of interest is $H_0 : \beta_4 = 0 \quad H_a : \beta_4 \neq 0$
and the test statistic is $t = .78$ with p-value $= .439$. H_0 is not rejected and we conclude that the quadratic term is not important in predicting mercury concentration for female dolphins.

f Answers will vary from student to student.

13.31 **a-b** The data is plotted below. It appears to be a curvilinear relationship, which could be described using the quadratic model $y = \beta_0 + \beta_1 x + \beta_2 x^2 + \varepsilon$.

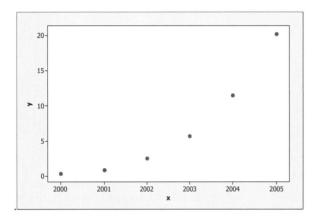

c The *Minitab* printout is shown below.

Regression Analysis: y versus x, x_sq
```
The regression equation is
y = 4114749 - 4113 x + 1.03 x_sq

Predictor      Coef    SE Coef       T       P
Constant    4114749     343582   11.98   0.001
x            -4113.4      343.2  -11.99   0.001
x_sq         1.02804    0.08568   12.00   0.001
S = 0.523521   R-Sq = 99.7%    R-Sq(adj) = 99.5%

Analysis of Variance
Source          DF        SS       MS        F       P
Regression       2    297.16   148.58   542.11   0.000
Residual Error   3      0.82     0.27
Total            5    297.98
```

d The hypothesis of interest is $H_0 : \beta_1 = \beta_2 = 0$
and the test statistic is $F = 542.11$ with p-value $= .000$. H_0 is rejected and we conclude that the model provides valuable information for the prediction of y.

e $R^2 = .997$. Hence, 99.7% of the total variation is accounted for by using x and x^2 in the model.

f The residual plots are shown below. There is no reason to doubt the validity of the regression assumptions.

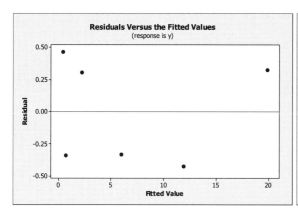
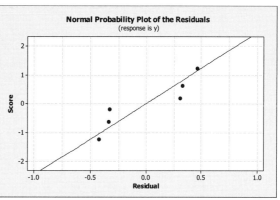

13.33 The hypothesis of interest is $H_0 : \beta_1 = \beta_2 = \beta_3 = 0$

and the test statistic is $F = 1.49$ with p-value $= .235$. H_0 is not rejected and we conclude that the model does not provide valuable information for the prediction of y. This matches the results of the analysis of variance F-test.

13.35 **a** $R^2 = .999$. Hence, 99.9% of the total variation is accounted for by using x and x^2 in the model.

b The hypothesis of interest is $H_0 : \beta_1 = \beta_2 = 0$

and the test statistic is $F = 1676.61$ with p-value $= .000$. H_0 is rejected and we conclude that the model provides valuable information for the prediction of y.

c The hypothesis of interest is $H_0 : \beta_1 = 0$

and the test statistic is $t = -2.65$ with p-value $= .045$. H_0 is rejected and we conclude that the linear regression coefficient is significant when x^2 is in the model.

d The hypothesis of interest is $H_0 : \beta_2 = 0$

and the test statistic is $t = 15.14$ with p-value $= .000$. H_0 is rejected and we conclude that the quadratic regression coefficient is significant when x is in the model.

e When the quadratic term is removed from the model, the value of R^2 decreases by $99.9 - 93.0 = 6.9\%$. This is the additional contribution of the quadratic term when it is included in the model.

f The clear pattern of a curve in the residual plot indicates that the quadratic term should be included in the model.

Case Study: "Buying Up the Canadian Economy" – Another Look

1 The three figures shown below represent the data with a linear, quadratic and cubic model fitted to them.

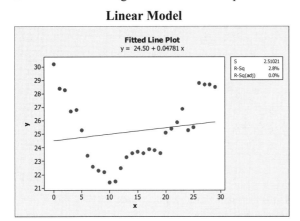

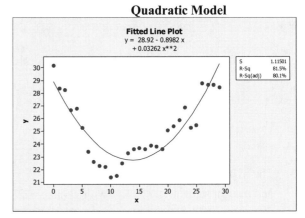

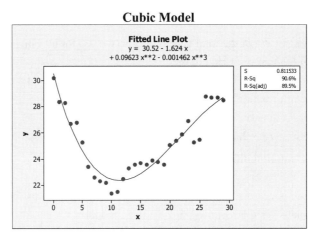

2 The residuals using the fitted linear regression model are found by calculating $y_i - \hat{y}_i$ for each of the 30 values of y. These values are plotted in the figure below. Notice the periodic trend to the residuals, indicating that a higher order, possibly quadratic or cubic model might be useful.

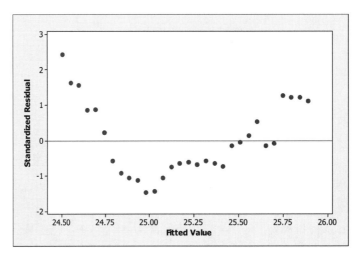

3 When the linear model is fit, $R^2 = 2.8\%$ while $R^2 = 81.5\%$ for the quadratic fit. The quadratic term is highly significant ($t = 7.45$ with p-value $= .000$) and the residuals from the quadratic fit are plotted in the figure below.

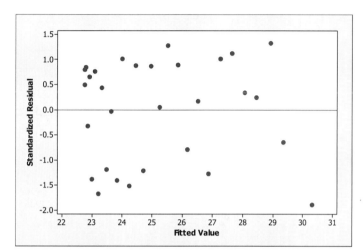

4 When the cubic model is fit, $R^2 = 90.6\%$ -- higher than R^2 for the quadratic model. Also, for the cubic model, $R^2(\text{adj}) = 89.5\%$ compared to $R^2(\text{adj}) = 80.1\%$ for the quadratic model. The cubic term is significant ($t = -5$ with p-value $= 0$). The cubic model is the best of the three choices.

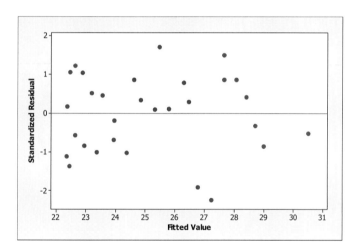

Project 13: Aspen Mixedwood Forests in Canada, Part 2

(i) Scatterplots of all the variables are shown below.

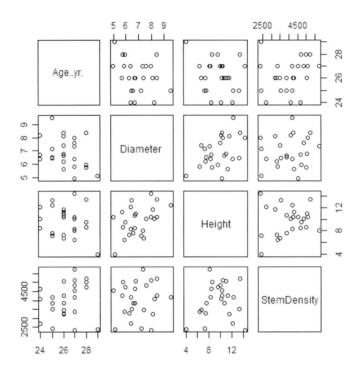

It is difficult to say whether or not linear regression would be useful in this case, based solely on these plots.

(ii) Next, we will model each variable separately.

Fit 1: Dependent: Age, Independent: Diameter

```
              Estimate Std. Error t value Pr(>|t|)
(Intercept)   28.5331    1.6368   17.432  9.44e-15
Diameter      -0.3311    0.2333   -1.419   0.169
Residual standard error: 1.304 on 23 degrees of freedom
Multiple R-squared: 0.08051,   Adjusted R-squared: 0.04054
F-statistic: 2.014 on 1 and 23 DF,  p-value: 0.1693
```

Fit 2: Dependent: Age, Independent: Height

```
              Estimate Std. Error t value Pr(>|t|)
(Intercept)   27.3704    1.0982   24.922  <2e-16
Height        -0.1160    0.1094   -1.061   0.3
Residual standard error: 1.328 on 23 degrees of freedom
Multiple R-squared: 0.04665,   Adjusted R-squared: 0.005196
F-statistic: 1.125 on 1 and 23 DF,  p-value: 0.2998
```

Fit 3: Dependent: Age, Independent: Stem Density

```
              Estimate  Std. Error  t value  Pr(>|t|)
(Intercept)  2.508e+01  1.175e+00    21.347   <2e-16
StemDensity  2.893e-04  2.855e-04     1.013    0.322
Residual standard error: 1.331 on 23 degrees of freedom
Multiple R-squared: 0.04272,    Adjusted R-squared: 0.001102
F-statistic: 1.026 on 1 and 23 DF,  p-value: 0.3215
```

In all three cases, the slope of the regression is not significant, judging from the high *p*-values. The *p*-values for the overall *F*-test are also all high. So, taking each variable separately, linear regression is not appropriate for modeling the dependent variable *y*.

(iii) A model with all possible interactions (including a 3-way interaction) is fit, and the results are below.

```
                             Estimate  Std. Error  t value  Pr(>|t|)
(Intercept)                 6.059e+01   2.025e+01    2.992    0.0082
Diameter                   -5.064e+00   2.946e+00   -1.719    0.1038
Height                     -4.342e+00   2.288e+00   -1.898    0.0748
StemDensity                -4.010e-03   6.654e-03   -0.603    0.5547
Diameter:Height             5.880e-01   3.079e-01    1.910    0.0732
Diameter:StemDensity        5.755e-04   9.257e-04    0.622    0.5424
Height:StemDensity          6.794e-04   7.062e-04    0.962    0.3495
Diameter:Height:StemDensity -8.667e-05  9.359e-05   -0.926    0.3674
Residual standard error: 1.068 on 17 degrees of freedom
Multiple R-squared: 0.544,    Adjusted R-squared: 0.3562
F-statistic: 2.897 on 7 and 17 DF,  p-value: 0.03462
```

The question asks that we use the overall *F*-test to assess the merits of the model. In this case, since the *p*-value for the *F*-statistic is 0.03462, and since the *p*-value is less than 5% significance, we can conclude that some of the parameters are relevant/significant.

(iv) We can conduct a separate *t*-test for each independent variable to determine if they are significant. The *p*-values are given in the table of the preceding question. For all 3 independent variables, the *p*-value is greater than 5% significance, and so we would not reject the hypothesis that their regression parameters are equal to 0. The same conclusion can be drawn if we look at the *p*-values for each independent variable separately, as we did in question (ii).

(v) In this case, we obtain:

```
                 Estimate  Std. Error  t value  Pr(>|t|)
(Intercept)      44.83589    5.96431     7.517   2.2e-07
Diameter         -2.74817    0.91615    -3.000   0.00683
Height           -1.61817    0.56865    -2.846   0.00968
Diameter:Height   0.23574    0.08436     2.794   0.01086
Residual standard error: 1.16 on 21 degrees of freedom
Multiple R-squared: 0.3364,   Adjusted R-squared: 0.2416
F-statistic: 3.549 on 3 and 21 DF,  p-value: 0.03199
```

This fit is better than the one from part (iii). The *p*-values for all terms are very low, and therefore significant. The residual standard error and the *p*-value for the *F*-Statistic are comparable to the fit in part (iii), but based on less independent variables. As a whole, the fit here is superior.

(vi) From part (v) above, $\beta_1 = -2.74817$, $\beta_2 = -1.61817$ and $\beta_3 = 0.23574$. In the absence of interaction terms, each individual parameter would represent the change in the response for a one unit increase in that independent variable, if all of the other variables were held constant. But with interaction terms, such an interpretation is not meaningful. Thus, we interpret them simply as parameters used for this particular model.

(vii) Yes, the quadratic term in part (v) seems to be very useful. The *p*-value for the *t*-test for β_3 is 0.01086, which implies that it is significant (it is less than 5%). The R^2 value is somewhat low, but reasonable. The residual standard error and the *p*-value for the *F*-Statistic are comparable to the regression from part (iii), and are decent.

(viii) The residual plots below indicate no serious violation of the regression model assumptions.

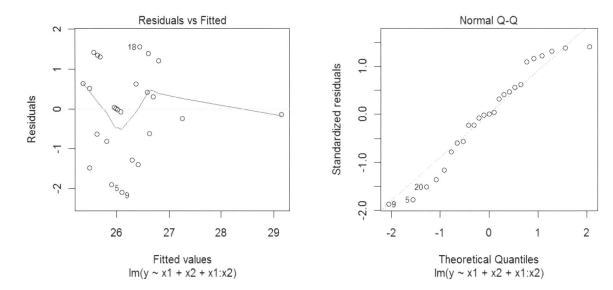

(ix) I would say Diameter at Breast Height makes the most significant contribution, since it has the lowest *p*-value for its *t*-test.

(x) Yes, the interaction term is certainly significant, judging from its low *p*-value (0.01086).

(xi) The results from the regression are:

```
                Estimate Std. Error t value Pr(>|t|)
(Intercept)     46.46127    6.69899   6.936 9.82e-07
Diameter        -3.36046    1.42120  -2.365   0.0283
Height          -1.46794    0.63521  -2.311   0.0316
I(Height^2)     -0.02775    0.04865  -0.570   0.5748
Diameter:Height  0.29320    0.13229   2.216   0.0384
Residual standard error: 1.179 on 20 degrees of freedom
Multiple R-squared: 0.347,    Adjusted R-squared: 0.2164
F-statistic: 2.657 on 4 and 20 DF,  p-value: 0.063
```

The additional quadratic term is not significant, as the *p*-value is 0.5748. This term should not be added to the model.

(xii) The best model is the one from part (v). Note that Stem Density is not an independent variable for this model, and so the value 4900 is ignored. Using statistical software, we find that a point prediction for *y* (Age) is 26.86701 when DBH is 8 and Canopy Height is 15. A 95% prediction interval is found to be (24.04376, 29.69027).

(xiii) Adding Stem Density produces the following fit:

```
                Estimate Std. Error t value Pr(>|t|)
(Intercept)    46.9099595  5.3251686   8.809 2.55e-08
Diameter       -3.4047892  0.8465425  -4.022 0.000668
Height         -2.0292136  0.5258319  -3.859 0.000978
StemDensity     0.0006109  0.0002320   2.633 0.015944
Diameter:Height 0.2943709  0.0777478   3.786 0.001159
Residual standard error: 1.024 on 20 degrees of freedom
Multiple R-squared: 0.5072,    Adjusted R-squared: 0.4087
F-statistic: 5.147 on 4 and 20 DF,  p-value: 0.005127

Analysis of Variance Table

Model 1: y ~ x1 + x2 + x1:x2
Model 2: y ~ x1 + x2 + x1:x2 + x3
  Res.Df    RSS Df Sum of Sq      F  Pr(>F)
1     21 28.243
2     20 20.973  1     7.270 6.9328 0.01594
With an F-value of 6.9328, the null hypothesis is rejected.
```

A partial *F*-test on x_3 yields a *p*-value of 0.01594, exactly the same as the *p*-value for the *t*-test for that variable (as it must when the reduced model is exactly one variable less). And so, at 5% significance, the reduced model is deemed to be inadequate.

Chapter 14: Analysis of Categorical Data

14.1 See Section 14.1 of the text.

14.3 For a test of specified cell probabilities, the degrees of freedom are $k-1$. Use Table 5, Appendix I:
    a $df = 6$; $\chi^2_{.05} = 12.59$; reject H_0 if $X^2 > 12.59$
    b $df = 9$; $\chi^2_{.01} = 21.666$; reject H_0 if $X^2 > 21.666$

14.5   **a**   Three hundred responses were each classified into one of five categories. The objective is to determine whether or not one category is preferred over another. To see if the five categories are equally likely to occur, the hypothesis of interest is

$$H_0 : p_1 = p_2 = p_3 = p_4 = p_5 = \frac{1}{5}$$

versus the alternative that at least one of the cell probabilities is different from 1/5.

    b   The number of degrees of freedom is equal to the number of cells, k, less one degree of freedom for each linearly independent restriction placed on $p_1, p_2, \ldots, p_k$. For this exercise, $k = 5$ and one degree of freedom is lost because of the restriction that

$$\Sigma p_i = 1$$

Hence, X^2 has $k - 1 = 4$ degrees of freedom.

    c   The rejection region for this test is located in the upper tail of the chi-square distribution with $df = 4$. From Table 5, the appropriate upper-tailed rejection region is $X^2 > \chi^2_{.05} = 9.4877$.

    d   The test statistic is

$$X^2 = \sum \frac{(O_i - E_i)^2}{E_i}$$

which, when n is large, possesses an approximate chi-square distribution in repeated sampling. The values of O_i are the actual counts *observed* in the experiment, and

$$E_i = np_i = 300(1/5) = 60.$$

A table of observed and expected cell counts follows:

Category	1	2	3	4	5
O_i	47	63	74	51	65
E_i	60	60	60	60	60

Then

$$X^2 = \frac{(47-60)^2}{60} + \frac{(63-60)^2}{60} + \frac{(74-60)^2}{60} + \frac{(51-60)^2}{60} + \frac{(65-60)^2}{60}$$
$$= \frac{480}{60} = 8.00$$

    e   Since the observed value of X^2 does not fall in the rejection region, we cannot conclude that there is a difference in the preference for the five categories.

14.7 One thousand cars were each classified according to the lane which they occupied (one through four). If no lane is preferred over another, the probability that a car will be driven in lane i, $i = 1, 2, 3, 4$ is ¼. The null hypothesis is then

$$H_0 : p_1 = p_2 = p_3 = p_4 = \frac{1}{4}$$

and the test statistic is

$$X^2 = \sum \frac{(O_i - E_i)^2}{E_i}$$

with $E_i = np_i = 1000(1/4) = 250$ for $i = 1, 2, 3, 4$. A table of observed and expected cell counts follows:

Lane	1	2	3	4
O_i	294	276	238	192
E_i	250	250	250	250

Then

$$X^2 = \frac{(294-250)^2}{250} + \frac{(276-250)^2}{250} + \frac{(238-250)^2}{250} + \frac{(192-250)^2}{250}$$

$$= \frac{6120}{250} = 24.48$$

The rejection region with $k - 1 = 3$ df is $X^2 > \chi^2_{.05} = 7.81$. Since the observed value of X^2 falls in the rejection region, we reject H_0. There is a difference in preference for the four lanes.

14.9 If the frequency of occurrence of a heart attack is the same for each day of the week, then when a heart attack occurs, the probability that it falls in one cell (day) is the same as for any other cell (day). Hence,

$$H_0 : p_1 = p_2 = \cdots = p_7 = \frac{1}{7}$$

versus H_a : at least one p_i is different from the others, or equivalently, $H_a : p_i \neq p_j$ for some pair $i \neq j$

Since $n = 200$, $E_i = np_i = 200(1/7) = 28.571429$ and the test statistic is

$$X^2 = \frac{(24 - 28.571429)^2}{28.571429} + \cdots + \frac{(29 - 28.571429)^2}{28.571429} = \frac{103.71429}{28.571429} = 3.63$$

The degrees of freedom for this test of specified cell probabilities is $k - 1 = 7 - 1 = 6$ and the upper tailed rejection region is

$$X^2 > \chi^2_{.05} = 12.59$$

H_0 is not rejected. There is insufficient evidence to indicate a difference in frequency of occurrence from day to day.

14.11 Similar to previous exercises. The hypothesis to be tested is $H_0 : p_1 = p_2 = \cdots = p_{12} = \frac{1}{12}$

versus H_a : at least one p_i is different from the others with $E_i = np_i = 400(1/12) = 33.333$.
The test statistic is

$$X^2 = \frac{(38 - 33.33)^2}{33.33} + \cdots + \frac{(35 - 33.33)^2}{33.33} = 13.58$$

The upper tailed rejection region is with $\alpha = .05$ and $k - 1 = 11$ df is $X^2 > \chi^2_{.05} = 19.675$. The null hypothesis is not rejected and we cannot conclude that the proportion of cases varies from month to month.

14.13 It is necessary to determine whether proportions given by the Mars Company are correct. The null hypothesis to be tested is $H_0: p_1 = .13; p_2 = .14; p_3 = .13; p_4 = .24; p_5 = .20; p_6 = .16$
against the alternative that at least one of these probabilities is incorrect. A table of observed and expected cell counts follows:

Colour	Brown	Yellow	Red	Blue	Orange	Green
O_i	70	72	61	118	108	85
E_i	66.82	71.96	66.82	123.36	102.80	82.24

The test statistic is
$$X^2 = \frac{(70-66.82)^2}{66.82} + \frac{(72-71.96)^2}{71.96} + \cdots + \frac{(85-82.24)^2}{82.24} = 1.247$$
The number of degrees of freedom is $k - 1 = 5$ and, since the observed value of $X^2 = 1.247$ is less than $\chi^2_{.10} = 9.24$, the p-value is greater than .10 and the results are not significant. We conclude that the proportions reported by the Mars Company are substantiated by our sample.

14.15 It is necessary to determine whether admission rates differ from the previously reported rates. A table of observed and expected cell counts follows:

	Unconditional	Trial	Refused	Totals
O_i	329	43	128	500
E_i	300	25	175	500

The null hypothesis to be tested is $H_0: p_1 = .60; p_2 = .05; p_3 = .35$
against the alternative that at least one of these probabilities is incorrect. The test statistic is
$$X^2 = \frac{(329-300)^2}{300} + \frac{(43-25)^2}{25} + \frac{(128-175)^2}{175} = 28.386$$
The number of degrees of freedom is $k - 1 = 2$ and the rejection region $X^2 > \chi^2_{.05} = 5.99$. The null hypothesis is rejected, and we conclude that there has been a departure from previous admission rates. Notice that the percentage of unconditional admissions has risen slightly, the number of conditional admissions has increased, and the percentage refused admission has decreased at the expense of the fist two categories.

14.17 Refer to Section 14.4 of the text. For a 3×5 contingency table with $r = 3$ and $c = 5$, there are $(r-1)(c-1) = (2)(4) = 8$ degrees of freedom.

14.19 The hypotheses to be tested are
H_0: opinion groups are independent of gender
H_a: opinion groups are dependent on gender
and the test statistic is the chi-square statistic given in the printout as $X^2 = 18.352$ with p-value $= .000$.
Because of the small p-value, the results are highly significant, and H_0 is rejected. There is evidence of a dependence between opinion group and gender. The conditional distributions of the three groups for males and females are shown in the following table.

	Group 1	Group 2	Group 3	Total
Males	$\frac{37}{158} = .23$	$\frac{49}{158} = .31$	$\frac{72}{158} = .46$	1.00
Females	$\frac{7}{88} = .08$	$\frac{50}{88} = .57$	$\frac{31}{88} = .35$	1.00

You can see that men tend to favor the group 3 opinion, while almost 60% of the women favor the group 2 opinion. The group 1 opinion contains a very small proportion of the women, but almost 25% of the men.

14.21 a The hypothesis of independence between attachment pattern and child care time is tested using the chi-square statistic. The contingency table, including column and row totals and the estimated expected cell counts, follows.

	Child Care			
Attachment	Low	Moderate	High	**Total**
Secure	24 (24.09)	35 (30.97)	5 (8.95)	64
Anxious	11 (10.91)	10 (14.03)	8 (4.05)	29
Total	111	51	297	459

The test statistic is

$$X^2 = \frac{(24-24.09)^2}{24.09} + \frac{(35-30.97)^2}{30.97} + \cdots + \frac{(8-4.05)^2}{4.05} = 7.267$$

and the rejection region is $X^2 > \chi^2_{.05} = 5.99$ with 2 df. H_0 is rejected. There is evidence of a dependence between attachment pattern and child care time.

b The value $X^2 = 7.267$ is between $\chi^2_{.05}$ and $\chi^2_{.025}$ so that $.025 < p\text{-value} < .05$. The results are significant.

14.23 a The hypothesis of independence between type of pharmacy and waiting time is tested using the chi-square statistic. The contingency table, including column and row totals and the estimated expected cell counts, follows.

	Waiting Time				
Pharmacy	15 minutes or less	16-20 minutes	More than 20 min	Other	**Total**
Supermarket	75 (83.14)	44 (27.86)	21 (24.86)	10 (14.14)	150
Drugstore	119 (110.86)	21 (37.14)	37 (33.14)	23 (18.86)	200
Total	194	65	58	33	350

The test statistic is

$$X^2 = \frac{(75-83.14)^2}{83.14} + \frac{(44-27.86)^2}{27.86} + \cdots + \frac{(23-18.86)^2}{18.86} = 20.937$$

and the rejection region is $X^2 > \chi^2_{.01} = 11.3449$ with 3 df. H_0 is rejected. There is evidence of a difference in waiting times between pharmacies in supermarkets and drugstores.

b The contingency table is collapsed to include only "more than 20 minutes" and "not more than 20 minutes". The table is shown below, along with the estimated expected cell counts.

	Waiting Time		
Pharmacy	More than 20 min	Other	**Total**
Superstore	21 (24.86)	129 (125.14)	150
Drugstore	37 (33.14)	163 (166.86)	200
Total	58	292	350

The test statistic is

$$X^2 = \frac{(21-24.86)^2}{24.86} + \cdots + \frac{(163-166.86)^2}{166.86} = 1.255$$

and the rejection region is $X^2 > \chi^2_{.01} = 6.63$ with 1 df. H_0 is not rejected. There is insufficient evidence to indicate a difference in waiting times between pharmacies in supermarkets and drugstores.

14.25 **a** The hypothesis of independence between salary and number of workdays at home is tested using the chi-square statistic. The contingency table, including column and row totals and the estimated expected cell counts, generated by ***Minitab*** follows.

Chi-Square Test: Less than one, At least one, not all, All at home
```
Expected counts are printed below observed counts
Chi-Square contributions are printed below expected counts

                 At
              least
        Less   one,    All
        than   not     at
         one   all    home   Total
    1     38    16     14      68
        36.27 21.08  10.65
        0.083  1.224  1.051

    2     54    26     12      92
        49.07 28.52  14.41
        0.496  0.223  0.404

    3     35    22      9      66
        35.20 20.46  10.34
        0.001  0.116  0.174

    4     33    29     12      74
        39.47 22.94  11.59
        1.060  1.601  0.014

Total   160    93     47     300

Chi-Sq = 6.447, DF = 6, P-Value = 0.375
```

The test statistic is

$$X^2 = \frac{(38-36.27)^2}{36.27} + \frac{(16-21.08)^2}{21.08} + \cdots + \frac{(12-11.59)^2}{11.59} = 6.447$$

and the rejection region with $\alpha = .05$ and $df = 3(2) = 6$ is $X^2 > \chi^2_{.05} = 12.59$ and the null hypothesis is not rejected. There is insufficient evidence to indicate that salary is dependent on the number of workdays spent at home.

b The observed value of the test statistic, $X^2 = 6.447$, is less than $\chi^2_{.10} = 10.6446$ so that the *p*-value is more than .10. This would confirm the non-rejection of the null hypothesis from part a.

14.27 Similar to previous exercises, except that the number of observations per row was selected prior to the experiment. The test procedure is identical to that used for an $r \times c$ contingency table. The contingency table, including column and row totals and the estimated expected cell counts, follows.

Population	Category 1	Category 2	Category 3	Total
1	108 (102.33)	52 (47.33)	40 (50.33)	200
2	87 (102.33)	51 (47.33)	62 (50.33)	200
3	112 (102.33)	39 (47.33)	49 (50.33)	200
Total	307	142	151	600

a The test statistic is

$$X^2 = \frac{(108-102.33)^2}{102.33} + \frac{(52-47.33)^2}{47.33} + \cdots + \frac{(49-50.33)^2}{50.33} = 10.597$$

using calculator accuracy.

b With $(r-1)(c-1) = 4$ df and $\alpha = .01$, the rejection region is $X^2 > 13.2767$.

c The null hypothesis is not rejected. There is insufficient evidence to indicate that the proportions depend upon the population from which they are drawn.

d Since the observed value, $X^2 = 10.597$, falls between $\chi^2_{.05}$ and $\chi^2_{.025}$,

$.025 < p\text{-value} < .05$.

14.29 Because a set number of Canadian children in each sub-population were each fixed at 1000, we have a contingency table with fixed rows. The table, with estimated expected cell counts appearing in parentheses, follows.

	Overweight/Obese	Not Overweight/Obese	Total
White	189 (200)	811 (800)	1000
Black	301 (200)	699 (800)	1000
Southeast/East Asian	117 (200)	883 (800)	1000
Off-reserves Aboriginal	231 (200)	769 (800)	1000
Other	162 (200)	838 (800)	1000
Total	1000	4000	5000

The test statistic is

$$X^2 = \frac{(189-200)^2}{200} + \frac{(301-200)^2}{200} + \cdots + \frac{(838-800)^2}{800} = 122.6$$

and the rejection region with 4 df is $X^2 > 13.2767$. H_0 is rejected and we conclude that the incidence of overweight/obese children is dependent on the ethnic origin.

14.31 **a** The number of people in each of the three income categories was chosen in advance; each of these populations represents a multinomial population in which we measure education levels.

b The 4×3 contingency table is analyzed as in previous exercises. The ***Minitab*** printout below shows the observed and estimated expected cell counts, the test statistic and its associated *p*-value.

Chi-Square Test: 70-99K, 100-249K, 250K or more
```
Expected counts are printed below observed counts
Chi-Square contributions are printed below expected counts

                          250K
                            or
         70-99K  100-249K  more   Total
    1        32        20    23      75
          25.00     25.00 25.00
          1.960     1.000 0.160

    2        13        16     1      30
          10.00     10.00 10.00
          0.900     3.600 8.100

    3        43        51    60     154
          51.33     51.33 51.33
          1.353     0.002 1.463

    4        12        13    16      41
          13.67     13.67 13.67
          0.203     0.033 0.398

Total       100       100   100     300

Chi-Sq = 19.172, DF = 6, P-Value = 0.004
```

The results are significant at the 1% level (*p*-value = .004) and we conclude that there is a difference in the level of wealth depending on educational attainment.

c Answers will vary from student to student.

14.33 The number of observations per column were selected prior to the experiment. The test procedure is identical to that used for an $r \times c$ contingency table. The contingency table, including column and row totals and the estimated expected cell counts, follows.

Family Members	Type			Total
	Apartment	Duplex	Single Residence	
1	8 (9.67)	20 (9.67)	1 (9.67)	29
2	16 (11)	8 (11)	9 (11)	33
3	10 (11.33)	10 (11.33)	14 (11.33)	34
4 or more	6 (8)	2 (8)	16 (8)	24
Total	40	40	40	120

The test statistic is

$$X^2 = \frac{(8-9.67)^2}{9.67} + \frac{(20-9.67)^2}{9.67} + \cdots + \frac{(16-8)^2}{8} = 36.499$$

using computer accuracy. With $(r-1)(c-1) = 6$ *df* and $\alpha = .01$, the rejection region is $X^2 > 16.8119$. The null hypothesis is rejected. There is sufficient evidence to indicate that family size is dependent on type of family residence. It appears that as the family size increases, it is more likely that people will live in single residences.

14.35 If the housekeeper actually has no preference, he or she has an equal chance of picking any of the five floor polishes. Hence, the null hypothesis to be tested is

$$H_0 : p_1 = p_2 = p_3 = p_4 = p_5 = \frac{1}{5}$$

The values of O_i are the actual counts observed in the experiment, and $E_i = np_i = 100(1/5) = 20$.

Polish	A	B	C	D	E
O_i	27	17	15	22	19
E_i	20	20	20	20	20

Then $X^2 = \dfrac{(27-20)^2}{20} + \dfrac{(17-20)^2}{20} + \cdots + \dfrac{(19-20)^2}{20} = 4.40$

The p-value with $df = k-1 = 4$ is greater than .10 and H_0 is not rejected. We cannot conclude that there is a difference in the preference for the five floor polishes. Even if this hypothesis **had** been rejected, the conclusion would be that at least one of the value of the p_i was significantly different from 1/6. However, this does not imply that p_i is necessarily greater than 1/6. Hence, we could not conclude that polish A is superior.

If the objective of the experiment is to show that polish A is superior, a better procedure would be to test an hypothesis as follows: $H_0 : p_1 = 1/6 \qquad H_a : p_1 > 1/6$

From a sample of $n = 100$ housewives, $x = 27$ are found to prefer polish A. A z-test can be performed on the single binomial parameter p_1.

14.37 The data is analyzed as a 2×3 contingency table with estimated expected cell counts shown in parentheses.

	Small	Medium	Large	Total
Fatal	67 (61.43)	26 (28.04)	16 (19.53)	109
Not Fatal	128 (133.57)	63 (60.96)	46 (42.47)	237
Total	195	89	62	346

The test statistic is

$$X^2 = \frac{(67-61.43)^2}{61.43} + \frac{(26-28.04)^2}{28.04} + \cdots + \frac{(46-42.47)^2}{42.47} = 1.885$$

The p-value with 2 df is greater than .10 and H_0 is not rejected. There is insufficient evidence to indicate that the frequency of fatal accidents is depending on the size of automobiles.

14.39-14.40 a Let p_1 be the probability that a prone-sleeping infant rolls over at the 4-month checkup and let p_2 be the probability that a supine or side-sleeping infant rolls over. Then the hypothesis to be tested is $\qquad H_0 : p_1 - p_2 = 0 \qquad H_a : p_1 - p_2 > 0$

Using the procedure described in Chapter 9 for testing an hypothesis about the difference between two binomial parameters, the following estimators are calculated:

$$\hat{p}_1 = \frac{x_1}{n_1} = \frac{93}{121} \qquad \hat{p}_2 = \frac{x_2}{n_2} = \frac{119}{199} \qquad \hat{p} = \frac{x_1 + x_2}{n_1 + n_2} = \frac{93+119}{320} = .6625$$

The test statistic is

$$z = \frac{\hat{p}_1 - \hat{p}_2 - 0}{\sqrt{\hat{p}\hat{q}(1/n_1 + 1/n_2)}} = \frac{.7686 - .59799}{\sqrt{.6625(.3375)\left(\dfrac{1}{121} + \dfrac{1}{199}\right)}} = 3.1297$$

with one-tailed p-value given as
$$p\text{-value} = P(z > 3.13) < (.5 - .4990) = .001$$

b Using a 2×2 contingency table, the number of infants who roll over and those who do not roll over are recorded in the 2×2 table. The Minitab printout shows the chi-square test ($X^2 = 9.795$) with p-value $= .002$.

Chi-Square Test: Prone, Supine
```
Expected counts are printed below observed counts
Chi-Square contributions are printed below expected counts

        Prone   Supine   Total
    1      93      119     212
         80.16   131.84
         2.056    1.250

    2      28       80     108
         40.84    67.16
         4.036    2.454

Total     121      199     320

Chi-Sq = 9.795, DF = 1, P-Value = 0.002
```

Remember that the chi-square test rejects H_0 in favor of a non-directional alternative using only large values of X^2. To obtain the directional test in part **a**, you must first check to see that the direction of the difference is as indicated by H_a ($\hat{p}_1 > \hat{p}_2$). The p-value for the test is then half the two-tailed p-value or

$$p\text{-value} = \frac{1}{2}(.002) = .001$$

as in part **a**. Also, the two test statistics are related as:
$$z^2 = (3.1297)^2 = 9.795 = X^2$$

14.41 Each of the three milestones are tested as in Exercise 14.41 as a 2×2 contingency table using the chi-square test or the two-sample z test for binomial proportions. The researchers are interested in detecting a difference in the proportions of infants achieving a particular milestone at the 4-month checkup. It is not clear from the table whether the tests are one- or two-tailed; however, the p-values for each of the three tests are large enough to indicate that none of the proportions are significantly different for the two groups of infants. The *Minitab* chi-square printouts below answer these questions.

Chi-Square Test: Prone, Supine
```
Expected counts are printed below observed counts
Chi-Square contributions are printed below expected counts
        Prone   Supine   Total
    1      79      144     223
         76.12   146.88
         0.109    0.056

    2       6       20      26
          8.88    17.12
         0.932    0.483

Total      85      164     249
Chi-Sq = 1.579, DF = 1, P-Value = 0.20
```

Chi-Square Test: Prone, Supine
```
Expected counts are printed below observed counts
Chi-Square contributions are printed below expected counts
        Prone   Supine  Total
    1     102      167    269
       103.46   165.54
        0.021    0.013

    2       3        1      4
         1.54     2.46
        1.388    0.868

Total     105      168    273
Chi-Sq = 2.290, DF = 1, P-Value = 0.130
2 cells with expected counts less than 5.
```

Chi-Square Test: Prone, Supine
```
Expected counts are printed below observed counts
Chi-Square contributions are printed below expected counts
        Prone   Supine  Total
    1     107      183    290
       107.05   182.95
        0.000    0.000

    2       3        5      8
         2.95     5.05
        0.001    0.000

Total     110      188    298
Chi-Sq = 0.001, DF = 1, P-Value = 0.972
1 cells with expected counts less than 5.
```

Notice that the researcher's *p*-values match the values given in the printout, indicating that the researchers are doing two-tailed tests. However, *Minitab* warns you that the assumption that E_i is greater than or equal to five for each cell has been violated. Since the effect of a small expected value is to inflate the value of X^2, you need not be too concerned (X^2 was not big enough to reject H_0 in any of the tests).

14.43 The 3×2 contingency table is analyzed as in previous exercises. The *Minitab* printout is shown below.

Chi-Square Test: Present, Absent
```
Expected counts are printed below observed counts
Chi-Square contributions are printed below expected counts
        Present  Absent  Total
    1       42       58    100
         29.00    71.00
         5.828    2.380

    2       23       77    100
         29.00    71.00
         1.241    0.507

    3       22       78    100
         29.00    71.00
         1.690    0.690

Total       87      213    300
Chi-Sq = 12.336, DF = 2, P-Value = 0.002
```

The test statistic is $X^2 = 12.336$ with *p*-value $= .002$ and the null hypothesis is rejected. There is evidence of a significant difference in the rate of salmonella contamination among these three processing plants. The first plant appears to have the most contamination.

14.45 The objective is to determine whether or not attitudes regarding parking policy are independent of status. The data is analyzed as a 2×3 contingency table with estimated expected cell counts shown in parentheses.

Party	Status			Total
	Student	Faculty	Administration	
Favor	252 (237.44)	107 (114.16)	43 (50.40)	402
Oppose	139 (153.56)	81 (73.84)	40 (32.60)	260
Total	391	188	83	662

The test statistic is

$$X^2 = \frac{(252-237.44)^2}{237.44} + \frac{(107-114.16)^2}{114.16} + \cdots + \frac{(40-32.60)^2}{32.60} = 6.18$$

With $df = 2$, the p-value is between .025 and .05 and H_0 is rejected. There is a difference in opinion due to status.

14.47 In order to perform a chi-square "goodness of fit" test on the given data, it is necessary that the values O_i and E_i are known for each of the five cells. The O_i (the number of measurements falling in the i-th cell) are given. However, $E_i = np_i$ must be calculated. Remember that p_i is the probability that a measurement falls in the i-th cell. The hypothesis to be tested is

H_0 : the experiment is binomial   versus H_a : the experiment is not binomial

Let x be the number of successes and p be the probability of success on a single trial. Then, assuming the null hypothesis to be true,

$$p_0 = P(x=0) = C_0^4 p^0 (1-p)^4 \qquad p_1 = P(x=1) = C_1^4 p^1 (1-p)^3$$
$$p_2 = P(x=2) = C_2^4 p^2 (1-p)^2 \qquad p_3 = P(x=3) = C_3^4 p^3 (1-p)^1$$
$$p_4 = P(x=4) = C_4^4 p^4 (1-p)^0$$

Hence, once an estimate for p is obtained, the expected cell frequencies can be calculated using the above probabilities. Note that each of the 100 experiments consists of four trials and hence the complete experiment involves a total of 400 trials.

The best estimator of p is $\hat{p} = x/n$ (as in Chapter 9). Then,

$$\hat{p} = \frac{x}{n} = \frac{\text{number of successes}}{\text{number of trials}} = \frac{0(11)+1(17)+2(42)+3(12)+4(9)}{400} = \frac{1}{2}$$

The experiment (consisting of four trials) was repeated 100 times. There are a total of 400 trials in which the result "no successes in four trials" was observed 11 times, the result "one success in four trials" was observed 17 times, and so on. Then

$$p_0 = C_0^4 (1/2)^0 (1/2)^4 = 1/16 \qquad p_1 = C_1^4 (1/2)^1 (1/2)^3 = 4/16$$
$$p_2 = C_2^4 (1/2)^2 (1/2)^2 = 6/16 \qquad p_3 = C_3^4 (1/2)^3 (1/2)^1 = 4/16$$
$$p_4 = C_4^4 (1/2)^4 (1/2)^0 = 1/16$$

The observed and expected cell frequencies are shown in the following table.

x	0	1	2	3	4
O_i	11	17	42	21	9
E_i	6.25	25.00	37.50	25.00	6.25

and the statistic is

$$X^2 = \frac{(11-6.25)^2}{6.25} + \frac{(17-25.00)^2}{25.00} + \cdots + \frac{(9-6.25)^2}{6.25} = 8.56$$

In order to bound the *p*-value or set up a rejection region, it is necessary to determine the appropriate degrees of freedom associated with the test statistic. Two degrees of freedom are lost because:
1. The cell probabilities are restricted by the fact that $\sum p_i = 1$.
2. The binomial parameter *p* is unknown and must be estimated before calculating the expected cell counts. The number of degrees of freedom is equal to $k-1-1 = k-2 = 3$. With $df = 3$, the *p*-value for $X^2 = 8.56$ is between .025 and .05 and the null hypothesis can be rejected at the 5% level of significance. We conclude that the experiment in question does not fulfill the requirements for a binomial experiment.

14.49 **a** The contingency table with estimated expected cell counts in parentheses follows.

	Participate	Not Participate	Total
Approve	73 (62)	51 (62)	124
Do not approve	27 (38)	49 (38)	76
Total	100	100	200

The test statistic is

$$X^2 = \frac{(73-62)^2}{62} + \frac{(51-62)^2}{62} + \cdots + \frac{(49-38)^2}{38} = 10.272$$

The rejection region with 1 *df* has a right-tail area of $2(.05) = .10$ or $X^2 > \chi^2_{2(.05)} = 2.706$ and H$_0$ is rejected. There is evidence that approval or disapproval depends on whether workers participate in decision making.

b The one-tailed *z* test was used to test the hypothesis $H_0: p_1 - p_2 = 0 \quad H_a: p_1 - p_2 > 0$

Calculate $\hat{p}_1 = \frac{73}{100} = .73$, $\hat{p}_2 = \frac{51}{100} = .51$, and $\hat{p} = \frac{73+51}{200} = .62$. The test statistic is then

$$z = \frac{\hat{p}_1 - \hat{p}_2}{\sqrt{\hat{p}\hat{q}\left(\frac{1}{n_1}+\frac{1}{n_2}\right)}} = \frac{.73-.51}{\sqrt{.62(.38)(2/100)}} = 3.205$$

The rejection region for $\alpha = .05$ is $z > 1.645$ and H$_0$ is rejected. There is evidence that workers with participative decision making more generally approve of the firm's decisions than those without.

14.51 **a-b** The *Minitab* printouts below are used to analyze the data for the two contingency tables. The observed values of the test statistics are $X^2 = 19.043$ and $X^2 = 60.139$, for faculty and student responses, respectively. The rejection region, with $(3)(2) = 6$ *df* is $X^2 > 16.81$ with $\alpha = .01$ and H_0 is rejected for both cases.

Chi-Square Test (a): High, Medium, Low
```
Expected counts are printed below observed counts
Chi-Square contributions are printed below expected counts
        High   Medium     Low   Total
   1       4        0       0       4
        1.53     1.47    1.00
       3.968    1.467   1.000

   2      15       12       3      30
       11.50    11.00    7.50
       1.065    0.091   2.700

   3       2        7       7      16
        6.13     5.87    4.00
       2.786    0.219   2.250

   4       2        3       5      10
        3.83     3.67    2.50
       0.877    0.121   2.500

Total     23       22      15      60

Chi-Sq = 19.043, DF = 6, P-Value = 0.004
7 cells with expected counts less than 5.
```

Chi-Square Test (b): High, Medium, Low
```
Expected counts are printed below observed counts
Chi-Square contributions are printed below expected counts
        High   Medium     Low   Total
   1      19        6       2      27
        6.88     9.56   10.57
       21.379   1.325   6.944

   2      19       41      27      87
       22.16    30.80   34.04
       0.449    3.377   1.457

   3       3        7      31      41
       10.44    14.52   16.04
       5.303    3.891  13.943

   4       0        3       3       6
        1.53     2.12    2.35
       1.528    0.361   0.181

Total     41       57      63     161

Chi-Sq = 60.139, DF = 6, P-Value = 0.000
3 cells with expected counts less than 5.
```

c Answers will vary.
d Notice that the computer printout in both tables warns that some cells have expected cell counts less than 5. This is a violation of the assumptions necessary for this test, and results should thus be viewed with caution.

14.53 **a** Since the percentages do not add to 100%, there is a category missing. We call this category "Other" colors, with expected percentage $100-(29.72+11.00+\cdots+9.01)=31.95$. Since there were 250 trucks and vans in the survey, the number falling in the "Other" category is $250-(82+22+\cdots+20)=78$.

b The null hypothesis to be tested is
$$H_0: p_1=.2972;\ p_2=.1100;\ p_3=.0924;\ p_4=.0908;\ p_5=.0901;\ p_6=.3195$$
A table of observed and expected cell counts follows:

Color	White	Dark red	Green	Red	Black	Other
O_i	82	22	27	21	20	78
E_i	74.3	27.5	23.1	22.7	22.525	79.875

Then the test statistic is
$$X^2=\frac{(82-74.3)^2}{74.3}+\frac{(22-27.5)^2}{27.5}+\cdots+\frac{(78-79.875)^2}{79.875}=3.01$$

Since the observed value of X^2 with $df=k-1=5$ is less than $\chi^2_{.10}$, the p-value is greater than .10. We cannot reject H_0. There is insufficient evidence to suggest a difference from the given percentages.

14.55 The *Minitab* printout for this 2×4 contingency table is shown below.

Chi-Square Test: Banana, Cherry, WildFruit, Straw_Ban

```
Expected counts are printed below observed counts
Chi-Square contributions are printed below expected counts
        Banana  Cherry  WildFruit  Straw_Ban  Total
    1       14      20          7          9     50
         12.86   24.29       5.00       7.86
         0.102   0.756      0.800      0.166

    2        4      14          0          2     20
          5.14    9.71       2.00       3.14
         0.254   1.891      2.000      0.416
Total       18      34          7         11     70

Chi-Sq = 6.384, DF = 3, P-Value = 0.094
2 cells with expected counts less than 5.
```

The test statistic for the test of independence of the two classifications of $X^2=6.384$ with p-value $=.094$ and H_0 is not rejected. There is insufficient evidence to indicate a difference in the perception of the best taste between adults and children. If the company intends to use a flavor as a marketing tool, the cherry flavor does not seem to provide an incentive to buy this product.

14.57   The 4×4 contingency table is analyzed using **Minitab** to test

H_0 : fast food choice is independent of age group

H_a : fast food choice is dependent on age group

Chi-Square Test: McDonalds, Burger King, Wendys, Other

```
Expected counts are printed below observed counts
Chi-Square contributions are printed below expected counts
              Burger
     McDonalds   King  Wendys  Other  Total
 1         75     34     10      6    125
        59.75  38.25  16.00  11.00
        3.892  0.472  2.250  2.273

 2         89     42     19     10    160
        76.48  48.96  20.48  14.08
        2.050  0.989  0.107  1.182

 3         54     52     28     18    152
        72.66  46.51  19.46  13.38
        4.790  0.648  3.752  1.598

 4         21     25      7     10     63
        30.11  19.28   8.06   5.54
        2.758  1.698  0.140  3.582

Total     239    153     64     44    500
Chi-Sq = 32.182, DF = 9, P-Value = 0.000
```

The **Minitab** printout shows $X^2 = 32.182$ with p-value $= .000$, so that H_0 is rejected. There is a dependence between age group and favorite fast food restaurant. The practical implications of this conclusion can be explored by looking at the conditional distribution of fast food choices for each of the four age groups. Each student will present slightly different conclusions.

14.59   The 3×4 contingency table is analyzed as in previous exercises. The **Minitab** printout below shows the observed and estimated expected cell counts, the test statistic and its associated *p*-value.

Chi-Square Test: A, B, C, D

```
Expected counts are printed below observed counts
Chi-Square contributions are printed below expected counts
         A      B      C      D   Total
 1      55     40     43     30    168
     45.39  43.03  43.03  36.55
     2.035  0.214  0.000  1.173

 2      15     25     18     22     80
     21.61  20.49  20.49  17.40
     2.024  0.992  0.303  1.214

 3       7      8     12     10     37
     10.00   9.48   9.48   8.05
     0.898  0.230  0.672  0.473
Total   77     73     73     62    285
Chi-Sq = 10.227, DF = 6, P-Value = 0.115
```

The results are not significant (p-value $= .115$) and we cannot conclude that number of arrests is dependent on the educational achievement of a criminal offender.

14.61 Use the **Chi-Square Probabilities** applet, entering the right-tailed probability in the box marked "Prob:" and using the slider on the right side to adjust the degrees of freedom.

a The critical value of χ^2 for $\alpha = .05$ is 25.0, shown in the figure below.

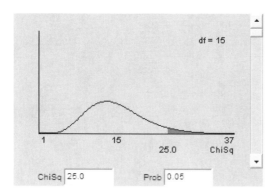

b For $\alpha = .01$ and 11 df, the critical value for χ^2 is 24.7.

14.63 Use the **Chi-Square Probabilities** applet. This time, enter the value of the test statistic in the box marked "ChiSq" and using the slider on the right side to adjust the degrees of freedom.

a p-value $= P(\chi^2 > .81) = .8471$ as shown in the figure below.

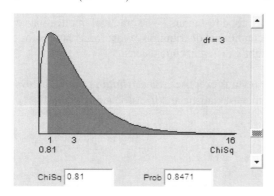

b p-value $= P(\chi^2 > 25.40) = .0204$.

14.65 Use the third **Goodness-of-Fit** applet, typing in the number of M&Ms of each color in the first row. The resulting test statistic should match the results in Exercise 14.13.

14.66-14.67 Use the **M&Ms** applet. Answers will vary from student to student.

14.69 The data is analyzed as a 2×3 contingency table with estimated expected cell counts shown in parentheses. Use the **Chi-Square Test of Independence** applet. Your results should agree with the hand calculations shown below.

	Manager			
	A	B	C	
Success (profit)	63 (63)	71 (63)	55 (63)	189
No profit	37 (37)	29 (37)	45 (37)	111
Total	100	100	100	300

The test statistic is

$$X^2 = \frac{(63-63)^2}{63} + \frac{(71-63)^2}{63} + \cdots + \frac{(45-37)^2}{37} = 5.491$$

With $(r-1)(c-1) = 2$ df, the p-value is bounded between .05 and .10 (the applet reports p-value = .0642). H_0 is not rejected and the results are declared not significant. There is not enough information to conclude that the proportion of successful purchases will differ among the managers.

Case Study:
Can a Marketing Approach Improve Library Services?

1. There are seven different 2×2 contingency tables to test the null hypothesis $H_0: p_1 = p_2$ for each of the seven questions. Using a Minitab computer package, the test statistics for the seven tests are shown below.

    $$X_1^2 = 10.342 \quad X_2^2 = 6.358 \quad X_3^2 = 0.661$$
    $$X_4^2 = 3.213 \quad X_5^2 = 9.654 \quad X_6^2 = 1.948$$
    $$X_7^2 = 0.040$$

    Since each test statistic has an approximate χ^2 distribution with 1 degree of freedom, the 3rd, 6th and 7th tests are not significant. The first test statistic has p-value $< .005$; the 2nd test statistic has $.025 < p$-value $< .01$; the 4th has $.05 < p$-value $< .10$; the 5th has p-value $< .005$. All of these values agree with the published results except for the 4th test statistic.

2. Notice that questions 11 and 13 (design) produce non-significant results, and that questions 5 and 6 (staff) produce one non-significant and one not too highly significant ($.05 < p$-value $< .10$) result. Questions 3, 4, and 7 are highly significant; that is, for questions about atmosphere, there is a significant difference in the responses for students and non-students.

3. No. The questions were asked of the same two groups of people and hence are dependent. They do not fit the normal criteria for a contingency table approach. There are techniques available, but they are not given in this text.

Project 14A: Child Safety Seat Survey, Part 3

a) A test of independence between age group and restraint type would best be tested using a contingency table and the χ2 statistical test of significance. Our null hypothesis would assume independence between the categories of age group and restraint type.

b) The sampling distribution of the test statistic is approximately chi-square with (r-1)(c-1) degrees of freedom. For this particular example, we have r=4 and c=4. Hence, our test statistic would be chi-square with 9 degrees of freedom.

c) i) The null and alternative hypotheses would be

H_0: Classification by age group and classification by restraint type are independent

H_a: The classification methods are dependent.

ii) The statistic used to test the hypotheses is a chi-square test on (4-1)(4-1)=9 degrees of freedom (i.e., our test statistic is approximately chi-square with 9 degrees of freedom). Specifically,

$$\chi^2_{test} = \sum_{i=1}^{r}\sum_{j=1}^{c} \frac{(O_{ij} - \hat{E}_{ij})^2}{\hat{E}_{ij}}$$

where

$$\hat{E}_{ij} = \frac{r_i c_j}{n}$$

and r_i and c_j are the row and column totals for row i and column j respectively.

iii) The main difficulty in calculating the test statistic is the fact that we have several cells with 0 counts. This may be a situation where insufficient individuals have been surveyed. However, one may also expect 0 counts within some of the cells. For example, infant seats are not typically used for children over 9 years of age. In a similar manner, one would not expect infants to be secured by seat belt only.

iv) One of the statistical issues associated with performing a formal test on the hypotheses described in part i) is the presence of low observation counts. Specifically, we have several cells that have fewer than 5 observations. It is possible that these small counts do not adequately reflect the true number of observations that should fall within these particular cells. As such, the marginal values might be lower than they should be. This could lead to inflated estimates of proportions in the other cells within any row (or column). Ultimately, the expected counts and observed counts will not be close, leading to inflated test statistics and an increased probability of rejecting the null hypothesis when it should not be rejected.

v) To alleviate the problems described above, one may wish to survey more individuals in order to potentially eliminate the zero counts. Another option would be to conduct the survey until k individuals are surveyed within each age group or within each restraint type. Of course, as previously mentioned, the counts may necessarily be 0 based on the standard use for each type of restraint specific to age group.

d) 1. The null and alternative hypotheses are

H_0: $p_A = p_B = p_C$. The type of restraint used is independent of the age group.

H_a: at least two of the restraint types differ.

2. Our particular study has been set up as a two-way classification with fixed column totals. In this case, our column totals are each 600. To test the null hypothesis, we use the χ^2 test for independence of row and column classifications. The test statistic is

$$\chi^2_{test} = \sum_{i=1}^{R}\sum_{j=1}^{C} \frac{(O_{ij} - \hat{E}_{ij})^2}{\hat{E}_{ij}}$$

The bracketed quantities in the table below represent the expected cell counts:

Age Group	A	B	C	Total
Todler	483 (261)	250 (261)	50 (261)	783
School	117 (339)	350 (339)	550 (339)	1017
Total	600	600	600	1800

Where, for example, the expected count for the school aged group who use a forward-facing infant seat (restraint type B) is

$$\hat{E}_{2,2} = \frac{1017 \times 600}{1800} = 339$$

Hence, our test statistic is

$$\chi^2_{test} = \sum_{i=1}^{R}\sum_{j=1}^{C} \frac{(O_{ij} - \hat{E}_{ij})^2}{\hat{E}_{ij}}$$
$$= \frac{(483-261)^2}{261} + \frac{(250-261)^2}{261} + \ldots + \frac{(550-339)^2}{261}$$
$$= 636.9376$$

3. The p-value for this test can be approximated using Table 5 in Appendix I. On (r-1)(c-1)=(2-1)(3-1)=2 degrees of freedom, we have $p(\chi^2 > 636.9376) < 0.005$.

Since our p-value is smaller than α=0.05, we reject the null hypothesis in favour of the alternative. There is sufficient evidence to suggest that at least two of the proportions are different. The three binomial proportions are

$$\hat{p}_A = \frac{483}{600} = 0.805$$
$$\hat{p}_B = \frac{250}{600} = 0.417$$
$$\hat{p}_C = \frac{50}{600} = 0.083$$

Note that these proportions are specific to the age group classified as toddler (1-4 years). For this age group there is a clear pattern of restraint use. That is, most toddlers use the forward-facing infant seat. Next in line are the booster seat, and finally seat belt only. This probably makes sense given the age and size of the children in question. If we calculate the proportions for the school age children (4-9 years), the pattern is reversed. The proportions using restraint type A, B and C are (0.195, 0.583, 0.917) respectively. Few school children fall under the forward-facing infant seat. Most use seat belts only.

We could also compare each of the three proportions listed above via pair-wise comparisons using a large sample z-test. Our null hypothesis would assume equality between the two proportions in question (i.e., $p_A=p_B$, or $p_A=p_C$, or $p_B=p_C$), while the alternative hypothesis would assume they were different.

The z-tests comparing p_A-p_B, p_A-p_C, and p_B-p_C, would be 13.795, 25.157 and 13.333 respectively. At a 95% level of significance, we would reject each of the null hypotheses. There is sufficient evidence to suggest that each of the proportions of toddlers using forward facing seats is significantly different than the proportion of toddlers using the booster seat, or seat belt only. Further, there is sufficient evidence to suggest that the proportion of toddlers using the booster seat is significantly different that the proportion of toddlers using a seat belt only.

4. If there are more than two row categories in a contingency table with fixed-total c columns, then the test of independence is equivalent to a test of the equality of c sets of multinomial proportions. Assume that there exist r row categories across each of the c columns. Note that within each column, we have a multinomial experiment. That is, we have a total of n trials (which is the same for each of the c columns) that are independently and identically assigned to each of the r rows. Specifically, O_1 observations are randomly assigned to the first row with probability p_1, O_2 observations are randomly assigned to the second row with probability p_2, and so forth. With a fixed column total k, we also have that $\sum O_i=k$. Since each of the columns represents a multinomial experiment, the test of independence is equivalent to testing whether or not c multinomial proportions are the same.

Project 14B: The Dating Strategies

a) To test if there has been a change from the online percentages in Quebec, we assume that the current responses should follow a 0.64:0.10:0.26 pattern. Our observed and expected counts are summarized below

Quebec	A	B	C	Total
Observed	190	50	55	**295**
Expected	188.8	29.5	76.7	**295**

The expected counts are determined by multiplying the observed row total by the online percentages provided. For example, 188.8=295*0.64.

The test statistic is

$$\chi^2_{test} = \sum_{i=1}^{r} \frac{(O_i - \hat{E}_i)^2}{\hat{E}_i}$$
$$= \frac{(190-188.8)^2}{188.8} + \frac{(50-29.5)^2}{29.5} + \frac{(55-76.7)^2}{76.7}$$
$$= 20.39276$$

The critical χ^2 value obtained from Table 5 in Appendix I with α=0.05 and (r-1)*(c-1)=(2-1)(3-1)=2 degrees of freedom is 5.99147. Since our test statistic is greater than 5.99147, we reject the null hypothesis. There is sufficient evidence to suggest that there has been a change from the online percentages in Quebec.

b) To test whether the distribution of the response to the question "does not matter at all" agrees with the corresponding distribution from the online survey, we use the following null and alternative hypotheses:

H_0: p_{AB}=0.13/0.70, p_{BC}=0.18/0.70, p_{ON}=0.13/0.70, p_{PQ}=0.26/0.70.

H_a: at least one of the probabilities is different.

Note that in this case we have divided the online survey results for column C by the grand total for column C (i.e., we have determined the conditional probabilities). This is to ensure that the observed proportions sum to 1.

The test statistic is determined using the observed and expected values indicated in the table below

Does not matter at all	AB	BC	ON	PQ	Total
Observed	40	30	55	55	**180**
Expected	33.43	46.29	33.43	66.86	**180**

The test statistic is

$$\chi^2_{test} = \sum_{i=1}^{r} \frac{(O_i - \hat{E}_i)^2}{\hat{E}_i}$$

$$= \frac{(40-33.43)^2}{33.43} + \frac{(30-46.29)^2}{46.29} + \frac{(55-33.43)^2}{33.43} + \frac{(55-66.86)^2}{66.86}$$

$$= 23.04487$$

The critical chi-square value on (2-1)(4-1)=3 degrees of freedom with α=0.01 is 11.3449. Since our test statistic is larger than the critical value, there is sufficient evidence to suggest that the distribution of the response "does not matter at all" does not agree with the distribution from the online survey. At least one of the proportions differs from the online survey.

c) To test whether or not the responses in category "A" follow the pattern of the online survey, we have the following observed and expected values:

Matter a bit	AB	BC	ON	PQ	Total
Observed	140	130	270	190	730
Expected	191.2546	180.4797	185.8672	172.3985	730

The test statistic is

$$\chi^2_{test} = 67.73468$$

Comparing this to the critical value (with α=0.025 and 3 degrees of freedom) of 9.34840, we would reject the null hypothesis. There is sufficient evidence to suggest that the proportions in the "A" category have changed from what was found from the online survey.

d) i) The null and alternative hypotheses are

H0: pAB=pBC=pON=pPQ=0.25.

Ha: at least one of the probabilities is different.

ii) The test statistic is based on the following observed and expected values

Does not matter at all	AB	BC	ON	PQ	Total
Observed	40	30	55	55	180
Expected	45	45	45	45	180

With this information, our test statistic is

$$\chi^2_{test} = 10$$

iii) We would reject the null hypothesis in favour of the alternative if our test statistic exceeds the critical chi-square value (using 3 degrees of freedom and α=0.05) of 7.81473.

iv) Since our test statistic falls in the rejection region, we reject the null hypothesis. There is sufficient evidence to suggest that the proportions in category "C" across the provinces are not all the same.

e) To test if there is a relationship between the responses to the question "Would financial status affect your decision as to whether or not you would be interested in pursuing a relationship with someone" and the province of residence of the respondent, we employ the chi-square test on the contingency table of responses. Our null hypothesis is that the response to the question and the province of residence are independent. Our alternative assumes a relationship between response and province. The observed and expected counts within each of the cells are presented below

Province	A	B	C	Total
Alberta	140 (135.81)	20 (30.70)	40 (33.49)	**200**
British Columbia	130 (129.02)	30 (29.16)	30 (31.81)	**190**
Ontario	270 (264.84)	65 (59.86)	55 (65.30)	**390**
Quebec	190 (200.33)	50 (45.28)	55 (49.40)	**295**
Total	**730**	**165**	**180**	**1075**

This gives us a chi-square test of

$$\chi^2_{test} = \sum_{i=1}^{r}\sum_{j=1}^{c} \frac{(O_{ij} - \hat{E}_{ij})^2}{\hat{E}_{ij}}$$

$$= \frac{(140 - 135.81)^2}{135.81} + \frac{(20 - 30.70)^2}{30.70} + \ldots + \frac{(55 - 49.40)^2}{49.40}$$

$$= 9.085627$$

The critical value ($\alpha=0.05$) on (r-1)(c-1)=(4-1)(3-1)=6 degrees of freedom is 12.5916. Since the test statistic is less than the critical value, we would not reject the null hypothesis. There is insufficient evidence to suggest that there is a relationship between the response to the question and the province of resident of the respondent. The p-value for this test is greater than 0.10. Since the province of residence is not related to the response, one may wish to pool the data to provide a better estimate of the proportion of individuals who would respond A, B, or C to the question.

Chapter 15: Nonparametric Statistics

15.1   **a**   If distribution 1 is shifted to the right of distribution 2, the rank sum for sample 1 (T_1) will tend to be large. The test statistic will be T_1^*, the rank sum for sample 1 if the observations had been ranked from large to small. The null hypothesis will be rejected if T_1^* is unusually small.

      b   From Table 7a with $n_1 = 6$, $n_2 = 8$ and $\alpha = .05$, H_0 will be rejected if $T_1^* \leq 31$.

      c   From Table 7c with $n_1 = 6$, $n_2 = 8$ and $\alpha = .01$, H_0 will be rejected if $T_1^* \leq 27$.

15.3   **a** H_0 : populations 1 and 2 are identical
          H_a : population 1 is shifted to the left of population 2

      b   The data, with ranks in parentheses, are given below.

Sample 1	Sample 2
1(1)	4(5)
3(3.5)	7(9)
2(2)	6(7.5)
3(3.5)	8(10)
5(6)	6(7.5)

Note that tied observations are given an average rank, the average of the ranks they would have received if they had not been tied. Then

$$T_1 = 1 + 3.5 + 2 + 3.5 + 6 = 16$$
$$T_1^* = n_1(n_1 + n_2 + 1) - T_1 = 5(10 + 1) - 16 = 39$$

      c   With $n_1 = n_2 = 5$, the one-tailed rejection region with $\alpha = .05$ is found in Table 7a to be $T_1 \leq 19$.

      d   The observed value, $T_1 = 16$, falls in the rejection region and H_0 is rejected. We conclude that population 1 is shifted to the right of population 2.

15.5   If H_a is true and population 1 lies to the right of population 2, then T_1 will be large and T_1^* will be small. Hence, the test statistic will be T_1^* and the large sample approximation can be used. Calculate

$$T_1^* = n_1(n_1 + n_2 + 1) - T_1 = 12(27) - 193 = 131$$
$$\mu_T = \frac{n_1(n_1 + n_2 + 1)}{2} = \frac{12(26 + 1)}{2} = 162$$
$$\sigma_T^2 = \frac{n_1 n_2(n_1 + n_2 + 1)}{12} = \frac{12(14)(27)}{12} = 378$$

The test statistic is

$$z = \frac{T_1 - \mu_T}{\sigma_T} = \frac{131 - 162}{\sqrt{378}} = -1.59$$

The rejection region with $\alpha = .05$ is $z < -1.645$ and H_0 is not rejected. There is insufficient evidence to indicate a difference in the two population distributions.

15.7 The hypothesis of interest is

H_0 : populations 1 and 2 are identical

versus H_a : population 2 is shifted to the right of population 1

The data, with ranks in parentheses, are given below.

20s	11(20)	7(11)	6(7.5)	8(14)	6(7.5)	9(16.5)	2(2)	10(18.5)	3(3.5)	6(7.5)
65-70s	1(1)	9(16.5)	6(7.5)	8(14)	7(11)	8(14)	5(5)	7(11)	10(18.5)	3(3.5)

Then

$$T_1 = 20 + 11 + \cdots + 7.5 = 108$$
$$T_1^* = n_1(n_1 + n_2 + 1) - T_1 = 10(20+1) - 108 = 102$$

The test statistic is

$$T = \min(T_1, T_1^*) = 102$$

With $n_1 = n_2 = 10$, the one-tailed rejection region with $\alpha = .05$ is found in Table 7a to be $T_1^* \leq 82$ and the observed value, $T = 102$, does not fall in the rejection region; H_0 is not rejected. We cannot conclude that this drug improves memory in mean aged 65 to 70 to that of 20 year olds.

15.9 Similar to previous exercises. The data, with corresponding ranks, are shown in the following table.

Deaf (1)	Hearing (2)
2.75 (15)	0.89 (1)
2.14 (11)	1.43 (7)
3.23 (18)	1.06 (4)
2.07 (10)	1.01 (3)
2.49 (14)	0.94 (2)
2.18 (12)	1.79 (8)
3.16 (17)	1.12 (5.5)
2.93 (16)	2.01 (9)
2.20 (13)	1.12 (5.5)
$T_1 = 126$	

Calculate

$$T_1 = 126$$
$$T_1^* = n_1(n_1 + n_2 + 1) - T_1 = 9(19) - 126 = 45$$

The test statistic is

$$T = \min(T_1, T_1^*) = 45$$

With $n_1 = n_2 = 9$, the two-tailed rejection region with $\alpha = .05$ is found in Table 7b to be $T_1^* \leq 62$. The observed value, $T = 45$, falls in the rejection region and H_0 is rejected. We conclude that the deaf children do differ from the hearing children in eye-movement rate.

15.11   The data, with corresponding ranks, are shown in the following table.

Lake 1	Lake 2
399.7 (12.5)	345.9 (2)
430.9 (16)	368.8 (6)
394.1 (10)	399.7 (12.5)
411.1 (14)	385.6 (7)
416.7 (15)	351.5 (3)
391.2 (8.5)	337.4 (1)
396.9 (11)	354.4 (4)
456.4 (18)	391.2 (8.5)
360.0 (5)	
433.7 (17)	
	$T_1 = 44$

Calculate

$$T_1 = 44$$
$$T_1^* = n_1(n_1 + n_2 + 1) - T_1 = 8(18 + 1) - 44 = 108$$

The test statistic is $T = \min(T_1, T_1^*) = 44$

With $n_1 = 8$ and $n_2 = 10$, the two-tailed rejection region with $\alpha = .05$ is found in Table 7b to be $T \leq 53$. The observed value, $T = 44$, falls in the rejection region and H_0 is rejected. We conclude that the distribution of weights for the tagged turtles exposed to the two lake environments were different.

15.13 **a**   If a paired difference experiment has been used and the sign test is one-tailed $(H_a : p > .5)$, then the experimenter would like to show that one population of measurements lies above the other population. An exact practical statement of the alternative hypothesis would depend on the experimental situation.

b   It is necessary that α (the probability of rejecting the null hypothesis when it is true) take values less than $\alpha = .15$. Assuming the null hypothesis to be true, the two populations are identical and consequently, $p = P(A \text{ exceeds } B \text{ for a given pair of observations})$ is 1/2. The binomial probability was discussed in Chapter 5. In particular, it was noted that the distribution of the random variable x is symmetrical about the mean np when $p = 1/2$. For example, with $n = 25$, $P(x = 0) = P(x = 25)$. Similarly, $P(x = 1) = P(x = 24)$ and so on. Hence, the lower tailed probabilities tabulated in Table 1, Appendix I will be identical to their upper tailed equivalent probabilities. The values of α available for this upper tailed test and the corresponding rejection regions are shown below.

Rejection Region	α
$x \geq 20$	.002
$x \geq 19$	.007
$x \geq 18$	.022
$x \geq 17$	.054
$x \geq 16$	.115

15.15 Similar to Exercise 15.14. The rejection regions and levels of α are given in the table for the three different values of n, and a one-tailed test.

$n = 10$	$n = 15$	$n = 20$
$x \le 0 \ \alpha = .001$	$x \le 2 \ \alpha = .004$	$x \le 3 \ \alpha = .001$
$x \le 1 \ \alpha = .011$	$x \le 3 \ \alpha = .018$	$x \le 4 \ \alpha = .006$
$x \le 2 \ \alpha = .055$	$x \le 4 \ \alpha = .059$	$x \le 5 \ \alpha = .021$
		$x \le 6 \ \alpha = .058$
		$x \le 7 \ \alpha = .132$

For the two-tailed test, the rejection regions with $.01 < \alpha < .15$ are shown below.

$n = 10$	$n = 15$	$n = 20$
$x \le 0; x \ge 10 \ \alpha = .002$	$x \le 2; x \ge 13 \ \alpha = .008$	$x \le 3; x \ge 17 \ \alpha = .002$
$x \le 1; x \ge 9 \ \alpha = .022$	$x \le 3; x \ge 12 \ \alpha = .036$	$x \le 4; x \ge 16 \ \alpha = .012$
$x \le 2; x \ge 8 \ \alpha = .110$	$x \le 4; x \ge 11 \ \alpha = .118$	$x \le 5; x \ge 15 \ \alpha = .042$
		$x \le 6; x \ge 14 \ \alpha = .116$

15.17 a If assessors A and B are equal in their property assessments, then p, the probability that A's assessment exceeds B's assessment for a given property, should equal $1/2$. If one of the assessors tends to be more conservative than the other, then either $p > 1/2$ or $p < 1/2$. Hence, we can test the equivalence of the two assessors by testing the hypothesis

$$H_0 : p = 1/2 \quad \text{versus} \quad H_a : p \ne 1/2$$

using the test statistic x, the number of times that assessor A exceeds assessor B for a particular property assessment. To find a two-tailed rejection region with α close to .05, use Table 1 with $n = 8$ and $p = .5$. For the rejection region $\{x = 0, x = 8\}$ the value of α is $.004 + .004 = .008$, while for the rejection region $\{x = 0, 1, 7, 8\}$ the value of α is $.035 + .035 = .070$ which is closer to .05. Hence, using the rejection region $\{x \le 1 \text{ or } x \ge 7\}$, the null hypothesis is not rejected, since x = number of properties for which A exceeds B = 6. The p-value for this two-tailed test is

$$p\text{-value} = 2P(x \ge 6) = 2(1 - .855) = .290$$

Since the p-value is greater than .10, the results are not significant; H_0 is not rejected (as with the critical value approach).

b The t statistic used in Exercise 10.45 allows the experimenter to reject H_0, while the sign test fails to reject H_0. This is because the sign test used less information and makes fewer assumptions than does the t test. If all normality assumptions are met, the t test is the more powerful test and can reject when the sign test cannot.

15.19 Similar to Exercise 15.18. The hypothesis to be tested is

$$H_0 : p = 1/2 \quad \text{versus} \quad H_a : p > 1/2$$

using the sign test with x, the number of "elevated" blood lead levels observed in $n = 17$ people, as the test statistic. Using the large sample approximation, the test statistic is

$$z = \frac{x - .5n}{.5\sqrt{n}} = \frac{15 - .5(17)}{.5\sqrt{17}} = 3.15$$

and the one-tailed rejection region with $\alpha = .05$ is $z > 1.645$. The null hypothesis is rejected and we conclude that the indoor firing range has the effect of increasing a person's blood lead level.

15.21 **a** H_0: population distributions 1 and 2 are identical
H_a: the distributions differ in location

b Since Table 8, Appendix I gives critical values for rejection in the lower tail of the distribution, we use the smaller of T^+ and T^- as the test statistic.

c From Table 8 with $n = 30$, $\alpha = .05$ and a two-tailed test, the rejection region is $T \leq 137$.

d Since $T^+ = 249$, we can calculate

$$T^- = \frac{n(n+1)}{2} - T^+ = \frac{30(31)}{2} - 249 = 216.$$

The test statistic is the smaller of T^+ and T^- or $T = 216$ and H_0 is not rejected. There is no evidence of a difference between the two distributions.

15.23 Since $n > 25$, the large sample approximation to the signed rank test can be used to test the hypothesis given in Exercise 15.21a. Calculate

$$E(T) = \frac{n(n+1)}{4} = \frac{30(31)}{4} = 232.5$$

$$\sigma_T^2 = \frac{n(n+1)(2n+1)}{24} = \frac{30(31)(61)}{24} = 2363.75$$

The test statistic is

$$z = \frac{T - E(T)}{\sigma_T} = \frac{216 - 232.5}{\sqrt{2363.75}} = -.34$$

The two-tailed rejection region with $\alpha = .05$ is $|z| > 1.96$ and H_0 is not rejected. The results agree with Exercise 15.21d.

15.25 **a** The hypothesis to be tested is

H_0: population distributions 1 and 2 are identical
H_a: the distributions differ in location

and the test statistic is T, the rank sum of the positive (or negative) differences. The ranks are obtained by ordering the differences according to their absolute value. Define d_i to be the difference between a pair in populations 1 and 2 (i.e., $x_{1i} - x_{2i}$). The differences, along with their ranks (according to absolute magnitude), are shown in the following table.

d_i	.1	.7	.3	–.1	.5	.2	.5
Rank $|d_i|$	1.5	7	4	1.5	5.5	3	5.5

The rank sum for positive differences is $T^+ = 26.5$ and the rank sum for negative differences is $T^- = 1.5$ with $n = 7$. Consider the smaller rank sum and determine the appropriate lower portion of the two-tailed rejection region. Indexing $n = 7$ and $\alpha = .05$ in Table 8, the rejection region is $T \leq 2$ and H_0 is rejected. There is a difference in the two population locations.

b The results do not agree with those obtained in Exercise 15.16. We are able to reject H_0 with the more powerful Wilcoxon test.

15.27 a Similar to Exercise 15.26. The Wilcoxon signed rank test is used, and the differences, along with their ranks (according to absolute magnitude), are shown in the following table.

d_i	−4	2	−2	−5	−3	0	1	1	−6
Rank $\lvert d_i \rvert$	6	3.5	3.5	7	5	--	1.5	1.5	8

The sixth pair has zero difference and is hence eliminated from consideration. Pairs 7 and 8, 2 and 3 are tied and receive an average rank. Then $T^+ = 6.5$ and $T^- = 29.5$ with $n = 8$. Indexing $n = 8$ and $\alpha = .05$ in Table 8, the lower portion of the two-tailed rejection region is $T \leq 4$ and H_0 is not rejected. There is a insufficient evidence to detect a difference in the two machines.

b If a machine continually breaks down, it will eventually be fixed, and the breakdown rate for the following month will decrease.

15.29 a The paired data are given in the exercise. The differences, along with their ranks (according to absolute magnitude), are shown in the following table.

d_i	1	2	−1	1	3	1	−1	3	−2	3	1	0
Rank $\lvert d_i \rvert$	3.5	7.5	3.5	3.5	10	3.5	3.5	10	7.5	10	2.5	--

Let $p = P(\text{A exceeds B for a given intersection})$ and x = number of intersections at which A exceeds B. The hypothesis to be tested is

$$H_0 : p = 1/2 \quad \text{versus} \quad H_a : p \neq 1/2$$

using the sign test with x as the test statistic.

Critical value approach: Various two tailed rejection regions are tried in order to find a region with $\alpha \approx .05$. These are shown in the following table.

Rejection Region	α
$x \leq 1; x \geq 10$	.012
$x \leq 2; x \geq 9$	.066
$x \leq 3; x \geq 8$	.226

We choose to reject H_0 if $x \leq 2$ or $x \geq 9$ with $\alpha = .066$. Since $x = 8$, H_0 is not rejected. There is insufficient evidence to indicate a difference between the two methods.

p-value approach: For the observed value $x = 8$, calculate the two-tailed p-value:

$$p\text{-value} = 2P(x \geq 8) = 2(1 - .887) = .226$$

Since the p-value is greater than .10, H_0 is not rejected.

b To use the Wilcoxon signed rank test, we use the ranks of the absolute differences shown in the table above. Then $T^+ = 51.5$ and $T^- = 14.5$ with $n = 11$. Indexing $n = 11$ and $\alpha = .05$ in Table 8, the lower portion of the two-tailed rejection region is $T \leq 11$ and H_0 is not rejected, as in part **a**.

15.31 a Since the experiment has been designed as a paired experiment, there are three tests available for testing the differences in the distributions with and without imagery – (1) the paired difference t test; (2) the sign test or (3) the Wilcoxon signed rank test. In order to use the paired difference t test, the scores must be approximately normal; since the number of words recalled has a binomial distribution with $n = 25$ and unknown recall probability, this distribution may not be approximately normal.

b Using the **sign test**, the hypothesis to be tested is
$$H_0 : p = 1/2 \quad \text{versus} \quad H_a : p > 1/2$$
For the observed value $x = 0$ we calculate the two-tailed p-value:
$$p\text{-value} = 2P(x \leq 0) = 2(.000) = .000$$
The results are highly significant; H_0 is rejected and we conclude there is a difference in the recall scores with and without imagery.

Using the **Wilcoxon signed-rank test**, the differences will all be positive ($x = 0$ for the sign test), so that and
$$T^+ = \frac{n(n+1)}{2} = \frac{20(21)}{2} = 210 \quad \text{and} \quad T^- = 210 - 210 = 0$$
Indexing $n = 20$ and $\alpha = .01$ in Table 8, the lower portion of the two-tailed rejection region is $T \leq 37$ and H_0 is rejected.

15.33 Similar to Exercise 15.32. The data with corresponding ranks in parentheses are shown below.

Treatment			
1	2	3	4
124 (9)	147 (20)	141 (17)	117 (4.5)
167 (26)	121 (7)	144 (18.5)	128 (10.5)
135 (14)	136 (15)	139 (16)	102 (1)
160 (24)	114 (3)	162 (25)	119 (6)
159 (23)	129 (12)	155 (22)	128 (10.5)
144 (18.5)	117 (4.5)	150 (21)	123 (8)
133 (13)	109 (2)		
$T_1 = 127.5$	$T_2 = 63.5$	$T_3 = 119.5$	$T_4 = 40.5$
$n_1 = 7$	$n_2 = 7$	$n_3 = 6$	$n_4 = 6$

The test statistic, based on the rank sums, is
$$H = \frac{12}{n(n+1)} \sum \frac{T_i^2}{n_i} - 3(n+1)$$
$$= \frac{12}{26(27)} \left[\frac{(127.5)^2}{7} + \frac{(63.5)^2}{7} + \frac{(119.5)^2}{6} + \frac{(40.5)^2}{6} \right] - 3(27) = 13.90$$

The rejection region with $\alpha = .05$ and $k - 1 = 3$ df is based on the chi-square distribution, or $H > \chi^2_{.05} = 7.81$. The null hypothesis is rejected and we conclude that there is a difference among the four treatments.

15.35 Similar to Exercise 15.32. The data with corresponding ranks in parentheses are shown below.

Age			
10 – 19	20 – 39	40 – 59	60 – 69
29 (21)	24 (8)	37 (39)	28 (18)
33 (29.5)	27 (15)	25 (10.5)	29 (21)
26 (12.5)	33 (29.5)	22 (5.5)	34 (34)
27 (15)	31 (24)	33 (29.5)	36 (37.5)
39 (40)	21 (3)	28 (18)	21 (3)
35 (36)	28 (18)	26 (12.5)	20 (1)
33 (29.5)	24 (8)	30 (23)	25 (10.5)
29 (21)	34 (34)	34 (34)	24 (8)
36 (37.5)	21 (3)	27 (15)	33 (29.5)
22 (5.5)	32 (25.5)	33 (29.5)	32 (25.5)
$T_1 = 247.5$	$T_2 = 168$	$T_3 = 216.5$	$T_4 = 188$
$n_1 = 10$	$n_2 = 10$	$n_3 = 10$	$n_4 = 10$

a The test statistic, based on the rank sums, is

$$H = \frac{12}{n(n+1)} \sum \frac{T_i^2}{n_i} - 3(n+1)$$

$$= \frac{12}{40(41)} \left[\frac{(247.5)^2}{10} + \frac{(168)^2}{10} + \frac{(216.5)^2}{10} + \frac{(188)^2}{10} \right] - 3(41) = 2.63$$

The rejection region with $\alpha = .01$ and $k - 1 = 3$ df is based on the chi-square distribution, or $H > \chi^2_{.01} = 11.35$. The null hypothesis is not rejected. There is no evidence of a difference in location.

b Since the observed value $H = 2.63$ is less than $\chi^2_{.10} = 6.25$, the p-value is greater than .10.

c-d From Exercise 11.60, $F = .87$ with 3 and 36 df. Again, the p-value is greater than .10 and the results are the same.

15.37 Similar to previous exercises. The ranks of the data are shown below.

Campaigns		
1	2	3
11.5	6	1.5
7	15	8
1.5	13	4
10	14	11.5
3	5	9
$T_1 = 33$	$T_2 = 53$	$T_3 = 34$
$n_1 = 5$	$n_2 = 5$	$n_3 = 5$

a The test statistic is

$$H = \frac{12}{n(n+1)} \sum \frac{T_i^2}{n_i} - 3(n+1)$$

$$= \frac{12}{15(16)} \left[\frac{(33)^2}{5} + \frac{(53)^2}{5} + \frac{(34)^2}{5} \right] - 3(16) = 2.54$$

With $k - 1 = 2\, df$, the observed value $H = 2.54$ is less than $\chi^2_{.10} = 4.61$, the p-value is greater than .10. The null hypothesis is not rejected and we cannot conclude that there is a difference in the three population distributions.

15.39 Similar to Exercise 15.38. The ranks of the data are shown below.

	Treatment			
Block	1	2	3	4
1	4	1	2	3
2	4	1.5	1.5	3
3	4	1	3	2
4	4	1	2	3
5	4	1	2.5	2.5
6	4	1	2	3
7	4	1	3	2
8	4	1	2	3
	$T_1 = 32$	$T_2 = 8.5$	$T_3 = 18$	$T_4 = 21.5$

a The test statistic is

$$F_r = \frac{12}{bk(k+1)} \sum T_i^2 - 3b(k+1)$$

$$= \frac{12}{8(4)(5)} \left[(32)^2 + (8.5)^2 + 18^2 + (21.5)^2 \right] - 3(8)(5) = 21.19$$

and the rejection region is $F_r > \chi^2_{.05} = 7.81$. Hence, H₀ is rejected and we conclude that there is a difference among the four treatments.

b The observed value, $F_r = 21.19$, exceeds $\chi^2_{.005}$, p-value < .005.

c-e The analysis of variance is performed as in Chapter 11. The ANOVA table is shown below.

Source	df	SS	MS	F
Treatments	3	198.34375	66.114583	75.43
Blocks	7	220.46875	31.495536	35.93
Error	21	18.40625	0.876488	
Total	31	437.40625		

The analysis of variance F test for treatments is $F = 75.43$ and the approximate p-value with 3 and 21 df is p-value < .005. The result is identical to the parametric result.

15.41 Similar to Exercise 15.38, with rats as blocks. The data are shown below along with corresponding ranks within blocks. Note that we have rearranged the data to eliminate the random order of presentation in the display.

	Treatment		
Rat	A	B	C
1	6 (3)	5 (2)	3 (1)
2	9 (2.5)	9 (2.5)	4 (1)
3	6 (2)	9 (3)	3 (1)
4	5 (1)	8 (3)	6 (2)
5	7 (1)	8 (2.5)	8 (2.5)
6	5 (1.5)	7 (3)	5 (1)
7	6 (2)	7 (3)	5 (1)
8	6 (1)	7 (2.5)	7 (2.5)
	$T_1 = 14$	$T_2 = 21.5$	$T_3 = 12.5$

a The test statistic is

$$F_r = \frac{12}{bk(k+1)} \sum T_i^2 - 3b(k+1)$$

$$= \frac{12}{8(3)(4)} \left[(14)^2 + (21.5)^2 + (12.5)^2 \right] - 3(8)(4) = 5.81$$

and the rejection region is $F_r > \chi^2_{.05} = 5.99$. Hence, H₀ is not rejected and we cannot conclude that there is a difference among the three treatments.

b The observed value, $F_r = 5.81$, falls between $\chi^2_{.05}$ and $\chi^2_{.10}$. Hence, $.05 < p\text{-value} < .10$.

15.43 Table 9, Appendix I gives critical values r_0 such that $P(r_s \geq r_0) = \alpha$. Hence, for an upper-tailed test, the critical value for rejection can be read directly from the table.
    a $r_s \geq .425$
    b $r_s \geq .601$

15.45 For a two-tailed test of correlation, the value of α given along the top of the table is doubled to obtain the **actual** value of α for the test.
    a To obtain $\alpha = .05$, index .025 and the rejection region is $|r_s| \geq .400$.
    b To obtain $\alpha = .01$, index .005 and the rejection region is $|r_s| \geq .526$.

15.47 a The two variables (rating and distance) are ranked from low to high, and the results are shown in the following table.

Voter	x	y	Voter	x	y
1	7.5	3	7	6	4
2	4	7	8	11	2
3	3	12	9	1	10
4	12	1	10	5	9
5	10	8	11	9	5.5
6	7.5	11	12	2	5.5

Calculate $\sum x_i y_i = 442.5 \quad \sum x_i^2 = 649.5 \quad \sum y_i^2 = 649.5$
$n = 12 \quad \sum x_i = 78 \quad \sum y_i = 78$

Then $S_{xy} = 422.5 - \frac{78^2}{12} = -84.5 \quad S_{xx} = 649.5 - \frac{78^2}{12} = 142.5 \quad S_{yy} = 649.5 - \frac{78^2}{12} = 142.5$

and $r_s = \frac{S_{xy}}{\sqrt{S_{xx}S_{yy}}} = \frac{-84.5}{142.5} = -.593$.

b The hypothesis of interest is H_0: no correlation versus H_a: negative correlation. Consulting Table 9 for $\alpha = .05$, the critical value of r_s, denoted by r_0 is $-.497$. Since the value of the test statistic is less than the critical value, the null hypothesis is rejected. There is evidence of a significant negative correlation between rating and distance.

15.49 a The data are ranked separately according to the variables x and y.

Rank x	7	6	5	4	1	12	8	3	2	11	10	9
Rank y	7	8	4	5	2	10	12	3	1	6	11	9

Since there were no tied observations, the simpler formula for r_s is used, and

$$r_s = 1 - \frac{6\sum d_i^2}{n(n^2-1)} = 1 - \frac{6\left[(0)^2 + (-2)^2 + \cdots + (0)^2\right]}{12(143)}$$
$$= 1 - \frac{6(54)}{1716} = .811$$

b To test for positive correlation with $\alpha = .05$, index .05 in Table 9 and the rejection region is $r_s \geq .497$. Hence, H_0 is rejected, there is a positive correlation between x and y.

15.51 Refer to Exercise 15.50. To test for positive correlation with $\alpha = .05$, index .05 in Table 9 and the rejection region is $r_s \geq .600$. We reject the null hypothesis of no association and conclude that a positive correlation exists between the teacher's ranks and the ranks of the IQs.

15.53 The ranks of the two variables are shown below.

Leaf	1	2	3	4	5	6	7	8	9	10	11	12
Rank x	10.5	5.5	7.5	7.5	4	9	2	5.5	1	12	10.5	3
Rank y	12	7.5	9	6	4.5	10	3	4.5	1	11	7.5	2

Calculate $\sum x_i y_i = 636.25$ $\sum x_i^2 = 648.5$ $\sum y_i^2 = 649$

$n = 12$ $\sum x_i = 78$ $\sum y_i = 78$

Then $S_{xy} = 636.25 - \frac{78^2}{12} = 129.25$ $S_{xx} = 648.5 - \frac{78^2}{12} = 141.5$ $S_{yy} = 649 - \frac{78^2}{12} = 142$

and $r_s = \frac{S_{xy}}{\sqrt{S_{xx}S_{yy}}} = \frac{129.25}{\sqrt{141.5(142)}} = .913$.

To test for correlation with $\alpha = .05$, index .025 in Table 9, and the rejection region is $|r_s| \geq .591$. The null hypothesis is rejected and we conclude that there is a correlation between the two variables.

15.55 **a** Define $p = P(\text{response for stimulus 1 exceeds that for stimulus 2})$ and x = number of times the response for stimulus 1 exceeds that for stimulus 2. The hypothesis to be tested is

$H_0 : p = 1/2$ versus $H_a : p \neq 1/2$

using the sign test with x as the test statistic. Notice that for this exercise $n = 9$, and the observed value of the test statistic is $x = 2$. Various two tailed rejection regions are tried in order to find a region with $\alpha \approx .05$. These are shown in the following table.

Rejection Region	α
$x = 0; x = 9$	.004
$x \leq 1; x \geq 8$	.040
$x \leq 2; x \geq 7$	.180

We choose to reject H_0 if $x \leq 1$ or $x \geq 8$ with $\alpha = .040$. Since $x = 2$, H_0 is not rejected. There is insufficient evidence to indicate a difference between the two stimuli.

b The experiment has been designed in a paired manner, and the paired difference test is used. The differences are shown below.

$d_i :$ $-.9$ -1.1 1.5 -2.6 -1.8 -2.9 -2.5 2.5 -1.4

The hypothesis to be tested is $H_0 : \mu_1 - \mu_2 = 0$ $H_a : \mu_1 - \mu_2 \neq 0$

Calculate $\bar{d} = \frac{\sum d_i}{n} = \frac{-9.2}{9} = -1.022$ $s_d^2 = \frac{\sum d_i^2 - \frac{(\sum d_i)^2}{n}}{n-1} = \frac{37.14 - 9.404}{8} = 3.467$

and the test statistic is

$$t = \frac{\bar{d}}{\sqrt{\frac{s_d^2}{n}}} = \frac{-1.022}{\sqrt{\frac{3.467}{9}}} = -1.646$$

The rejection region with $\alpha = .05$ and 8 df is $|t| > 2.306$ and H_0 is not rejected.

15.57 **a** Define $p = P(\text{school A exceeds school B in test score for a pair of twins})$ and $x =$ number of times the score for school A exceeds the score for school B. The hypothesis to be tested is
$$H_0 : p = 1/2 \quad \text{versus} \quad H_a : p \neq 1/2$$
using the sign test with x as the test statistic. Notice that for this exercise $n = 10$, and the observed value of the test statistic is $x = 7$.

Critical value approach: Various two tailed rejection regions are tried in order to find a region with $\alpha \approx .05$. These are shown in the following table.

Rejection Region	α
$x = 0; x = 10$	.002
$x \leq 1; x \geq 9$	.022
$x \leq 2; x \geq 8$	.110

We choose to reject H_0 if $x \leq 1$ or $x \geq 9$ with $\alpha = .022$. Since $x = 7$, H_0 is not rejected. There is insufficient evidence to indicate a difference between the two schools.

p-value approach: For the observed value $x = 7$, calculate the two-tailed p-value:
$$p\text{-value} = 2P(x \geq 7) = 2(1-.828) = .344$$
and H_0 is not rejected. There is insufficient evidence to indicate a difference between the two schools.

b Consider the one-tailed test of hypothesis as follows:
$$H_0 : p = 1/2 \quad \text{versus} \quad H_a : p > 1/2$$
This alternative will imply that school A is superior to school B. From Table 1, the one-tailed rejection region with $\alpha \approx .05$ is $x \geq 8$ with $\alpha = .055$. The null hypothesis is still not rejected, since $x = 7$. (The one-tailed p-value $= .172$.)

15.59 The data, with corresponding ranks, are shown in the following table.

A (1)	B (2)
6.1 (1)	9.1 (16)
9.2 (17)	8.2 (8)
8.7 (12)	8.6 (11)
8.9 (13.5)	6.9 (2)
7.6 (5)	7.5 (4)
7.1 (3)	7.9 (7)
9.5 (18)	8.3 (9.5)
8.3 (9.5)	7.8 (6)
9.0 (1.5)	8.9 (13.5)
$T_1 = 94$	

The difference in the brightness levels using the two processes can be tested using the nonparametric Wilcoxon rank sum test, or the parametric two-sample t test.

1 To test the null hypothesis that the two population distributions are identical, calculate
$$T_1 = 1 + 17 + \cdots + 1.5 = 94$$
$$T_1^* = n_1(n_1 + n_2 + 1) - T_1 = 9(18+1) - 94 = 77$$
The test statistic is $T = \min(T_1, T_1^*) = 77$.

With $n_1 = n_2 = 9$, the two-tailed rejection region with $\alpha = .05$ is found in Table 7b to be $T_1^* \leq 62$. The observed value, $T = 77$, does not fall in the rejection region and H_0 is not rejected. We cannot conclude that the distributions of brightness measurements are different for the two processes.

2 To test the null hypothesis that the two population means are identical, calculate

$$\bar{x}_1 = \frac{\sum x_{1j}}{n_1} = \frac{74.4}{9} = 8.2667$$

$$\bar{x}_2 = \frac{\sum x_{2j}}{n_2} = \frac{73.2}{9} = 8.1333$$

$$s^2 = \frac{(n_1-1)s_1^2 + (n_2-1)s_2^2}{n_1+n_2-2} = \frac{625.06 - \frac{(74.4)^2}{9} + 599.22 - \frac{(73.2)^2}{9}}{16} = .8675$$

and the test statistic is $t = \dfrac{\bar{x}_1 - \bar{x}_2}{\sqrt{s^2\left(\frac{1}{n_1}+\frac{1}{n_2}\right)}} = \dfrac{8.27 - 8.13}{\sqrt{.8675\left(\frac{2}{9}\right)}} = .304$

The rejection region with $\alpha = .05$ and 16 degrees of freedom is $|t| > 1.746$ and H_0 is not rejected. There is insufficient evidence to indicate a difference in the average brightness measurements for the two processes.
Notice that the nonparametric and parametric tests reach the same conclusions.

15.61 Since this is a paired experiment, you can choose either the sign test, the Wilcoxon signed rank test, or the parametric paired t test. Since the tenderizers have been scored on a scale of 1 to 10, the parametric test is not applicable. Start by using the easiest of the two nonparametric tests – the sign test. Define
$p = P(\text{tenderizer A exceeds B for a given cut})$ and x = number of times that A exceeds B. The hypothesis to be tested is

$$H_0: p = 1/2 \quad \text{versus} \quad H_a: p \neq 1/2$$

using the sign test with x as the test statistic. Notice that for this exercise $n = 8$ (there are two ties), and the observed value of the test statistic is $x = 2$.

p-value approach: For the observed value $x = 2$, calculate

$$p\text{-value} = 2P(x \le 2) = 2(.145) = .290$$

Since the p-value is greater than .10, H_0 is not rejected. There is insufficient evidence to indicate a difference between the two tenderizers.
If you use the Wilcoxon signed rank test, you will find $T^+ = 7$ and $T^- = 29$ which will not allow rejection of H_0 at the 5% level of significance. The results are the same.
Student's t may not be appropriate here as the data are on a 10 point scale.

15.63 To test for negative correlation with $\alpha = .05$, index .05 in Table 9, and the rejection region is $r_s \le -.564$.
The null hypothesis is rejected and we conclude that there is a negative correlation between the two variables.

15.65 The hypothesis to be tested is

H₀: population distributions 1 and 2 are identical
Hₐ: the distributions differ in location

and the test statistic is T, the rank sum of the positive (or negative) differences. The ranks are obtained by ordering the differences according to their absolute value. Define d_i to be the difference between a pair in populations 1 and 2 (i.e., $x_{1i} - x_{2i}$). The differences, along with their ranks (according to absolute magnitude), are shown in the following table.

d_i	−31	−31	−6	−11	−9	−7	7
Rank $\mid d_i \mid$	14.5	14.5	4.5	12.5	10.5	7	7

d_i	−11	7	−9	−2	−8	−1	−6	−3
Rank $\mid d_i \mid$	12.5	7	10.5	2	9	1	4.5	3

The rank sum for positive differences is $T^+ = 14$ and the rank sum for negative differences is $T^- = 106$ with $n = 15$. Consider the smaller rank sum and determine the appropriate lower portion of the two-tailed rejection region. Indexing $n = 15$ and $\alpha = .05$ in Table 8, the rejection region is $T \le 25$ and H₀ is rejected. We conclude that there is a difference between math and art scores.

15.67 Similar to Exercise 15.38. The ranks within each block are shown below.

Varieties	1	2	3	4	5	6	T_i
A	4	2	3	3	3	3	18
B	1	3	2	2	2	2	12
C	5	4	4	4	4	5	26
D	3	5	5	5	5	4	27
E	2	1	1	1	1	1	7

a Use the Friedman F_r statistic calculated as

$$F_r = \frac{12}{30(6)}\left[18^2 + 12^2 + \cdots + 7^2\right] - 3(36) = 20.13$$

and the rejection region is $F_r > \chi^2_{.05} = 9.49$. Hence, H₀ is rejected and we conclude that there is a difference in the levels of yield for the five varieties of wheat.

b From Exercise 11.68, $F = 18.61$ and H₀ is rejected. The results are the same.

15.69 **a-b** Since the experiment is a completely randomized design, the Kruskal Wallis H test is used. The combined ranks are shown below.

Plant	Ranks					T_i
A	9	12	5	1	7	34
B	11	15	4	19	14	63
C	3	13	2	9	6	33
D	20	17	9	16	18	80

The test statistic, based on the rank sums, is

$$H = \frac{12}{n(n+1)} \sum \frac{T_i^2}{n_i} - 3(n+1)$$

$$= \frac{12}{20(21)}\left[\frac{(34)^2}{5} + \frac{(63)^2}{5} + \frac{(33)^2}{5} + \frac{(80)^2}{5}\right] - 3(21) = 9.08$$

With $df = k - 1 = 3$, the observed value $H = 9.08$ is between $\chi^2_{.025}$ and $\chi^2_{.05}$ so that $.025 < p\text{-value} < .05$. The null hypothesis is rejected and we conclude that there is a difference among the four plants.

c From Exercise 11.66, $F = 5.20$, and H_0 is rejected. The results are the same.

15.71 **a** Neither of the two plots follow the general patterns for normal populations with equal variances.

b Use the Friedman F_r test for a randomized block design. The **Minitab** printout follows.

Friedman Test: Cadmium versus Harvest blocked by Rate
```
S = 10.33   DF = 2    P = 0.006

                        Sum
                        of
Harvest  N  Est Median  Ranks
1        6     202.29   11.0
2        6     201.21    7.0
3        6     300.73   18.0
Grand median = 234.74
```

Since the p-value is .006, the results are highly significant. There is evidence of a difference among the responses to the three rates of application.

15.73 The data are already in rank form. The "substantial experience" sample is designated as sample 1, and $n_1 = 5, n_2 = 7$. Calculate

$$T_1 = 19$$
$$T_1^* = n_1(n_1 + n_2 + 1) - T_1 = 5(13) - 19 = 46$$

The test statistic is $T = \min(T_1, T_1^*) = 19$

With $n_1 = n_2 = 12$, the one-tailed rejection region with $\alpha = .05$ is found in Table 7a to be $T_1 \leq 21$. The observed value, $T = 19$, falls in the rejection region and H_0 is rejected. There is sufficient evidence to indicate that the review board considers experience a prime factor in the selection of the best candidates.

15.75 Define $p = P(\text{student exhibits increased productivity after the installation})$ and x = number of students who exhibit increased productivity. The hypothesis to be tested is $H_0: p = 1/2$ versus $H_a: p > 1/2$ using the sign test with x as the test statistic. Using the large sample approximation, the test statistic is

$$z = \frac{x - .5n}{.5\sqrt{n}} = \frac{21 - .5(35)}{.5\sqrt{35}} = 1.18$$

and the one-tailed rejection region with $\alpha = .05$ is $z > 1.645$. The null hypothesis is rejected and we conclude that the new lighting was effective in increasing student productivity.

15.77 Similar to Exercise 15.32. The data with corresponding ranks in parentheses are shown below.

Training Periods (hours)			
.5	1.0	1.5	2.0
8 (9.5)	9 (11.5)	4 (1.5)	4 (1.5)
14 (14)	7 (7)	6 (5)	7 (7)
9 (11.5)	5 (3.5)	7 (7)	5 (3.5)
12 (13)		8 (9.5)	
$T_1 = 48$	$T_2 = 22$	$T_3 = 23$	$T_4 = 12$
$n_1 = 4$	$n_2 = 3$	$n_3 = 4$	$n_4 = 3$

The test statistic, based on the rank sums, is

$$H = \frac{12}{n(n+1)} \sum \frac{T_i^2}{n_i} - 3(n+1)$$

$$= \frac{12}{14(15)} \left[\frac{(48)^2}{4} + \frac{(22)^2}{3} + \frac{(23)^2}{4} + \frac{(12)^2}{3} \right] - 3(15) = 7.4333$$

The rejection region with $\alpha = .01$ and $k - 1 = 3$ df is based on the chi-square distribution, or $H > \chi^2_{.01} = 11.34$. The null hypothesis is not rejected and we conclude that there is insufficient evidence to indicate a difference in the distribution of times for the four groups.

15.79 a The batsmen have already been ranked. Let x be the true ranking and let y be my ranking. Since no ties exist in either the x or y rankings.

$$r_s = 1 - \frac{6 \sum d_i^2}{n(n^2 - 1)} = 1 - \frac{6\left[(-2)^2 + (1)^2 + \cdots + (1)^2\right]}{8(63)} = 1 - \frac{6(22)}{504} = .738$$

To test for positive correlation with $\alpha = .05$, index .05 in Table 9 and the rejection region is $r_s \geq .643$. We reject the null hypothesis of no association and conclude that a positive association exists.

Introduction to Probability and Statistics, 2ce

Case Study:
How's Your Cholesterol Level?

1. The design is a randomized block design, with tasters as blocks and the three types of eggs as treatments.

2. Since the judges are only asked to provide a rating between 0 and 20, the resulting data are not likely to be normally distributed. The parametric assumptions are probably not satisfied in this case, and a non-parametric test should be used.

3. The Friedman F_r test should be used. The ranks within each block are shown below.

Tasters	H	S	E
Bowe	3	2	1
Carroll	3	1	2
Katzl	2	3	1
O'Connell	2	3	1
Passot	3	2	1
Totals	14	10	6

The Friedman F_r statistic is

$$F_r = \frac{12}{15(4)} = \left[14^2 + 10^2 + 6^2\right] - 3(20) = 66.4 - 60 = 6.4$$

and the rejection region is $F_r > \chi^2_{.05} = 5.99$. Hence, H_0 is rejected and we conclude that there is a difference in the average scores for the three brands of egg substitutes.

Project 15A: Air Conditional Makes You Gain Weight

a   For this particular study, we wish to determine if the distribution of basal metabolic rate for females who use air conditioning is the same as for females who do not use air conditioning. Our null and alternative hypotheses are

H_0: the distribution of basal metabolic rate for females from Vancouver who use air conditioning during the hours of night sleep is the same as for females from Vancouver who do not use air conditioning during the hours of night sleep.

H_a: the distributions differ.

The Wilcoxon rank sum test begins by ordering all of the observations from smallest to largest. The observations are ranked, where the smallest observation is assigned a rank of 1, the second smallest is assigned a rank of 2, and so forth. Additionally, we label the observations based on whether they belong to Group A, or Group B. We thus obtain the following list of ordered basal metabolic rates

950 (1, A), 987 (2, B), 998 (3, B), 1011 (4, B), 1027 (5, A), 1086 (6, B), 1088 (7, A), 1090 (8, A), 1099 (9, B), 1101 (10, B), 1125 (11, B), 1145 (12, A), 1150 (13.5, A), 1150 (13.5, A), 1199 (15, A)

Since the value 1150 appears twice, we have a tie. We accommodate the tie by finding the average of the ranks these observations would have been assigned (i.e., 13 and 14). In this case, each of these observations are assigned a rank of (13+14)/2=13.5.

Since Group A has 8 observations, and Group B has 7 observations, we set n1=7 and n2=8. Thus, when calculating T1, we sum the ranks assigned to observations from Group B. Hence

T1=2+3+4+6+9+10+11=45
T1*=n1(n1+n2+1)-T1=7(7+8+1)-45=102-45=57

Our test statistic is the minimum of T1 and T1*. In this case, we set Ttest=min(45,57)=45. Note that since this is a two tailed test, we need to look up a value associated with α/2=0.05/2=0.025.
The critical value when α/2=0.025 (found in Table 7b in Appendix I) is Tcrit=38.

Our decision is to reject the null hypothesis in favour of the alternative if our test statistic is less than or equal to the critical value. In this case, since Ttest>Tcrit, we do not reject the null hypothesis. There is insufficient evidence to suggest that the distributions are different.

b To test whether or not the metabolic rates for Group A are significantly higher than those for Group B, we update our null and alternative hypotheses.

H_0: the distribution of basal metabolic rate for females from Vancouver who use air conditioning during the hours of night sleep is the same as for females from Vancouver who do not use air conditioning during the hours of night sleep.

H_a: the distribution of basal metabolic rate for females from Vancouver who use air conditioning during the hours of night sleep is greater than for females from Vancouver who do not use air conditioning during the hours of night sleep.

While our alternative hypothesis has changed, the ranks and test statistic remain the same. We need only look up the appropriate critical value. Since our alternative is one-sided, we look up the critical value when $\alpha=0.05$. In this case, we find Tcrit=41.

Our decision is to reject the null hypothesis in favour of the alternative if our test statistic is less than or equal to 41. Since 45>41, there is insufficient evidence to suggest that the basal metabolic rate for females who use air conditioning during the hours of night sleep is higher than for females who do not use air conditioning during the hours of night sleep.

c The normal approximation is generally valid when the sample sizes for both groups is 10 or larger. In this case, we have both sample sizes smaller than 10. For this reason, we do not use the normal approximation.

If we assume the data come from a normal distribution, we could use a t-test to compare the mean basal metabolic rates. Of course, we would have to check whether or not we could assume that the group variances were equivalent in order to select the most appropriate t-test statistic.

Project 15B: Does Drinking Water Increase Metabolism?

a Our null and alternative hypotheses are

H_0: the distribution of basal metabolic rate for females who drink the appropriate amount of water (A) is identical to the distribution of basal metabolic rate for females who do not drink the appropriate amount of water (B), and p(A>B)=p=0.5.

H_a: the distribution of basal metabolic rate for females who drink the appropriate amount of water is shifted to the right of the distribution of basal metabolic rate for females who do not drink the appropriate amount of water, and p>0.5.

b To determine the sign test statistic, we first determine the differences in our observations (before-after = x-y). The sign of the differences are identified (as indicated within the brackets).

-45 (-), -11 (-), -70 (-), -38 (-), -31 (-), -46 (-), -54 (-), -43 (-), -48 (-), -10 (-)

The test statistic is the number of positives observed. In this particular case, T=0, since there are no positive differences.

c Since our data are dependent (i.e., observations before and after treatment are related), the appropriate test for this situation is the sign test for a paired experiment. Here we assume under the null hypothesis that the probability that *after* values exceed *before* values is 0.5. The number of observational pairs where the difference is positive represents a binomial random variable.

d Based on the alternative hypothesis, this is a one-tailed test. We reject the null hypothesis when $\alpha=0.10$ if our observed test statistic is less than 3. This can be determined reviewing the cumulative binomial probabilities when n=10 and p=0.5. Note that $p(x \leq 3)=0.172$ and $p(x \leq 2)=0.055$. Hence, if we observed x<3 we would reject the null hypothesis in favour of the alternative.

e Since our test statistic falls within the rejection region, we can conclude at the $\alpha=0.10$ level that drinking an appropriate amount of water was helpful in increasing metabolism.

f The null hypothesis would be rejected if $\alpha>0.000976525$. This value represents the p-value for our test statistic. That is, the probability of observing an observation as extreme or more extreme than that which was observed. In our case, the p-value is calculated using

$$\text{p-value}=p(x \leq 0)={}_{10}C_0(0.5)^0(0.5)^{10}=0.5^{10}=0.000976525$$

g Our large-sample z statistic for testing the hypothesis in part (a), and given a sample size n=10 and x=0, is

$$\begin{aligned} z &= \frac{x-0.5n}{0.5\sqrt{n}} \\ &= \frac{0-0.5(10)}{0.5\sqrt{10}} \\ &= -3.162278 \end{aligned}$$

The critical value is $z_{crit}=-1.282$. Since our test statistic is less than our critical value, we reject the null hypothesis in favour of the alternative.

h Our decisions are the same for both (e) and (g). That is, there is sufficient evidence to suggest that an appropriate amount of water will increase the basal metabolic rate.

i The null and alternative hypotheses for the Wilcoxon Signed-Rank test are

H₀: the relative frequency distribution of basal metabolic rate for females who drink the appropriate amount of water is identical to the relative frequency distribution of basal metabolic rate for females who do not drink the appropriate amount of water.

H$_a$: the relative frequency distribution of basal metabolic rate for females who drink the appropriate amount of water is shifted to the right of the relative frequency distribution of basal metabolic rate for females who do not drink the appropriate amount of water.

j   To determine the test statistic for the Wilcoxon Signed-Rank test, we begin by calculating the differences (before-after) and then ranking the absolute values of the differences from smallest to largest as indicated by the values in the brackets.

-45 (6), -11 (2), -70 (10), -38 (4), -31 (3), -46 (7), -54 (9), -43 (5), -48 (8), -10 (1)

To determine the test statistic, we need to calculate T^- and T^+. T^- represents the sum of all ranks associated with a negative difference, while T^+ represents the sum of all ranks associated with a positive difference. Here we have

$$T^- = 1+2+3+4+5+6+7+8+9+10 = 55$$
$$T^+ = 0$$

Since our alternative hypothesis is one sided (where our *after* distribution is hypothesized to fall to the right of the *before* distribution), we select T^+ as our test statistic.

k   The rejection region for our test can be found by looking up our critical value (Table 8 in Appendix I). For our particular study and given a one-tailed alternative hypothesis with α=0.05, we obtain a critical value of 11. That is, we reject the null hypothesis in favour of the alternative if our test statistic is less than or equal to 11.

l   Since our test statistic T^+=0 falls within the rejection region, we reject the null hypothesis in favour of the alternative. That is, there is sufficient evidence to suggest that drinking the appropriate amount of water will increase the basal metabolic rate.

m   The p-value for the Wilcoxon Signed-Rank test can be approximated by examining the values in Table 8 of Appendix I. Our observed test statistic of 0 given n=10 for a one-tailed alternative hypothesis would be rejected even if α=0.005 (which would coincide with a rejection region such that you would reject the null hypothesis if $T^+ \leq 3$). Hence, we can state that the p-value for this test is p-value≤0.005.

n   The large-sample Wilcoxon Signed-Rank test for testing the hypothesis in (i) is calculated as

$$z = \frac{T^+ - E(T^+)}{\sigma_{T^+}}$$
$$= \frac{0 - 27.5}{\sqrt{96.25}}$$
$$= -2.803060$$

where E(T^+)=n(n+1)/4=10(11)/4=27.5, and σ^2=n(n+1)(2n+1)/24=96.25. Our critical value for a one-tailed alternative hypothesis (with α=0.05) is -1.282. We would reject the null hypothesis if our test statistic is less than -1.282. Since our test statistic is -2.803060, we reject the null hypothesis in favour of the alternative. There is sufficient evidence to suggest that drinking the appropriate amount of water will increase the basal metabolic rate.

o   The results from (e), (l) and (n) all agree. That is, they all suggest that drinking the appropriate amount of water will increase basal metabolic rate.

p   To determine Spearman's rank correlation coefficient, we first rank each group, and then determine Sxx, Sxy and Syy, as shown below.

Before (x): 1105(7), 1077(5), 1020 (4), 989 (3), 1120 (8), 1099 (6), 1145 (10), 907 (1), 1131 (9), 988 (2)

After (y): 1150(7), 1088(4), 1090 (5), 1027 (3), 1151 (8), 1145 (6), 1199 (10), 950 (1), 1179 (9), 998 (2)

$$S_{xx} = \sum_{i=1}^{n} (x_i - \bar{x})^2$$
$$= (7 - 5.5)^2 + (5 - 5.5)^2 + \ldots + (2 - 5.5)^2$$
$$= 82.5$$
$$S_{xy} = \sum_{i=1}^{n} (x_i - \bar{x})(y_i - \bar{y})$$
$$= (7 - 5.5)(7 - 5.5) + (5 - 5.5)(4 - 5.5) + \ldots + (2 - 5.5)(2 - 5.5)$$
$$= 81.5$$
$$S_{yy} = \sum_{i=1}^{n} (y_i - \bar{y})^2$$
$$= (7 - 5.5)^2 + (4 - 5.5)^2 + \ldots + (2 - 5.5)^2$$
$$= 82.5$$

The Spearman Rank Correlation coefficient is calculated as

$$r_s = \frac{S_{xy}}{\sqrt{S_{xx} S_{yy}}}$$
$$= \frac{81.5}{\sqrt{82.5 \cdot 82.5}}$$
$$= 0.9878788$$

q   The critical value for testing a correlation between the data given α/2=0.025 (for a two tailed hypothesis) is obtained from Table 9 in Appendix I. For our test, with n=10, we have a critical value of 0.648. We will reject the null hypothesis if our test statistic exceeds this value. Since our test statistic is 0.9878788, we reject the null hypothesis in favour of the alternative. There is sufficient evidence to suggest that there is a relationship between *x* and *y*.

Project 15C: Increase Your Overall Muscle Mass and Boost Your Metabolism

a To use the Kruskal-Wallis *H* statistic, we first rank the observations from smallest to largest. Numbers inside the brackets represent ranks, S1 through S5 represent the strategy. We thus have

877 (1, S3)	890 (2, S3)	905 (3, S5)	955 (4, S2)	975 (5, S1)
981 (6, S4)	987 (7, S2)	993 (8, S5)	995 (9, S4)	998 (10, S1)
1001 (11, S2)	1009 (12, S5)	1012 (13, S4)	1034 (14, S4)	1055 (15, S1)
1074 (16, S1)	1089 (17, S3)	1098 (18, S4)	1099 (19, S2)	1101 (20, S1)
1123 (21, S1)	1129 (22, S4)	1134 (23, S2)	1143 (24, S5)	1151 (25, S3)
1159 (26, S5)	1173 (27, S3)	1188 (28, S3)	1191 (29, S5)	1195 (30, S2)

We sum the ranks for each of the treatments (S1 through S5). This gives us $T_1=87$, $T_2=94$, $T_3=100$, $T_4=82$ and $T_5=102$.

Our test statistic is $H_{test}=12/(30*31)*(1/6)*(87^2+94^2+100^2+82^2+102^2)-3(30+1)=0.6193548$.

Using $\alpha=0.05$, we would obtain a critical χ^2 value (on k-1=5-1=4 degrees of freedom) of 9.48773. We would reject the null hypothesis if our test statistic H exceeded 9.48773. Since our test statistic is less, we do not reject the null hypothesis. That is, there is insufficient evidence to suggest that there is a difference in location for at least two of the population distributions.

b Using Table 5 in Appendix I, with 4 degrees of freedom, we can say that $p(\chi^2>0.6193548)>0.10$.

c Since our p-value is larger than α, we would fail to reject the null hypothesis.

d Using the statistical package R (http://www.r-project.org/) we obtain the following output for Analysis of Variance

	Df	Sum Sq	Mean Sq	F value	Pr(>F)
factor(strat)	4	2297	574	0.0587	0.9932
Residuals	25	244734	9789		

The approximate p-value for the *F*-statistic in testing equality of means is 0.9932. At the $\alpha=0.05$ level, we would not reject the null hypothesis. There is insufficient evidence to suggest that any of the means differ from the others.

e   While the p-values are different, they both support the null hypothesis.

For the second experiment we have the following solutions:

a   The experimental design is a Randomized Complete Block Design.

b   The appropriate method for the analysis of an RCBD is the Friedman F_r test for Randomized Block Designs. To calculate the F_r test statistic, we begin by ranking the observations within each block.

Age	S1	S2	S3	S4	S5
15-19	1187 (4)	1103 (3)	997 (1)	1029 (2)	1207 (5)
20-24	1310 (5)	1278 (3)	1251 (2)	1198 (1)	1305 (4)
25-29	1287 (4.5)	1191 (2)	1287 (4.5)	1121 (1)	1256 (3)
30-34	1223 (3)	1167 (2)	1234 (4)	1109 (1)	1245 (5)
35-39	1250 (4)	1289 (5)	1223 (1)	1244 (3)	1230 (2)
40-44	1312 (4)	11324 (5)	1288 (2)	1295 (3)	1159 (1)

Summing the ranks for each treatment we have

T1=4+5+4.5+3+4+4=24.5
T2=20
T3=14.5
T4=11
T5=20

Our F_r test statistic is calculated as

$$F_r = \frac{12}{bk(k+1)} \sum_{i=1}^{k} T_i^2 - 3b(k+1)$$
$$= \frac{12}{180} \left(24.5^2 + 20^2 + 14.5^2 + 11^2 + 20^2 \right) - 108$$
$$= 7.433333$$

c   Since our F_r test = 7.433333, which is smaller than our critical χ^2 value of 9.48773, we do not reject the null hypothesis. We have sufficient evidence to suggest that the strategies are equally effective.

d   The observed significance level of this test can be determined as 1-p-value. Here our p-value exceeds 0.10, since $p(\chi^2 > 7.433333) > 0.10$ on 4 degrees of freedom. Thus, our significance level is less than 90%.

e   Using the statistical package R (http://www.r-project.org/) we obtain the following output for Analysis of Variance

	Df	Sum Sq	Mean Sq	F value	Pr(>F)
factor(strat)	4	13470054	3367514	0.9964	0.4325
factor(age)	5	17880117	3576023	1.0581	0.4123
Residuals	20	67594925	3379746		

The approximate p-value for the F test associated with the strategies is 0.4325. At the 0.05 level, we would fail to reject the null hypothesis. There is sufficient evidence to suggest that all of the strategies result in similar BMR values.

f   While the p-values are different, they both support the null hypothesis.